INVERTEBRATE AND FISH TISSUE CULTURE

INVERTEBRATE AND FISH TISSUE CULTURE

Proceedings of the Seventh International Conference on Invertebrate and Fish Tissue Culture, Japan, 1987

Edited by

Yukiaki Kuroda
National Institute of Genetics, Japan

Edouard Kurstak
University of Montreal, Canada

and

Karl Maramorosch
Cook College Rutgers, The State University, U.S.A.

With 207 Figures and 75 Tables

JAPAN SCIENTIFIC SOCIETIES PRESS Tokyo
SPRINGER-VERLAG Berlin Heidelberg New York London Paris Tokyo

Softcover reprint of the hardcover 1st edition 1988

Published jointly by:
Japan Scientific Societies Press Tokyo

and
Springer-Verlag Berlin Heidelberg New York London Paris Tokyo

ISBN-13: 978-3-642-73628-5 e-ISBN-13: 978-3-642-73626-1
DOI: 10.1007/978-3-642-73626-1

Sole distribution rights outside Japan granted to
Springer-Verlag Berlin Heidelberg New York London Paris Tokyo

Preface

This volume contains the edited proceedings of the VII International Conference on Invertebrate and Fish Tissue Culture, held May 10–16, 1987 at Ohito Hotel, Ohito, near Shizuoka, Izu Peninsula, Japan. It seems appropriate to present here briefly the history of this conference series, which started 25 years ago at a time when invertebrate tissue culture was still in its infancy.

The first conference was at Montpellier, France, on October 22–24, 1962, organized under the auspices of the Centre National de la Recherche Scientifique, the Pasteur Institute, the Institut National de la Recherche Agronomique of France, and the University of Montpellier. Prof. C. Vago, former director of the Saint Christol les Ales, France, Station de Recherches Cytopathologiques was the conference organizer and the papers presented at this meeting appeared as a special volume of the Annales des Epiphyties, Paris, in 1963.

The second conference, under the sponsorship of the Italian Academy of Sciences and the Istituto Lombardo Baselli Foundation, was held at Villa Carlotta in Tremezzo on Lago di Como, Italy, on September 9–10, 1967. This time Prof. Claudio Barigozzi of the Istituto di Genetica, Milano University, was responsible for the organization of the conference and the proceedings were published by the Istituto Lombardo in Milano in 1968.

The third conference was held in Smolenice near Bratislava, Czechoslovakia, June 22–25, 1971, under the auspices of the Slovak Academy of Sciences and the Institute of Virology at Bratislava. Its director, Prof. D. Blaškovič, and the staff of the Virology Institute were responsible for the organization and the proceedings appeared in 1973 as a hard cover volume of the Slovak Academy of Sciences Publishing House.

The fourth conference was held at Mont Gabriel, Quebec, Canada on June 5–8, 1975. The meeting was sponsored by the University of Montreal, the U.S.A. National Institutes of Health, the National Research Council of Canada, the Medical Research Council of Canada, and the Faculty of Medicine of the University of Montreal. Professors E. Kurstak and K. Maramorosch, who were responsible for organizational matters, edited the proceedings that were published in 1976 by Academic Press in a volume entitled "Invertebrate Tissue Culture: Applications in Medicine, Biology and Agriculture."

The fifth conference was held at Rigi-Kaltbad near Luzerne, Switzerland, April 23–27, 1979. The Swiss National Science Foundation and the Swiss Academy of Sciences, the Welcome Trust, London, the University of Zürich and several Swiss industrial and pharmaceutical companies contributed financially and the proceedings were published by Elsevier North Holland Biomedical Press in 1980, under the editorship of E. Kurstak, K. Maramorosch and A. Dübendorfer.

The sixth conference at St. Augustine, Florida, from June 5 to 10, 1983, was chaired by E. Kurstak, H. Oberlander and K. Maramorosch. Twenty-one edited keynote papers presented at this conference and updated by the authors were published by Elsevier Biomedical as a hard cover volume, entitled "In Vitro Invertebrate Hormones and Genes," edited by E. Kurstak and H. Oberlander.

Starting with the fourth conference, the administrative and organizational matters were expertly handled by Prof. E. Kurstak's Conference International Office at the University of Montreal. This permitted the flawless exchange of information, notification of subsequent meeting venues, announcements, and last, but not least, the uniform preparation of manuscripts for publication, as well as the scientific program organization.

The seventh International Conference on Invertebrate and Fish Tissue Culture was broadened, as indicated by its name, by the addition of fish tissue culture. This decision was taken because of the considerable interest in this subject in Japan, the United States, Canada and several other countries. The conference was chaired by E. Kurstak, Y. Kuroda, and K. Maramorosch. The International Advisory Committee included W. Ahne (F.R.G.), D.P. Anderson (U.S.A.), E. Berger (U.S.A.), M. Best-Belpomme (France), R. Cavalloro (Italy), A. Dübendorfer (Switzerland), Gao Shang Yin (China), R.H. Goodwin (U.S.A.), W.F. Hink (U.S.A.), A. Igarashi (Japan), V.I. Kakpakov (U.S.S.R.), D.C. Kelly (U.K.), H. Laufer (U.S.A.), D.E. Lynn (U.S.A.), H.G. Miltenburger (F.R.G.), J. Mitsuhashi (Japan), B.L. Nicholson (U.S.A.), H. Oberlander (U.S.A.), J. Peleg (Israel), J.M. Quiot (France), J. Řeháček (Czechoslovakia), T. Sano (Japan), S.S. Sohi (Canada) and K. Wolf (U.S.A.). The Japanese Organizing Committee was composed of Y. Kuroda, Chairman, T. Ohtaki, fund raising, J. Mitsuhashi, program, T. Miyake, social events, and A. Machii, exhibition. Financial support came from several sources, primarily from the Commemorative Association for the Japan World Exposition, the Japanese Foundation for Advancement of International Science, the Japanese Tissue Culture Association, the Japan Society for Cell Biology and several Japanese industrial concerns.

The sessions dealt with new cultures and techniques, physiology and endocrinology of cells *in vitro*, production of specific proteins *in vitro*, cell differentiation and gene expression, application of cell cultures to pathology and microbiology, fish cell culture development and application, and new cell lines and their characterization.

Among the highlights in invertebrate tissue culture development and application were results presented on molecular mechanisms of gene regulation using DNA mediated gene transfer techniques, on ecdysterone regulatory hormone influence in differentiation of embryonic cells, and on metabolism of cells in culture leading to identification of functions for individual ribosomal proteins. Interesting data were reported on the establishment of new cell lines, their simplified culture media, and on the use of cells for evaluation of cytotoxicity, accumulation, and metabolization of insecticides. Important data also were reported on production of retrovirus-like particles by transposons in insect genome and on the isolation of a new retrotransposon, in fly genome of an intracellular life cycle, similar to the retroviruses. Very encouraging results were obtained in production of a large quantity of Japanese encephalitis and dengue viruses in suspension culture of mosquito cells. A laboratory scale mass production system of a high virus yield using mosquito cells on microcarriers is of importance for the production of viral vaccines.

At this conference several experts presented data of a very high standard on fish tissue culture and on their applications. The advent of biotechnology to classical cell culture changes the perspective of this important science. No longer are the culturists bound to merely offering nutrient gifts to their fastidious, unpredictable, cell cultures, for now, the cells can be kicked and tricked into performing for their masters by summoning new growth factors, conditioned media, and hormones and maximizing equipment and techniques.

The route to this new era has heretofore been directed by the cells themselves; those that will grow have been chosen, expanded and used for specific functions. The main purposes of fish tissue culture have been for diagnosis of viral fish pathogens, as the rainbow trout line, RTG-II, became the first unofficial standard amongst laboratories for screening hatchery and fish populations for viral agents causing diseases such as pancreatic necrosis, and infectious hematopoietic necrosis. Now there are over 75 fish cell lines available, which are recommended for growth and/or identification of specific fish pathogens. Indeed, at present, still the most important role of the fish tissue culturist is the maintenance of cell lines for use in diagnosis of viral fish disease agents. The development of plaques resulting from the virus growth and destruction of the cells remains the primary indicator of viral infections. If specific antisera against suspected pathogens is available, it is used for a neutralization assay, further identifying the viral agent. Recently, new sensitive adaptations of the immunoassays include immunoblot, the identification of the viral antigen-antibody complexes on a support matrix using enzyme immunosorbent indicators, and Western blotting, using a transfer of the specific antigens or antibody after electrophoretic separation.

These tedious assays of identification of viral agents by plaquing and neutralization are being replaced, however, the more sophisticated immunoassays require high quality support systems furnishing the biological assay materials. There is hope that this need will be fulfilled by the commercial companies as the demand for these biologicals increases, and that a successful market will be formed in the future.

Further exciting new areas for the application of fish tissue culture are developing. We tend to look at fishes as more primitive vertebrates, but the forces of evolution do not halt in any species. The tremendous variation in the morphological and biochemical traits of fishes demonstrates that each has refined its physiological niche and in doing so, may offer unique characteristics which can be exploited for biological studies. Basic traits include the ability of the lines to be held at room temperatures, without CO_2 supplement and without frequent media changes. These special characteristics make them candidates for participation in the next generation of biotechnology, as *in vitro* produced cellular products are becoming important. For instance, monoclonal antibodies from hybridomas resulting from specific antibody-producing cells and lymphoma cell lines can be used for more specific immunodiagnosis, and hold promise for therapy. Likewise, interferon, lymphokines, and hormones may be produced in much the same manner with fish cell lines, as variations in molecular structures between fish and mammal cell products may offer unusual applications. Fish cell lines will be important in environmental toxicology by demonstrating physiological changes as a result of exposure to pollutants, contaminants or drugs. The point-sensitive-action of the chemicals on the cellular biochemical cycles must be recognized. *In vitro* fish tissue culture can assist the demonstration of where the specific chemicals interact or interfere with the genome, physiology or structure of the cells.

The VII International Conference on Invertebrate and Fish Tissue Culture addressed these problems in the formal program and lengthy personal discussions continued throughout the meeting. The participants wish to thank the Japanese scientific community for sponsoring this timely conference.

The editors are deeply indebted to the contributors for the effort and care with which they prepared their manuscripts. Thanks also are addressed to the Japan Scientific Societies Press for cooperation and assistance in the production of this volume.

It should be noted that on the third timely day of the conference all participants enjoyed a full-day bus excursion to the Fuji-Hakone-Izu National Park. This trip also provided an excellent opportunity for extensive informal discussions in a splendid environment at the foot of Mt. Fuji. The accommodation at the 5-star Ohito Hotel with its beautiful gardens and the view of the snow-capped Mt. Fuji, as well as the natural hot spring pools added to the fine atmosphere of this truly international gathering.

The reception and sumptuous banquet, with musicians performing on old Japanese instruments, and the outstanding drummers left a deep and lasting impression. At the closing ceremony it was announced that the eighth International Conference on Invertebrate and Fish Tissue Culture will be held in Montreal, Canada.

Edouard KURSTAK
Yukiaki KURODA
Karl MARAMOROSCH
Douglas P. ANDERSON

Contents

PART 2 PHYSIOLOGY AND ENDOCRINOLOGY OF CELLS *IN VITRO*

PART 3 PRODUCTION OF SPECIFIC PROTEINS BY CULTURED CELLS *IN VITRO*

PART 4 CELL DIFFERENTIATION AND GENE EXPRESSION

PART 5 APPLICATION OF CELL CULTURES TO PATHOLOGY AND MICROBIOLOGY

PART 6 FISH CELL CULTURES

PART 7 NEW CELL LINES AND THEIR CHARACTERIZATION

Part 1

NEW CULTURES AND TECHNIQUES

PRIMARY CULTURES OF OVARIAN TISSUE FROM LOCUSTS, *Locusta migratoria* AND *Schistocerca gregaria* (ORTHOPTERA : ACRIDIDAE)

S.K. Raina and A.M. Khurad

Biological Pest Management, Department of Zoology,
Entomology Division, Nagpur University, Nagpur-440 010,
India

The primary cell cultures were initiated using ovarian tissue from IV and V instars of *Locusta migratoria* and *Schistocerca gregaria.* These cultures were set up in primaria flask using RK-86 medium supplemented with 20% foetal bovine serum. The tissue fragments of *L. migratoria* adhered to the surface and epithelial like cells started migrating after 8-10 days of initiation of cultures. The cells attached to the bottom of the flask and in 20-30 days grew considerably, some even to a giant size. These cells had been in culture for over 60 days. The explants of *S. gregaria* did not show any sign of attachment and the liberated cells remained suspended in the medium. After 30 days the cell number increased considerably even though they had remained static even after 180 days.

The cultured cells of insects provide an ideal system for studying insect pathogens and their propagation to use them against insect pests. Most of the continuous cell lines are derived from lepidopteran and dipteran insects (1,2,3). However, a few are available from orthopteroid group particularly grasshoppers and locusts. Vago *et al.* (4) achieved a long-term cellular outgrowths using dorsal vessels of *S. gregaria* and *L. migratoria.* Marks (5) succeeded in culturing cells of embryonic tissues of *Melanoplus bivittatus* for year. Raina and Ewen (6) tried to develop fat body cell cultures from *M. sanguinipes.* To date, only two cell lines from embryo of *M. sanguinipes* have been established using UMN-M-20 growth medium (7). The ovarian cell cultures have not been tried in orthopterans so far and these are available mainly in lepidopterans (3,8,9,10, 11). In the present work, an attempt has been made to develop the cell cultures from ovaries of *L. migratoria* and *S. gregaria.*

Materials and Methods

The required developmental stages of *L. migratoria* and *S. gregaria* were obtained from the stock cultures in an insectary developed for the Biological Pest Management Project. Early IV and V instars of these insects were surface sterilized with 70% ethyl alcohol and ovaries of both species were dissected out in Hank's balance salt solution (HBSS). About 10 ovaries of each species were pooled together, chopped into pieces and washed twice with HBSS. The chopped tissue was centrifuged and supernatant was discarded. The pellet was resuspended in 5 ml of newly formulated RK-86 medium (Table 1) supplemented with 20% FBS and transferred to 30 ml Falcon Polyestyrene disposable flasks (Becton and Dickinson). The pH of the medium was adjusted to 6.7 by adding 0.1 M KOH. The osmolality (460 mOsm/kg) was made to mimic that of the haemolymph of *L. migratoria* using Wescor INC. Osmometer with -5120 vapour pressure. The cultures were maintained at 27°C. The medium was replenished every 10-15 days.

Table 1. Composition of RK-86 medium

Component (mg/100 ml)	Component (mg/100 ml)
NaCl (450)	L-Tyrosine (7.5)
KCl (120)	L-Valine (30)
$NaHCO_3$(35)	Ornithine (2.5)
$MgSO_4$(126)	Taurine (2.5)
$MgCl_2$ (20)	L-Hydroxy proline (25)
NaH_2PO_4 (40)	Malic acid (50)
$CaCl_2$ (7.5)	α - Ketoglutaric acid (30)
L-Asparagine (50)	Fumaric acid (20)
L-Aspartic acid (35)	Succinic acid (10)
L-Glutamic acid (100)	Citric acid (10)
L-Alanine (10)	Glucose (1500)
β-Alanine (20)	Fructose (500)
L-Arginine (80)	Sucrose(300)
L-Cysteine (20)	Thiamine hydrochloride (0.002)
L-Glutamine (100)	Riboflavin (0.002)
L-Glycine (60)	Ca-pantothenate (0.002)
L-Histidine (25)	Pyridoxine hydrochloride (0.002)
L-Isoleucine (5)	p-Aminobenzoic acid (0.002)
L-Leucine (25)	Folic acid (0.001)
L-Lysine (20)	Niacin (0.003)
L-Methionine (30)	Choline chloride (0.03)
L-Phenylalanine (15)	Carnitine (0.002)
L-Proline (70)	Albumin, bovine (200)
L-Serine (30)	Yeast extract (300)
L-Threonine (10)	Lactalbumin hydrolysate (300)
L-Tryptophan (10)	Dl-Glycerophosphate (100)

Results and Discussion

Ten primary cultures of ovarian tissues from L. migratoria and five from S. gregaria were seeded initially. The tissue fragments and cells of L. migratoria adhered to the bottom of the vessel and the cells started migrating from the fragments after 8-10 days of initial setting of cultures. The epithelial like attached cells were small in size in early stage of growth and spread around the fragments with the help of their cytoplasmic processes forming patches. These patches joined together into a thin sheet of cells(Fig. 1). This type of epithelial like cells was observed in some primary cultures (11,12). The cytoplasm of these cells was granular and nuclei were dense with condensed chromatin. These cells in the initial period of growth showed mitotic figures (Fig. 2). A few of them even grew to a giant size displaying uniformly granulated cytoplasm, however, they did not survive longer. The growing patches were covered with small globules which appeared refractive under phase contrast optics. After 60 days of initial setting, the growth of the cells retarded and a few patches even deteriorated.

In S. gregaria cultures, the cells and explants remained freely suspended in the medium. The liberated cells were spherical having prominent nuclei with dense chromatin and a comparatively thinner cytoplasm. They looked partially refractive under phase contrast microscope. The cells seemed to multiply by mitoses and the new ones showed the same features, although their size differed considerably (Fig. 3). Initially, the cells increased in numbers but after a few weeks the growth rate was retarded. The cells had been alive in culture for more than 180 days. Such type of cells was reported in ovarian cultures of lepidopterans using various types of growth medium (8,9,10,11). In the present study efforts will be continued in developing the growth medium that gives better growth and multiplication of cells in order to develop a continuous line.

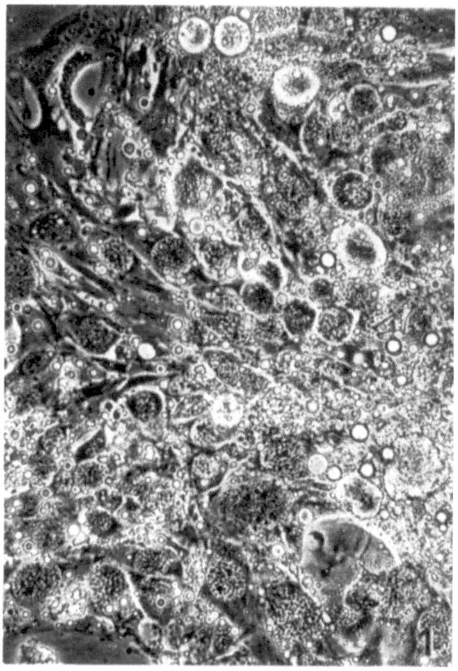

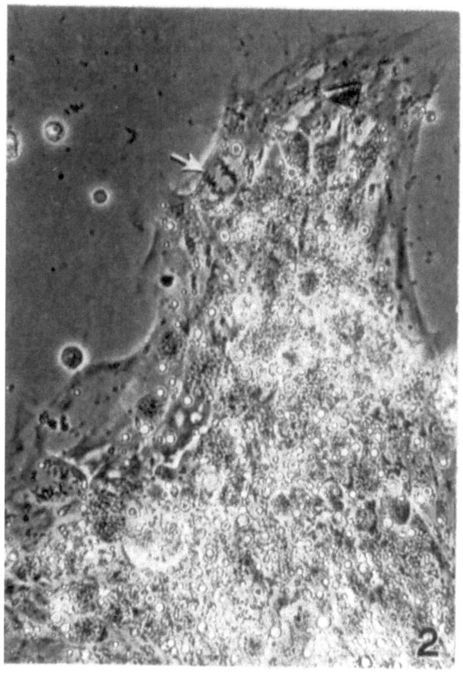

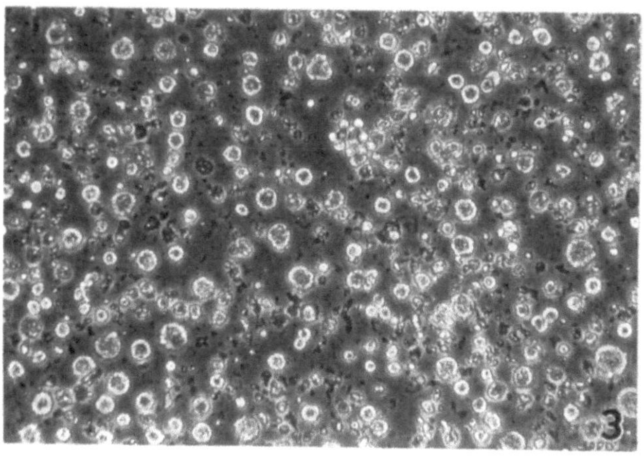

Fig. 1. Primary culture of ovarian tissue from L. migratoria after 50 days of initial setting. Note monolayer of epithelial like attached cells. Bright phase contrast. x 900.

Fig. 2. A patch of growing cells of L. migratoria showing mitotic figure (arrow). Bright phase contrast. x 900

Fig. 3. Primary culture of ovarian tissue from S. gregaria after 40 days of initial setting showing suspended spherical cells. Bright phase contrast. x 600.

Acknowledgements

The authors are grateful to IDRC, Canada for providing funds to undertake this work. The efforts made by Dr. Ken Riley are gratefully acknowledged. They are also thankful for technical help by Mr. T. Pandhripande and photographic assistance by Mr. Chandrakant Sarodey.

References

1. Hink, W.F. Invertebrate tissue culture : Research Applications (Maramorosch, K., ed.), pp. 319-369, Academic Press. 1976.
2. Hink, W.F. Invertebrate Systems in vitro (Kurstak, E., Maramorosch, K. and Dubendorfer, A., eds.), pp. 553-578, Elsevier Biomedical. 1980.
3. Hink, W.F. The 1983 compilation of invertebrate cell lines (unpublished).
4. Vago, C., Quiot, J.M. and Luciani, J. Proc. 2nd Int. Colloq. Invert. Tissue Cult. Italy. 110-118, 1967.
5. Marks, E.P. Tissue culture : Methods and Applications (Kruse, P.F. Jr. and Patterson, M.K. Jr., eds.), pp. 153-156, Academic Press. 1973.
6. Raina, S.K. and Ewen. A.B. Proc. 2nd Trienn. Meet. Pan Amer. Acridol. Soc. USA. 203, 1983.
7. Tsang, K.R., Freeman, F.A., Kurtti, T.J., Brooks, M.A. and Henry, J.E. Acrida. 10: 105-111, 1981
8. Vaughn, J.L., Goodwin, R.H., Tompkins, G.J. and McCawley, P. In Vitro 13 : 213-217, 1977.
9. Lui. Q., Hu. Y, and Shen, L. Contrib. Shanghai. Inst. Entomol. 2 : 128, 1981.
10. McIntosh, A.H., Andrews, P.A. and Ignoffo, C.M. In Vitro. 17 : 649-650, 1981.
11. Goodwin, R.H., Tompkins, G.J., Gettig, R.R. and Adams, J.F. In Vitro 18 : 843-850, 1982
12. Mitsuhashi, J. Zoological Science. 1 : 415-419, 1984.

CULTURE OF INSECT CELL LINES ORIGINATED FROM *MAMESTRA BRASSICAE* WITH AUTOCLAVED SERUM-FREE MEDIUM

M. Koike and K. Sato

Hokko Chemical Industries Co.Ltd., Central Research Laboratories, Toda 2165, Atsugi, Kanagawa 243, Japan

Some cell lines originated from *M.brassicae* are able to be subcultured with an autoclaved serum-free medium. The medium was modified to omit components of sodium hydrogen carbonate, sodium dihydrogen phosphate and serum. These cell lines were SES-MaBr-1,-2, -3,-4 (from fat cells), NIAS-MaBr-92,-93 (from hemolymph), NIAS-MaBr-32(from ovary). An inoculum of $2x10^5$ cells/ml of each cell line produced more than a concentration of 10^6 cells/ml in stationary cultures(5ml). In a spinner flask culture of 500ml, MaBr-93 grew to $1.0x10^6$ cells/ml from $2x10^5$ cells/ml. In a jar fermenter culture of 40 liters, it grew to $6.5x10^5$ cells/ml from $1.5x10^5$ cells/ml.

In agriculture fields, pest control has mainly been accomplished by the use of chemicals. However, appearance of resistance to chemicals and environment pollution requires the application of biological control.

Baculoviruses, which are pathogens of agriculturally important insects, are used for insect control.

The purpose of this work was to culture some cell lines derived from *Mamestra brassicae* in serum free medium.

We used 9 cell lines ; 7 cell lines originated from *M.brassicae*, SES-MaBr-1, SES-MaBr-2, SES-MaBr-3, SES-MaBr-4 (from fat cell), NIAS-MB-32 (from ovary), NIAS-MaBr-92, NIAS-MaBr-93 (from hemolymph), a cell line, NIAS-SpSe-1, originated from ovary of *Spilosuma seriatopuntata* (tiger moth), and a cell line, NIAS-PX-64, from ovary of *Papilio xuthus* (Swallowtail butterfly). These cell lines were established(1,2,3,4,5) and supplied by Dr.J.Mitsuhashi. They were subcultured in the plastic flask(size of $25cm^2$) at 25°C at intervals of 5-7 days using a split ratio of 1:4. Cell concentrations were calculated using a hemacytometer. Concentrations of glucose were measured using Iatro-Chrom GLU-LQ(Iatron), and concentrations of lactic acid, glutamic acid, glutamine, urea and ammonia were measured using a food analysis kit (Boehringer).

We attempted to culture MaBr-32 and MaBr-93 with serum-free medium as listed in Table 1.

Both cell lines were subcultured more than 10 times in serum-free Mitsuhashi and Maramorosch medium(5) sterilized by either filtration (MMSF) or autoclave(MMSFH). In another medium, culturing failed before the 5th passage. Mitsuhashi cultured some cell lines continuously in MMSF(5). We cultured 5 cell lines from armyworm, a cell line from tiger moth and a cell line from butterfly in MMSF and MMSFH. The cell lines from armyworm were subcultured continuosly, but not the cell lines from tiger moth and butterfly. We cultured 7 cell lines from fat cell, hemolymph, ovary of armyworm continuously in serum free medium, MMSF and MMSFH. Their growth was slower than in MM medium containing 3% FBS. The cell line MaBr-4 grew in MMSFH up to a concentration of $1.5x10^6$ cells/ml (Fig.1). MaBr-1,-2,-3,-93 grew similarly to MaBr-4, while MaBr-32 and -92 grew slower than MaBr-4. Using the medium MMSFH, we were able to sterilize by autoclaving rather than by filtration.

Table 1. SERUM FREE MEDIA #

1. Grace's insect medium	10. Leibovitz's L-15 *
2. Eagle's MEM with Hank's salts *	11. Dulbecco's modified Eagle
3. with Earle's salts*	12. Mitsuhashi & Maramorosch
4. RPMI 1640 *	
5. Ham's F10 *	13. Nutrient broth
6. F12 *	14. Heart infusion broth
7. Medium 199 with Hank's salts *	15. Brain heart infusion broth
8. McCoy's 5A	16. Tryptose phosphate broth
9. Waymouth's MB752/1 *	17. Mitsuhashi & Maramorosch

* Added with L-glutamine.
# Sterilization method: 1-12: Filtration 13-17: Autoclave

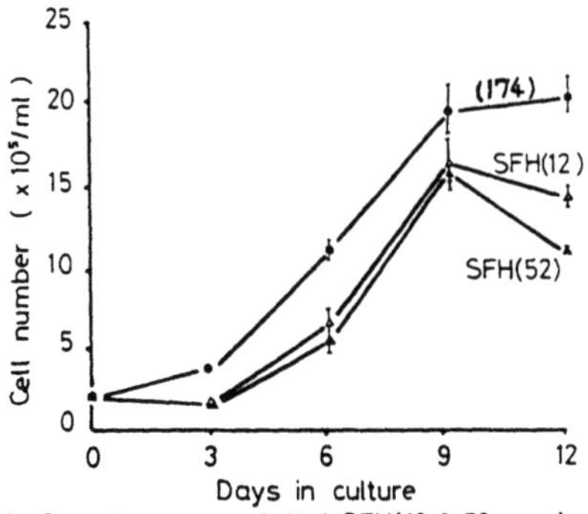

Fig.1. Growth curves of No4-SFH(12 & 52pass) and SES-MaBr-4(174 pass)

We studied the change of the concentration of some substances in the MMSFH cultured MaBr-93. The concentration of glucose decreased and that of lactic acid increased and the pH decreased according to the cell growth (Fig.2). The concentration of glutamic acid decreased. The concentration of glutamine increased during early culture but it decreased later. Ammonia decreased and urea changed only slightly (Fig.3). The change in the concentration of these substances was similar to changes that occur in MM containing 3% FBS. The increase of the concentration of ammonia in the MM medium at later stage was thought to be caused by the termination of cell growth. Mitsuhashi studied the consumption and the production of amino acids in the medium culturing many insect cell lines (1,3,6,7). He reported that the hemocyte cell line increased or decreased the concentration of some amino acids in a characteristic manner. Concentration of glutamine increased and that of glutamic acid decreased when the cells were cultured in MM medium containing 3% FBS. In addition, the concentration of ammonia decreased(3). We had similar results using the same cell line cultured in autoclaved serum free medium(MMSFH). The results suggest that MaBr-93 did not change the metabolism of tested substances when cells were cultured in MM medium containing 3% FBS and MMSFH. Mitsuhashi reported that glucose and lactic acid were consumed by the butterfly cell line. The pH increased in MGM-443 medium(8) but decreased in MGM-431(7). During culture of MaBr-93 in MMSFH, a concentration of lactic acid increased and pH value decreased. The results suggest that metabolism of lactic acid of MaBr-93 is different from that of butterfly cells.

When MMSFH was autoclaved, sometimes precipitates appeared and pH value increased to alkali side. We modified the medium to prevent precipitation (Table 2). Cell lines subcultured in MMSFH grew in No8, No10, No15. Growth of MaBr-93 is shown in Fig. 4. As shown, No10 and No15 media promoted cell growth. These media were contained twice the weight of lactalbumin hydrolysate that MMSFH did but its effect is not yet clear.

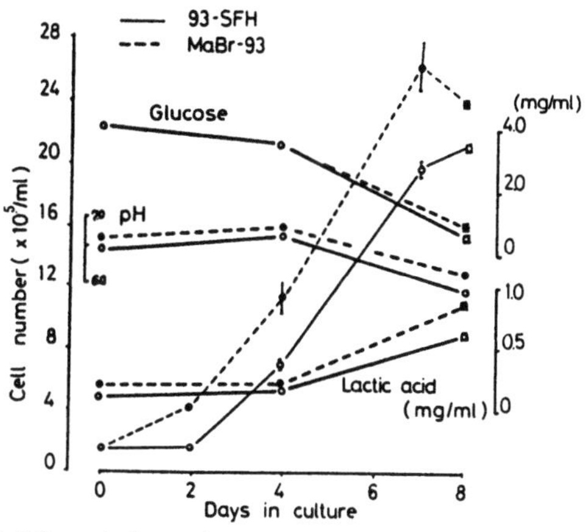

Fig.2. Concentrations of glucose and lactic acid

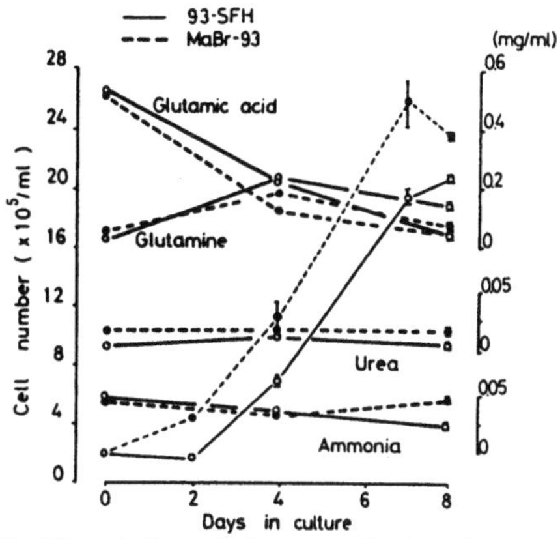

Fig.3. Concentrations of glutamine, glutamic acid, ammonia and urea

Table 2. MODIFIED MEDIA

Components(g/l)	M M	No8	No10	No15
NaCl	7.0	7.0	7.0	7.0
KCl	0.2	0.2	0.2	0.2
$NaOH_2PO_4H_2O$	0.2	-	-	-
$NaHCO_3$	0.12	-	-	-
$MgCl_2 6H_2O$	0.1	0.1	0.1	0.1
$CaCl_2 2H_2O$	0.2	0.2	0.1	0.1
Glucose	4.0	4.0	2.0	2.0
Lactalbumin hydrolyzate	6.5	6.5	13.0	13.0
Yeastolate	5.0	5.0	5.0	3.0

* pH adjusted to 6.5

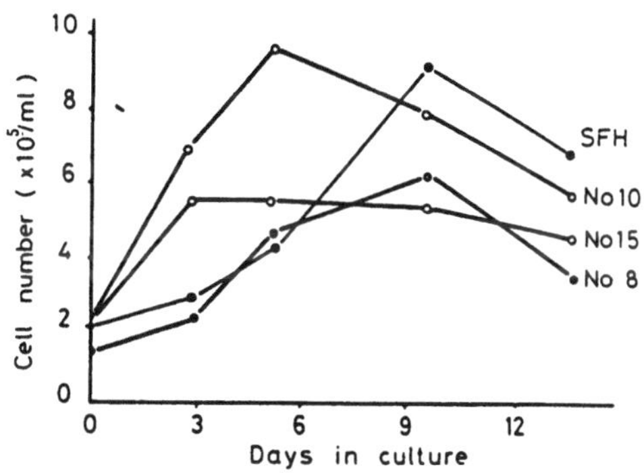

Fig.4. Growth curves in modified medium

We attempted large scale culture of MaBr-93 in No8. We tried to use an air lifter (Shibata Hario 11 fermenter). Cells clustered on the surface of floated bubbles and died within 3 days of culture. We thought that air bubbles injured the cells as Tramper and coworkers reported(9). They reported that air bubbles damaged a cultured cell line of Spodoptera frugiperda. MaBr-93 was cultured in roller bottles(50ml) rolling 1/5 rpm (Toyo Kagaku, Multi incubator). It attained a concentration of $1.8x10^6$ cells/ml from $2.0x10^5$ cells/ml in 12 days. Using a spinner system (Shibata Kagaku), MaBr-93 was cultured in 100ml bottles with bar type impellers at a speed of 60 rpm. It grew to $1.2x10^6$ cells/ml in 8 days(Fig.5). At a size of 500ml, the cells were cultured with a combination of bar and spun plate impellers at 40 rpm. This grew to $1.0x10^6$ cells/ml in 12 days (Fig.5). Using the spinner system, cell growth in No8 was similar to the growth in MM medium containing 3% FBS.

Using a small jar fermenter (Iwashiya Seibutsukagaku MB-5), we cultured MaBr-93 in a 1 liter bottle with marine type impeller at 150 rpm and in 3 liter bottles with turbine (upper) and marine (lower) impellers at 200 rpm (Fig.6). It grew to $1.0x10^6$ cells/ml in 1 liter culture in 7 days and $1.1x10^6$ cells/ml in 3 liter culture in 12 days.

Using a jar fermenter (Komatsugawa Kakoki 65 liter), we cultured MaBr-93 in 40 liter bottles with marine type impeller at 150 rpm (Fig.7). It grew to $6.5x10^5$ cells/ml in 7 days.

We cultured insect cells in autoclaved serum free medium in a jar fermenter with baffle plates removed from a bacteria culture system set. Our system is simple: jar

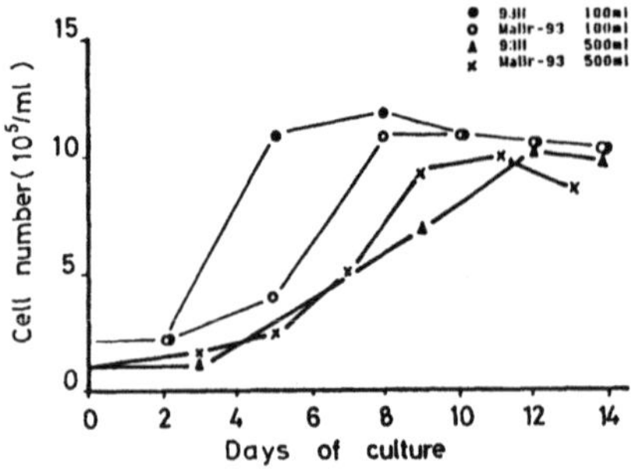

Fig.5. Growth curves of 93H & NIAS-MaBr-93 with spinner culture

93H : cultured with No8
MaBr-93 : cultured with MM-3%FBS

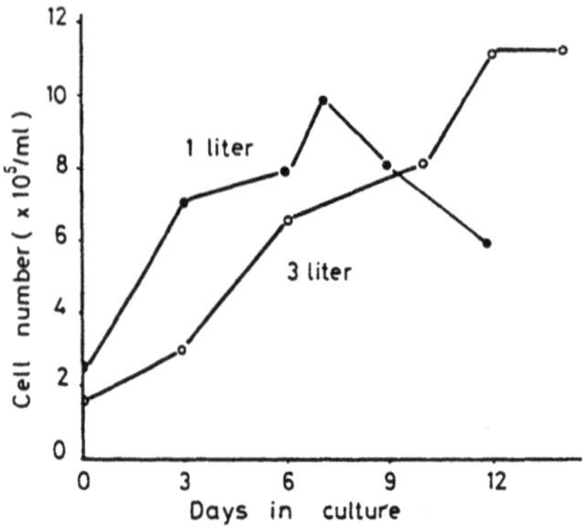

Fig.6. Growth curves in small size jar(1 and 3 lit.)

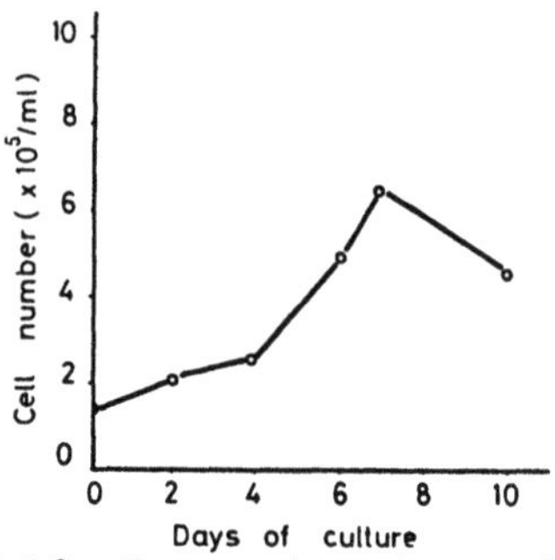

Fig.7. Growth curve in jar fermenter (40 lit.)

fermenter of a standard type for bacteria culture is available and needs no special apparatus nor modification. The medium is composed of simple components without serum and it can be sterilized by a steam system set jar fermenter, requiring no filtration system. So we culture these insect cells on a large scale at very low cost, like bacteria.

Acknowledgements

We thank Dr.J.Mitsuhashi for providing insect cell lines and suggestions.
This work was supported by funds obtained from the Research Association for Biotechnology of Agricultural Chemicals.

References

1. Inoue,H. and Mitsuhashi,J. Appl.Ent.Zool. 20:496-498,1985.
2. Mitsuhashi,J. Develop. Growth and Differ. 19:337-344,1977.
3. Mitsuhashi,J. and Shozawa,A. ibid. 27:599-606,1985.
4. Mitsuhashi,J. Zool.Sci. 1:415-419,1984.
5. Mitsuhashi,J. Advances in Cell Cultures. Vol.2 (Maramorosch,K. ed.), pp.133-196, Academic Press N.Y. 1982.
6. Mitsuhashi,J. Invertebrate Cell Culture Applications (Maramorosch,K. and Mitsuhashi,J. eds.), pp.9-52, Academic Press N.Y. 1982.
7. Mitsuhashi,J. Invertebrate Tissue Culture Research Applications (Maramorosch,K. ed.),pp.13-38, Academic Press N.Y. 1976.
8. Mitsuhashi,J. "Seibutsu no Seisankino no Kaihatsu" Reports of Special Study of Department Education,pp.364-382, 1982 (in Japanese).
9. Tramper,J. Williams,J.B. and Joustra,D. Enzyme Microb. Tachnol. 8:33-36,1986.

SOME CHARACTERS OF CELLS OF THE MIDGUT GLAND AND CHYTRIDS FROM PRIMARY CULTURES OF THE PRAWN PENAEUS JAPONICUS

A. Machii, K.T. Wada, M. Awaji, H.K. Nakamura, and S.J. Townsley*

National Research Institute of Aquaculture, Nansei, Mie 516-01, Japan
and *University of Hawaii, Honolulu, U.S.A.

Fundamental cell culture methods were developed for the midgut gland of the Kuruma prawn Penaeus japonicus and a chytrid parasite. In the first 5 (Pjmg 1-5) of 15 primary cultures, vigorous proliferations of a chytrid (Protista) parasite occurred. More than ten passage cultures of the chytrid have been maintained. In the last ten cultures(Pjmg 6-15), the chytrid cells did not proliferate. In the last six cultures (Pjmg 10-15), observations on chromosomes and esterase isozymes revealed that the proliferated cells were from the midgut gland of the prawn and not from the chytrid parasite.

In 1971, Peponnet and Quiot(1) reported that cultures of cardiac and stomach tissues from the marine crab Pachygrapsus marmoratus were maintained at least for four months. The next year, organotypic culture of invertebrates other than insects was described by Gomot(2). In recent studies on the in vitro culture of midgut gland from the prawn Paraemon serratus, Malcoste et al.(3) reported preliminary results on immunocytochemical detection of enzyme activity and the role of nucleic acid precursors in the cells. Chen et al.(4) described experiments on cultures of various tissues from the grass prawn Penaeus monodon. Although they performed four consecutive cultures from a gonad, they were not successful in getting cell proliferation in the midgut gland cultures.

This study was undertaken to establish culture methods for elucidation of the function of midgut-gland cells and the causes of necrotic disease in young shrimp. Observations on cultures of a chytrid parasite from the midgut gland of the prawn and an account of cell culture from the midgut gland are presented. Some characteristics of the midgut gland and chytrid cells determined by morphology, isozyme and chromosomal studies are also described.

Materials and Methods

Midgut glands of the Kuruma prawn Penaeus japonicus of 1 to 40 grams in body weight were used. Fifteen trials of primary-culture from the midgut gland are outlined in Table 1.

Prior to dissection, the animals were maintained for 18 to 96 hours in running sea water sterilized by ultraviolet light and millipore(0.45 μm) filtration. Prior to dissection an animal was put in a beaker with Dakin's fluid for 30 to 60 seconds. After this the animal was washed 3 times with balanced salt solution for the prawn(Table 2). The midgut gland was removed from the body through an opening made by cutting the carapace on the dorsal median line. After removal, it was minced with knives. The midgut-gland pieces were trypsinized(0.05 % trypsin in CMF) for 20 minutes in calcium- and magnesium-free fluid(Table 2) at room temperature. The cells were washed two times in a balanced salt solution with 10 percent fetal calf serum or medium Pj-2(Table 3) with 0.012 µg/ml of aprotinin. Next, the cells were centrifuged at 800 rpm for 7 minutes and the supernate was removed. A few drops of medium was added to the pellet, and the cells were then suspended by gentle pipetting and inoculated into glass TD15 flasks, plastic T25 flasks or 24-well plastic multiplates. Culture media used were Pf35(5) developed for the pearl oyster, and Pj1, Pj2 and Pj3 developed for the prawn. Cultures were kept at 25°C and pH 7.4.

Isozymes from the midgut gland of the adult prawn, from primary cell cultures, and from cultures after the fourth passage(Pjmg 1) were examined by starch gel electrophoresis(6). Isozyme and buffered solutions used are shown in Table 4. Chromosome preparations were made by the ordinary air-drying method, including treatment with the colchicine (5 µg/ml, 25°C, 15 hrs) and the hypotonic solution of 0.075M KCl, fixation with Carnoy solution and staining with Giemsa solution. Nuclear DNA of cultured cells fixed in 70 % methanol was stained with solutions of PI(propidium iodide) or DAPI(4',6-diamidino-2-phenylindole dihydrochloride). The cells on the third passage of Pjmg 3 were observed by taking 16 mm time-lapse micro-cinematography with intervals of 1, 5, and 10 minutes for 3.5 days at 18 to 25°C.

Results and Discussion

Inoculum cells were epithelial like in shape, 8 to 25 µm in diameter and had a large number of cytoplasmic granules. Differences were observed between cultures Pjmg 1 to 6 and Pjmg 6 to 15. In most of the 5th to 12th day cultures of Pjmg 1 to 5, the cytoplasm of some cells fragmented, and several small spherical cells appeared. The small cells increased rapidly in number, forming a confluent cell sheet of small and large spherical cells. The large ones were 8 to 15 µm, and the

Table 1. Conditions and results of the primary cultures from the prawn *Penaeus japonicus*.

No. of Cultures (Pjmg)	Date of culture conducted	Size of prawn (g)	Number of prawns	Place of collection of prawn	Medium used for cultures	Chytrids prolife-ration	Notes
1	10 Jan.'86	2	6	Y[1]	Pf35	+	
		40	3	Y	Pf35	+	
2	14 Jan.'86	2	4	Y	Pf35	+	
		40	2	Y	Pf35	+	
3	2 Oct.'86	1	11	A	Pf35	+	
4	8 Oct.'86	15	1	G	Pf35	+	Isozyme
5	25 Nov.'86	20	1	G	Pf35	+	
6	20 Jan.'87	15	1	Y	Pf35		
7	23 Jan.'87	15	1	Y	Pf35[2]	-	
8	30 Jan.'87	15	1	Y	Pf35	-	
9	18 Feb.'87	15	2	Y	Pj1[3]	-	
10	24 Feb.'87	15	1	Y	Pj1	-	M.[4] 2nd day
11	6 Mar.'87	15	1	Y	Pj2	-	M. 3rd day
12	6 Mar.'87	15	1	Y	Pj2	-	M. 3rd day
13	17 Mar.'87	15	1	Y	Pj3	-	M. 1st day
14	17 Mar.'87	15	1	Y	Pj3	-	M. 1st day
15	24 Mar.'87	15	1	A	Pj2	-	M. 1st day Isozyme

1) Y:Yamaguchi City, A:Ago Bay, G:Gokashyo Bay, 2)0.12 µg/ml aprotinin added, 3)30% FCS 4)M.:Metaphase observed on the day as shown in each.

Table 2. Balanced salt solution for the prawn (PJBSS) and calcium- and magnesium-free fluid (CMF) (mg/l)

Components	PJBSS	CMF
NaCl	23070	26220
KCl	950	1080
$MgSO_4$	2800	-
$MgCl_2$	1940	-
$CaCl_2$	990	-
NaH_2CO_3	300	-
NaH_2PO_4	44	44
K_2PO_4	-	600
Glucose	300	300

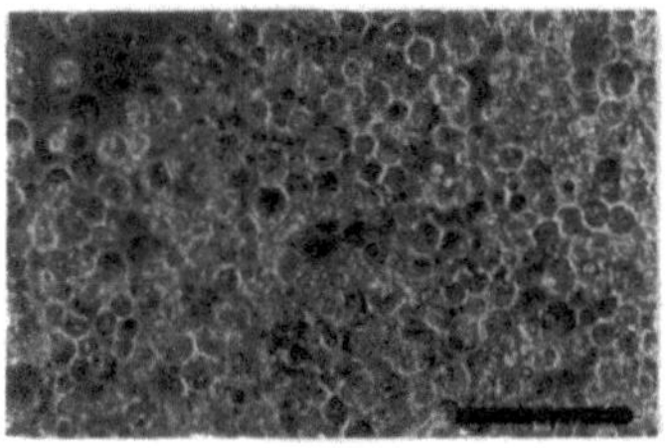

Figure 1. Cells on the 24 hour culture of the midgut gland. Bar = 10 µm.

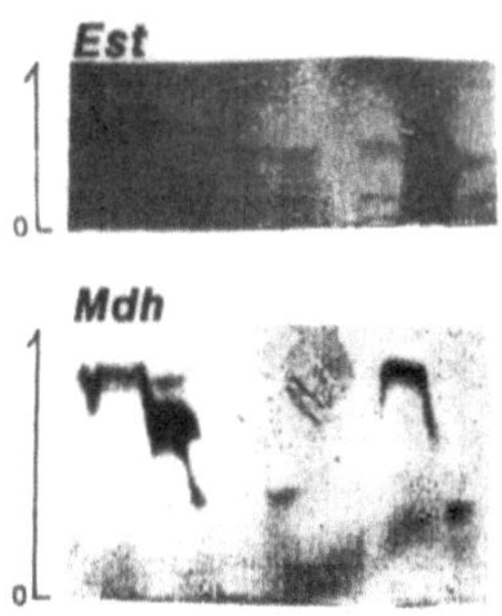

Figure 2. Zymogram pattern of esterase and malate dehydrogenase in the midgut gland, muscle and cultured midgut-gland cells on the third and sixth days.

Table 3. Composition of the culture medium(Pj 2)

Components	Concentration (mg/liter)	Components	Concentration (mg/liter)
Inorganic salts		Tyrosine	6.7
NaCl	18,880	Phenylalanine	147.7
KCl	780	β-Alanine	7.1
$MgSO_4$	2230	β-Amino-iso-butyric acid	2.2
$MgCl_2$	1540	γ-Aminobutyric acid	2.9
$CaCl_2$	800	Ornithine	20.5
$NaHCO_3$	300	Tryptophan	2.8
NaH_2PO_4	44	Lysine	320.8
		Histidine	100.5
Amino acids		3-Methylhistidine	2.3
Phosphoserine	13.3	Arginine	18.7
Taurine	53.8	Asparagine	27.3
Aspartic acid	10.7	Proline	47.7
Threonine	49.9		
Serine	34.1	Antibiotics	
Glutamic acid	12.8	Penicillin	200
Glutamine	55.3	Kanamycin	100
Glycine	80.8		
Alanine	53.8	Serum	
Citrulline	0.5	Fetal calf serum	200(ml)
α-Amino-n-butyric acid	0.6		
Valine	29.5	Other components*	
Cystine	2.9	Glucose	300
Methionine	13.2	Lactalbumin hydrolysate	100
Cystathionine	3.3		
Isoleucine	35.3	Phenol red	5.0
Leucine	46.9		

* Concentration of vitamins, nucleic acid precursors etc. are the same as those of medium 199, except ascorbic acid (40.0 mg/liter).

Table 4. Results of electrophoretic analysis of the cultured cells from Penaeus japonicus and buffer systems used.

Enzyme or protein	Pjmg-1 4th passage (Chytrid?)	Pjmg-15 3 days	Pjmg-15 6 days	Buffer
Alkaline phosphatase(Alp)	-	-	-	A
Esterase(Est)	x	+	+	B
Leucine aminopeptidase(Lap)	-			B
Malate dehydrogenase(Mdh)	x	-	x	A
Phosphoglucose isomerase(Pgi)	x	-	-	A
Phosphoglucomutase(Pgm)	x	-	-	B
Muscle protein(Pt)	-	-	-	A

-: not detected, +: same band detected as midgut gland, x: different band detected from midgut gland, A: Poulik discontinuous(pH 8.6), B: Tris-Boric-EDTA(pH 9.1)

Table 5. Some characteristics of the cultured cells.

Characteristics	Pjmg 6-15	Pjmg 1-5(Chytrid?)
Shape and size	epithelial like 8-15 μm	spherical, 2-15 μm rhizoid like filament
Movement	none	flagellate zoospores swim rapidly
Chromosome	2n = 86	could not be observed

small ones 2 to 4 μm in diameter. The cell sheet formed on the bottom of the culture was not a monolayer. On the surface of the culture medium were observed small floating sheets composed of small cells attached to each by rhizoid filaments.

First subcultures were taken a few days after confluent cell sheets were formed. Sixteen to 48 hours after subculture, numerous flagellated, rapidly swimming zoospores were observed in most cultures. In these cultured cells, no chromosomes were demonstrated with the standard Giemsa staining. Multinuclei were observed under fluorescent microscopy in the medium-size and large spherical cells stained with PI and DAPI. Different band patterns were observed in four enzymes in electrophoretic analysis of cells from fourth passage of Pjmg 1(Table 4). Observations on the Pjmg 2 and 3 culture by 16 mm time-lapse micro cinematography showed that the large spherical cells exploded releasing numerous zoospores which swam away rapidly. This shows that the large spherical cells are not prawn cells.

In cultures from Pjmg 6 to 15, cells inoculated were epithelial like in shape, and more than half of them were floating in the medium (Fig. 1). None of the small spherical cells or zoospores seen in the Pjmg 1 to 5 cultures were observed in Pjmg 6 to 15. After treatment with colchicine, numerous metaphase(2n=86)-stage cells were observed in the 24-, 48- and 72-hour cultures. In electrophoretic analysis the same esterase bands were observed in midgut gland before culture and in cultured cells on the third and the sixth days(Table 4, Fig. 2).

These results are summarized in Tables 1 and 5. On the basis of these results, differences between cultured midgut-gland cells and chytrid cells are evident. The authors presume that the cultured chytrid cells originated from chytrids previously parasitic in the hosts' tissue. Although midgut-gland cell cultures of 24, 48 and 72 hours revealed evident mitotic activity, they have not yet been subcultured.

Judging from these results the following conclusions can be drawn: 1) a fundamental method for chytrid-cell culture in vitro is established; 2) cell cultures of midgut gland from the prawn may be developed in the future if a method for eliminating the chytrid parasite from the cultures is found.

Acknowledgements

The authors express their sincere gratitude to Drs. T. Kawarabata of Kyushu University and K. Hatai of Nippon Veterinary and Zootechnical College for their assistance in identification of the contaminant cells. Thanks are also extended to Dr. M. G. Hadfield of The University of Hawaii for reading the manuscript.

References

1. Peponnet, F. and Quiot, J.M. Invertebrate Tissue Culture, Vol.1 (Vago, C. ed.), pp. 341-360, Academic Press, 1971.
2. Gomot, L. Invertebrate Tissue Culture, Vol. 2 (Vago, C. ed.), pp.41-136, Academic Press, 1972.
3. Malcoste, R., Wormhoudt, A. Van and Bellon-Humbert, C. Comt. Rend. HEBD. Seances Acad. Sci.(III), 296:597-602, 1983.
4. Chen, S.N., Chi, S.C., Kou, G.H. and Liao, I.C. Fish Pathology (Japan), 21:161-166, 1986.
5. Machii, A. Tissue Cult. Res. Comm., 4: 18-20, 1985.
6. De Matthaeis, E., Allegruci, G., Caccone, A., Cesaroni, D., Sbordoni M.C. and Sbordoni, V. Mar. Ecol. Prog. Ser., 12:191-197, 1983.

SIMPLIFICATION OF MEDIA AND UTILIZATION OF SUGARS BY INSECT CELLS IN CULTURES

J. Mitsuhashi

Division of Forest Protection,
Forestry and Forest Products Research Institute,
Matsunosato, Kukizaki-cho, Inashiki-gun, Ibaraki 305,
Japan

Culture media were modified with a view to simplification and cost reduction. The new medium (MTCM-1601) consisted of sea water of $\frac{1}{4}$ strength, lactalbumin hydrolysate, TC-yeastolate and table sugar. This medium could be sterilized by autoclaving. Most of the cell lines tested could be passed continuously in this medium, although growth was somewhat retarded in most cases. Using a culture with this medium, the ability of insect cells to utilize sucrose in the medium was proved, and biochemical analyses demonstrated the presence of α-glucosidase in the cells as well as in the media used for culturing cells.

Various types of media have hitherto been used for insect tissue cultures (1). This is due to people's continuous efforts to develop and improve culture media. There are three main objects of improving culture media : The first is to develop a medium with totipotency, in which any type of tissues or cells can be cultured successfully. Such a medium has been wanted for long time, but no one has so far been able to formulate a medium of this type. Therefore, there are still many species of insects from which even primary cultures cannot be started. The second is to obtain chemically defined media. This type of medium is extremely useful for studies of the biochemistry of cultured cells. Such a medium has been reported by Wilkie et al. (2). However, the use of this medium is limited to some continuous cell lines. The third is to formulate simple and low cost media. When one attempts to obtain large amounts of some substances including viruses from the cultured cells, large scale cultures of cells become necessary. For this purpose, media which can be prepared easily and are inexpensive are desirable. Presently used insect cell culture media contain fetal bovine serum in most cases and as long as people use such media containing expensive ingredients, large scale cultures of insect cells will never become practical. Recently, however, it has become possible to culture a considerable number of insect cell lines in serum-free culture media. Among such serum-free media, Mitsuhashi and Maramorosch's medium without serum (MM-SF) is the most simple and inexpensive (3). In the present study, attempts were made to modify the MM-SF medium so as to obtain more simple and inexpensive media. The mixture of inorganic salts in the MM-SF medium was successfully substituted with diluted sea water, and the glucose with table sugar. Since the table sugar was then the sole source of carbohydrates in the medium, the utilization of sucrose by insect cell lines was suspected, and analyses of culture media proved that insect cell lines possess α - glucosidase and can utilize sucrose in media.

Materials and Methods

Cell Lines : Lepidopteran and dipteran cell lines were used. They are listed in

Table 1. All the cells had been continuously cultured for many passages in the MM-SF medium.

Formulation of New Medium : The MM-SF medium was modified by substituting its salts and glucose with four-times diluted sea water and table sugar respectively, leaving the other components unchanged. The rate of dilution of sea water was determined with reference to the data of chemical analysis on sea water (4). The new medium was designated as MTCM-1601. The medium consisted of 25 ml sea water, 0.65g lactalbumin hydrolysate, 0.5g TC-yeastolate, 0.8g table sugar (obtained from food stores or supermarkets) and distilled water added to total volume of 100 ml. The above solution had a pH of about 6.5, and it was not necessary to adjust the pH. The medium could be sterilized by autoclaving for 15 min.

Sea Water : Sea water was obtained from Ago Bay in Mie Prefecture (courtesy of Dr. A. Machii, Nat. Inst. Aquacult.), Hitachi Harbor and Tokyo Harbor. The samples were sterilized by passing through membrane filters, and stored at room temperature for more than one year without deterioration.

Adaptation of Cells to New Medium : When the cells were subcultured, half of the cell suspension was discarded and an equal amount of the MTCM-1601 medium was added. When the cells reached the plateau level of their growth curves, half of the cell suspension was discarded and an equal amount of the MTCM-1601 medium was added. After subculturing in this manner five times, the original MM-SF medium should be reduced to one thirty second of its original strength. Then the cells were washed and cultured in a complete MTCM-1601 medium. Cells which continued to multiply in the MTCM-1601 medium for more than ten passages were thought to have adapted to the MTCM-1601 medium.

Growth Curves : The cells in the MTCM-1601 and MM-SF media were cultured for 7 days at 25°C. Cell enumeration was made at the beginning, middle and end of the cultures by counting the cells in a 10 square (50x10 μm) area of an occular micrometer.

Analysis of Sugar : The unused MTCM-1601 medium and the ones used for culturing NIAS-MB-32, NIAS-MaBr-93 and SES-MaBr-4 respectively were freed of cells by centrifugation. A 25 ml aliquot of each sample was analysed for sucrose concentration with a sugar analyser YSI model 27 (Yellow Springs Instrument Co., Inc., Yellow Springs, U.S.A.).

Detection of Cellular Enzyme : Cells equivalent to 6.0 mg protein were collected from each of the cell lines, washed and spun down. The cell pellet was sonicated in 2.0 ml distilled water and then centrifuged. 0.08 ml of the supernatant was distributed to each hole of the APIZYM-System (API SYSTEM, S.A., Montalieu, France). After 4 hr incubation at 37°C, APIZYM reagents were added, and the activity of the enzyme was evaluated from the color by comparison with the APIZYM reading scale.

Examination of α-Glucosidase Activity : Cells and their used culture media were examined for α-glucosidase activity. The NIAS-MB-32, NIAS-MaBr-93 and SES-MaBr-3 cell lines were used. About $2x10^8$ cells were harvested. The cells were sedimented by centrifugation. From the supernatants, proteins were precipitated by dropwise addition of three-times volume of saturated ammonium sulfate in a cold bath. The precipitate was collected by centrifugation at 15,000 rpm for 30 min. The supernatant was discarded, and the pellet was dissolved in 2.0 ml of phosphate buffer (0.1M, pH 7.0). Meanwhile, the cell pellet was dispersed in 2.0 ml of the phosphate buffer, and sonicated for 30 sec in a cold bath. These solutions in phosphate buffer were used as enzyme solutions. Half of each solution was boiled and used as a control. For the substrate of α-glucosidase, 0.03% aquous solution of 4-nitophenyl-α-D-glucopyranosid was used. The enzyme solution (0.2 ml) and acetate buffer (0.1M, pH 5.0, 0.2 ml) were mixed. The mixture and the substrate solution (0.2 ml) were incubated at 37°C separately for 5 min. Then both solutions were mixed and incubated for the time previously scheduled. After the incubation, the reaction was stopped by adding 5.0 ml 0.1N sodium hydroxide, and the optical density of the solution at 420 nm was measured with a Beckman DB-GT spectrophotometer.

Results

Many cell lines derived from Lepidoptera and Diptera were able to grow in the MTCM-1601 medium continuously for more than ten passages. These are listed in Table 1. The growth of cells in this medium was retarded to some extent compared with that in the MM-SF medium in most cases, whereas some cell lines such as SES-MaBr-3 grew better in the MTCM-1601 than the MM-SF medium (Fig. 1). The sea water collected from different districts showed no difference in the ability to support cell growth.

When the MTCM-1601 media which cultured some insect cell lines were analysed for their sucrose contents before and after the cultures, a marked decrease of sucrose concentration was observed after the cultures (Table 2). However, the degree of decrease fluctuated among the cell lines cultured.

Since table sugar of the MTCM-1601 is the sole carbohydrate present in the medium, it was supposed that insect cells utilized sucrose in media. To prove this, α-glucosidase activity in cells and culture media was examined. By semi-quantitative examination of cellular enzyme with an APIZYM-System, all the cells tested showed α-glucosidase activity, although the activity fluctuated among cell lines (Table 1). The presence of α-glucosidase in the cells as well as the media used for culturing cells was further ascertained. Only the results obtained from the NIAS-MB-32 cell line are shown (Fig. 2). Both the cells and the media showed α-glucosidase activity, although the activity fluctuated among cell lines examined. Fresh MTCM-1601 medium did not show any α-glucosidase activity.

Table 1. List of Cell Lines Used, with Growth in MTCM-1601 Medium and Activity of α-Glucosidase

Cell Lines	Culture in MTCM-1601*	Activity of α-Glucosidase**
NIAS-PX-58(1)	-	3
NIAS-PX-64(1)	-	2
NIAS-MB-25(2)	+	2
NIAS-MB-32(2)	+	3
NIAS-MaBr-92(3)	+	2
NIAS-MaBr-93(3)	+	5
NIAS-MaBr-85(4)	+	2
SES-MaBr-1(4)	+	3
SES-MaBr-2(4)	+	3
SES-MaBr-3(4)	+	4
SES-MaBr-4(4)	+	2
SES-MaBr-5(4)	+	4
NIAS-LeSe-11(5)	+	4
NIAS-SpSe-1(6)	+	2
FRI-SpIm-1229(7)	+	-
NIAS-AeAl-2(8)	+	5
NIH-SaPe-4(9)	+	3

(1) *Papilio xuthus* ovaries; (2) *Mamestra brassicae* ovaries; (3) *M. brassicae* hemocytes; (4) *M. brassicae* fat bodies; (5) *Leucania separata* fat bodies; (6) *Spilarctia seriatopunctata* fat bodies; (7) *Spirosoma imparilis* fat bodies; (8) *Aedes albopictus* neonate larvae; (9) *Sarcophaga peregrina* embryos.
* Cultured more than (+) or less than (-) ten passages.
** 5: equal to or more than 40 nmol.; 4: 30 nmol.; 3: 20 nmol.; 2: 10 nmol.

Discussion

In the early stages of invertebrate tissue culture development, Lewis had used diluted sea water as a base for a culture medium for grasshoppers (5). Although her result was not satisfactory from the viewpoint of cell cultures, it encouraged me to use sea water for my culture media. On the other hand, recent information on insect cell culture has shown that insect cells can grow in a wide range of pHs, osmotic pressures and Na/K ratios (1,6,7). This also suggests the possibility of using sea water as a base for culture media. Furthermore, the report that sea water contains all the inorganic salts necessary for insect cells in sufficient amounts (4) supported the idea of using sea water for the medium.

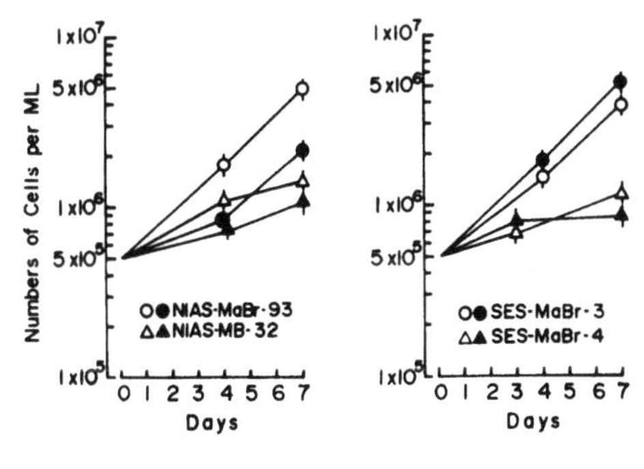

Fig. 1. Growth curves of four insect cell lines in MM-SF (O Δ) and MTCM-1601 (● ▲) media.

The chemical composition and degree of pollution of sea water may be different in different districts. Sea water from Ago Bay was supposedly the cleanest while that from Tokyo Harbor looked dirty, with an oily film on the surface. However, sea water from all districts could be used without deleterious effects.

Table 2. Sucrose in Media (mg/100 ml)

Media	Duration of Culture (Days)	Sucrose Concentration
MTCM-1601 Unused	0	673.3 ± 6.1
MTCM-1601 Used for NIAS-MB-32 Culture	16	163.5 ± 0.7
MTCM-1601 Used for NIAS-MaBr-93 Culture	10	644.5 ± 0.7
MTCM-1601 Used for SES-MaBr-4 Culture	13	595.5 ± 3.5

The fact that many cell lines could be cultured in the MTCM-1601 medium suggested the ability of insect cell lines to utilize sucrose in media, because table sugar consists mostly of sucrose. Grace and Brzostowski (8) have shown that *Antheraea eucalypti* cells utilized sucrose. However, there are also some reports claiming the inability of insect cells to utilize sucrose (9-11). Sucrose, if it is utilized, will be digested to glucose and fructose by enzymes before uptake by cells. The enzyme which mediates this process is α-glucosidase. Therefore the presence of α-glucosidase in the cells as well as the media used for cell cultures was examined. When the enzyme preparations from cells and media were given with a substrate, 4-nitrophenyl-α-D-glucopyranosid, reaction products increased with reaction time as well as with concentration of enzyme preparation, showing that both cells and media contain α-glucosidase. Probably, α-glucosidase is liberated from the cells into the culture media. Also, semi-quantitative analyses showed that all the insect cell lines examined possessed α-glucosidase to some extent.

In conclusion, based on the MM-SF medium, a simple medium MTCM-1601, consisting of sea water, lactalbumin hydrolysate, TC-yeastolate and table sugar was formulated. The medium could be sterilized by autoclaving. Most of the insect cell lines tested could be cultured continuously in the medium. Since table sugar was the only carbohydrate present in the MTCM-1601 medium, it became evident that insect cells can utilize sucrose, and the presence of α-glucosidase in cells and media was demonstrated.

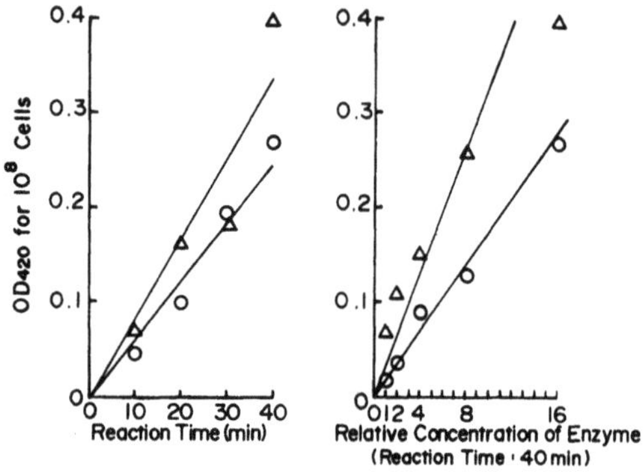

Fig. 2. Activity of α-glucosidase in NIAS-MB-32 cells (○) and the media (△) used for cultivation of the cells.

References

1. Mitsuhashi, J. Advances in Cell Culture (Maramorosch, K. ed.), pp. 133-196, Academic Press, 1982.
2. Wilkie, G.E.I., Stockdale, H. and Pirt, S.V. Develop. Biol. Standard., 46: 20-37, 1980.
3. Mitsuhashi, J. and Maramorosch, K. Contrib. Boyce Thompson Inst.,22:435-460, 1964.
4. Kawai, S. Rep. on "Development of Tissue Culture Techniques for Shellfish", pp. 5-11, Mie Univ., 1982. (in Japanese)
5. Lewis, M.R. Anat. Rec., 10: 287-299, 1916.
6. Mitsuhashi, J. Tissue Culture, 12: 213-218, 1986. (in Japanese)
7. Hink, W.F., Ralph, D.A. and Joplin, K.H. Comparative Insect Physiology, Biochemistry and Pharmacology (Kerkut, G.A. and Gilbert, L.I., eds.), pp. 547-470, Pergamon Press, 1985.
8. Grace, T.D.C. and Brzostowski, H.W. J. Insect Physiol., 12: 625-633, 1966.
9. Gottschewski, G.H.M. Roux' Arch. Entwicklungsmech. Organ., 152: 204-229, 1960.
10. Berridge, M.J. J. Insect Physiol., 12: 1523-1538, 1966.
11. Stockdale, H. and Gardiner, G.R. J. Invert. Pathol., 30: 330-336, 1977.

HORMONE AND GROWTH FACTOR EFFECTS ON THE PROLIFERATION OF DIPTERAN CELL LINES IN A DEFINED MEDIUM

C. Wyss

Dental Institute, University of Zürich, Plattenstr. 11, CH-8028 Zürich, Switzerland

Several cell lines of *Drosophila melanogaster* and one line each of *Chironomus tentans* and *Aedes albopictus*, originally established in widely differing, complex media, can be grown in supplemented ZW synthetic medium. For some lines *Drosophila* low molecular weight cationic growth factor, CalGF is sufficient, while others require CalGF, insulin and *Drosophila* fly extract, FX. All respond to ecdysterone and juvenile hormone. In contrast to *Drosophila* imaginal disc cell cultures, ecdysterone is not required and fetal calf serum is not toxic and can replace FX.

The following study was undertaken with the explicit aim to develop in vitro systems capable of reproducibly culturing specific normal cells, in particular of imaginal discs of *Drosophila*, controlled by autologous growth factors. Many aspects of insect biology will only be amenable to a functional analysis at the biochemical and genetic level by in vitro cell and organ culture, which optimally should be applicable to all organismic mutants available. Since it is unlikely that permanent cell lines display growth requirements not important in vivo, it seemed justified to use these technically simple models to advance our information on insect cell growth and differentiation. The great number of permanent cell lines established represent only a small section of all insect taxa and more importantly, only very few of them even approximately maintain the coordinated functions of a specific tissue at a particular developmental stage. Nevertheless, the functions which they do perform will be part of the animal's physiological possibilities. While nothing definite is known about the cellular processes leading from primary cultures to the development of permanently proliferating cell lines, the selection of nutritionally less demanding mutants generated by retrotransposon activity is a likely possibility. Experience with a number of Dipteran cell lines, all established in the absence of autologous growth factors, indicated that they differ in their nutritional requirements. Therefore, a comparative analysis might eventually yield a catalogue of growth factors relevant also for normal cells in vitro and possibly also in vivo.

Materials and Methods

Culture media

The chemically defined media ZO, ZR and ZW have been described earlier (1, 2, 3). The high molecular weight components of yeast extract (HYEC) were prepared as described (4) and used at a concentration sufficient for the soft agar cloning of line KcHP (5). Adult Drosophila fly extract (FX) and the low molecular weight cationic growth factor (CalGF) from Drosophila adults were prepared and used as described (2, 3). Heat inactivated fetal calf serum (FCS) and horse serum (HS) were from GIBCO-Biocult, yeast extract from Difco Laboratories, ecdysone from Fluka, ecdysterone from Rhoto and juvenile hormone III from Ecco Control. All other chemicals were of the highest purity available from Merck, Serva and Sigma. The water used throughout was from a Millipore Q water purification system. Media sterilization was with detergent free Gelman TCM metricel filters.

Cell lines

The following Drosophila cell lines were used:
Line KcHP was maintained as described (2); Dübendorfer's line D1 and Schneiders line S3 have been previously used, exploiting their more demanding physiology for the selection of somatic cell hybrids (6); more recently they were maintained in medium ZO with HYEC and 2% FCS; Line Bl-3-1 was derived from Gateffs line l(3)mbn(described in 7), kindly provided by E. Gateff, Freiburg, by cloning in soft agar (5) with medium ZO containing HYEC and 2% FCS. The same medium in liquid form was used for further maintainance; line AD-6-4 was cloned from a diploid derivative of Debecs haploid line 6 (8), kindly provided by A. Debec, Paris, as described for line Bl-3-1. The epithelial cell line of Chironomus tentans has been described (1).The Aedes albopictus cell line of Singh (9) was kindly provided by H. Koblet, Bern, in Mitsuhashi-Maramorosch (10) medium with 20% FCS; it was further maintained with reduced growth rate and lower final cell densities in medium ZO containing HYEC, 0.1 mg/l bovine insulin and 2% FCS.
All cells were grown at 25°C in humidified air in petri dishes or 24 well multidishes (Falcon). For experiments, cells were washed three times in phosphate buffered saline and counted in a Coulter counter.

Results and Discussion

To satisfy the nutritional requirements for proliferation of animal cells in vitro usually a synthetic basal medium is supplemented with complex components. Obviously the composition of the basal medium will influence the need for further growth factors. This is illustrated here for the two Drosophila cell lines S3 and D1. High molecular weight yeast extract components (HYEC) and mammalian serum were not sufficient to promote S3 and D1 proliferation in medium ZH, but did so in the more refined media ZO and ZR. Obviously it would be attractive to replace the heterologous, complex components HYEC and fetal calf serum (FCS) by defined and autologous defined growth factors. Table 1 shows the results of a factorial analysis of the effects of various growth factors on these two cell lines. Note that in the presence of FCS or FX for line S3 the addition of CalGF alone can replace HYEC, while line D1 additionally requires insulin. As shown previously for

Drosophila primary cultures (3), this indicates that FX contains no detectable insulin-like activity. It has also been shown that the activity of bovine insulin was blocked by antibodies to insulin and could not be replaced by human insulin like growth factor I (11). The results also demonstrate that both cell lines respond to CalGF, but it is not clear whether FX activity in these lines is through CalGF.

Table 1 Proliferation of lines S3 and D1 in minimal medium ZR depending on the supplement

	line S3		line D1	
	-CalGF	+CalGF	-CalGF	+CalGF
Supplement				
-	-	-	-	-
HYEC	-	-	-	-
FCS (2%)	-	+	-	-
HYEC + FCS	+	+	+	+
insulin (100 ng/ml)	-	-	-	-
HYEC + insulin	-	-	-	-
FCS + insulin	-	+	-	+
HYEC + FCS + insulin	+	+	+	+
FX (2%)	nd	+	nd	-
FX + insulin	nd	+	nd	+

Cells were washed three times in medium ZR and inoculated to the test media at 50,000 cells/ml in duplicate. -: no proliferation; +: proliferation; nd: test not done.

In similar experiments the requirements of two other Drosophila cell lines, Bl-3-1 and AD-6-4, both previously cloned in soft agar, the C.tentans epithelial cell line and Singh's A.albopictus cell line were determined. The results are summarized in Table 2, which also contains the data for Drosophila embryo cells and imaginal disc

Table 2. Requirements for proliferation of different cell types in medium ZO:

	CalGF	insulin	ecdysterone	FX	FCS/FX
D. melanogaster					
embryo cells	+	+	-	+	-
imaginal disc cells	+	+	+	+	-
KcHP	+	-	-	-	-
Bl-3-1	+	-	-	-	-
S-3	+	-	-	+	+
D-1	+	+	-	+	+
AD-6-4	+	-	-	+	+
C. tentans	+	+	-	+	+
A. albopictus	+	+	-	+	+

Compound required: +; FCS/FX + : FX replaceable by FCS.

cells (3). KcHP and Bl-3-1 in medium ZO require only CalGF (or HYEC) whereas S3 and AD-6-4 additionally need FX or FCS. Even more demanding are line D1 and the C.tentans and A.albopictus lines, which furthermore require insulin. Again it may be noted that specific requirements depend on the milieu as a whole, since all these lines were established and maintained in the absence of CalGF and insulin. A similar point can be made about FCS, which as a supplement to ZO or ZW proved toxic to Drosophila primary cultures (3). Yet, all Drosophila cell lines studied here originated from primary cultures set up and maintained in the presence of FCS. Interestingly then, for all the cell lines studied here FX could be replaced by FCS. It remains to be determined whether this apparent difference reflects only a higher resistance of established cell lines to toxic effects of mammalian serum, or also higher requirements for autologous growth factors in FX by primary cultures. In contrast to Drosophila imaginal disc cells in primary culture, none of the other cell systems studied here showed a requirement for ecdysterone. However, all of them proved to be responsive to this steroid, in that all proliferation was inhibited in the presence of 100 ng/ml of ecdysterone. Also, in all of them the inhibition by ecdysterone was counteracted by juvenile hormone. As regards the effects of insulin, the data in Table 2 refer to strict requirements. Previously it has been reported, that line KcHP is growth inhibited by (and thus responsive to) insulin (11). Line S3 in contrast, while not depending on insulin, is strongly stimulated by 100 ng/ml of insulin. Clearly, much further work is needed, to define the nature of the autologous growth factors supplied to Drosophila cell cultures either in the form of animal extracts (3, 12, 13) or as cell conditioned medium (14 - 16).

Acknowledgements

The results presented were obtained in 1981 at the Institute for Cell Biology of the ETH, Zürich, Switzerland, with the support of a ETH grant to C.W. and H. Eppenberger.

References

1. Wyss, C. Exp. Cell Res. 139: 309-319, 1982.
2. Wyss, C. Insect Biochem. 12: 515-522, 1982.
3. Wyss, C. Exp. Cell Res. 139: 297-307, 1982.
4. Wyss, C. J. Insect Physiol. 23: 739-747, 1977.
5. Wyss, C. Somatic Cell Genet. 5: 23-28, 1979.
6. Wyss, C. Somatic Cell Genet. 5: 29-37, 1979.
7. Hink, F. Invertebr. Systems in vitro (Kurstak, E., Maramorosch, K. and Dübendorfer, A., eds.) p 553-578, Elsevier. 1980.
8. Debec, A. Nature 274: 255-256, 1978.
9. Singh, K. R. P. Current Sci. 36: 506-508, 1967.
10. Mitsuhashi, J. and Maramorosch, K. Contr. Boyce Thompson Inst. 222, 435-460, 1964.
11. Wyss, C. Experientia 37: 664, 1981.
12. Kakpakov, V. T., Gvozdev, V. A., Platova, T. P. and Polukarova, L. G. Genetika 5: 67-75, 1969.
13. Dewhurst, S. and Sang, J. H. In Vitro 13: 305-310, 1977.
14. Davis, K. T. and Shearn, A. Science 196: 438-440, 1976.
15. Nakajima, S. and Miyake, T. Invertebrate Tissue Culture (Kurstak, E. and Maramorosch, K., eds.)pp. 279-287, Academic Press, 1976.
16. Richard-Molard, C. and Ohanessian, A. Wilhelm Roux Archives 181: 135-149: 1977.

NEW CULTURE MEDIA FOR INSECT CELLS

J. C. Landureau[1] and J. J. Lenoir - Rousseaux[2]

1) Unité de Biotechnologie, Université Pierre et Marie Curie, 75230 Paris Cedex 05, Unite de Virologie (Service de Pr C. Chany), Hopital Saint Vincent de Paul, 75674 Paris Cedex 14, France
2) Laboratoire de Biologie Animale, ufr Sciences, Université de Paris XII Val de Marne, Creteil Cedex, France

Insect cell and tissue cultures have been performed in media reproducing hemolymph characteristics. These culture conditions have selected cells able to develop hypoxic metabolism. Consequently, a new technology has been devised to improve long term culture of normoxic tissues (medium Bi6020)and mass production from established Diptera or Lepidoptera cell lines (medium Bi6030).

During the 4th meeting on Invertebrate Tissue Culture (Mont Gabriel, June 1975), E. P. Marks and the author compared the results of their respective in vitro approaches concerning cockroaches. It appeared that the use of M14-20 media only yielded positive results in tissue culture (1), whereas that protraction of medium S20 regularly led to the establishment of cell lines (2). The latter only occurred in primary cultures derived from embryos after hemocyte differentiation. These embryonic cell lines (EPa) were especially rich in hemocytes. Moreover hemocyte lines (HPa) were directly obtained from hemolymph and also (OPa) from imaginal ovarian sheath cultures (3).

Further experiments have shown that all hemocyte lines can be grown under drastic hypoxic conditions (4). We suggest that cultivating cockroach cells in medium S20 selects cells able to survive under hypoxia. The above mentioned hypoxic abilities of cockroach hemocytes may be extended to other insect cell lines: Drosophila Kc cells can be maintained under nitrogen and Aedes cells can easily be cloned under the agar layer. Moreover, all the studied insect cell lines manifest similar nutritive requirements (5). They need rather high concentrations of carbohydrates and amino acids, perform important transamination reactions and release in the culture media the typical end products of invertebrate anaerobic metabolism (4).

As insect hemolymph does not ensure respiratory gas exchanges, Carrel's dogma that closely links the physico-chemical characteristics of blood and culture medium does not apply for insects.

It was therefore necessary to design a new technology for cultivating normoxic tissues and improving mass cell production.

Design of Medium Bi6020 for Tissue Culture

The main modifications to the originally described medium S20 concern the decrease or removal of both reducing components and transamination substrates (Table 1).

These culture media, including balanced salt solutions, have been tested in separate laboratories. Partial results have already been published (6-10), but there is a great deal of scope for further applications.

Compiling the preliminary data, the reader can verify the efficiency of the new technology :

- The usual fetal calf serum (FCS) supplementation is reduced to 5% (6-8), 2% (10)
- The tested media support a long term culture of normoxic tissues : 3-14 days for prepupal _Pieris_ wing discs (9), 2-25 days for pupal _Tenebrio_ integuments (6) and male accessory glands (7), 15 days for adult _Periplaneta_ corpora allata (10) and 2 months for nymphal _Periplaneta_ fat bodies and ovaries (10).
- The cultured cells maintain their functional abilities and the target tissues differentiate _in vitro_ in response to hormonal induction. Nevertheless two weeks are necessary to deplete the primary effect of the hemolymph before tissue explantation (10, 12). Ecdysteroids induce imaginal differentiation of scales(9) and integuments (6), stimulate mitoses _in vitro_ (7, 8) and start vitellogenesis in fat bodies and oocytes (10,12). Co-cultured with corpora allata, the fat bodies acquire cell characteristics of the end of previtellogenesis.
- Cell proliferation occurs in primary cultures and will probably lead to the establishment of cell lines of well defined tissue origin :
Adipocyte mitoses have been observed for the first time within fat bodies (10).
The follicular cells actively divide beyond a 15 day culture (12).
Cultures of myoblasts and epithelial cells are regularly initiated from male accessory glands (8).
- The successive modifications reported in Table 1 improve the efficiency of culture media (Figs 1-2). They confirm our working assumption.

Table 1. From S20 to Bi6020 culture media (concentrations in moles x l^{-1}).

	Initial S20 (2)	Modified S20(6-8)	A217	020(10)	Bi6020(11)
ascorbic acid	0.01	0	0	0	0
asparagine, H_2O	0	0	3.3	3.3	1.3
aspartic acid	3.75	3.75	2.6	3.75	0.75
inorganic buffer	10 (PO_3)	10 (PO_3)	7.5 (PO_3)	5 (PO_4)	4.3 (PO_4)
organic buffer	0	0	2.5 (Pipes)	5 (Tris)	5 (Hepes)
cysteine, (HCl)	2.25	2.25	0	0.1	0
cystine, $(HCl)_2$	0	0	0.1	0.83	0.1
diholoside	0	17.5	14.6	14.6	11.7-29.2
glucose	22.2	5.6	22.2	2.8	0
glutamic acid	10.2	10.2	4.1	3.4	3.4
glutamine	0.5	1.7	3.4	5.6	5
glutathione reduced	0.05	0	0	0	0

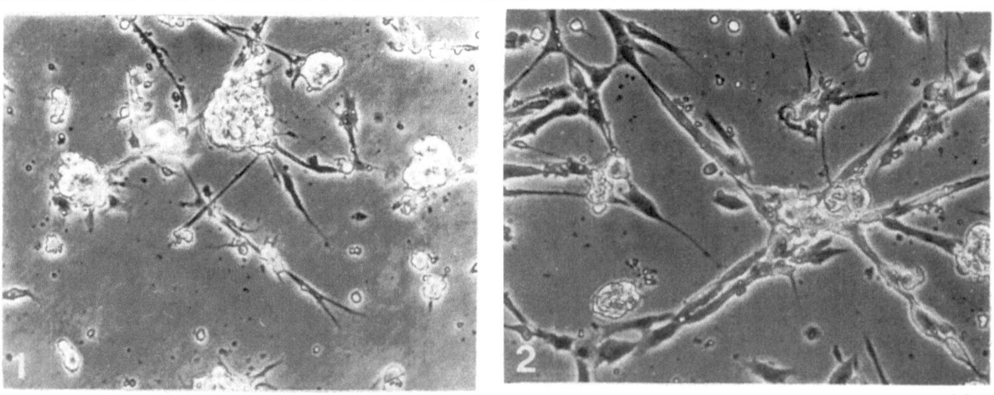

Figs. 1-2 Simultaneous 15 day cultures of _Tenebrio_ myoblasts in modified S20 and B16020 Media. (Courtesy of J.J. L-R. and Dr. G. M. Happ)

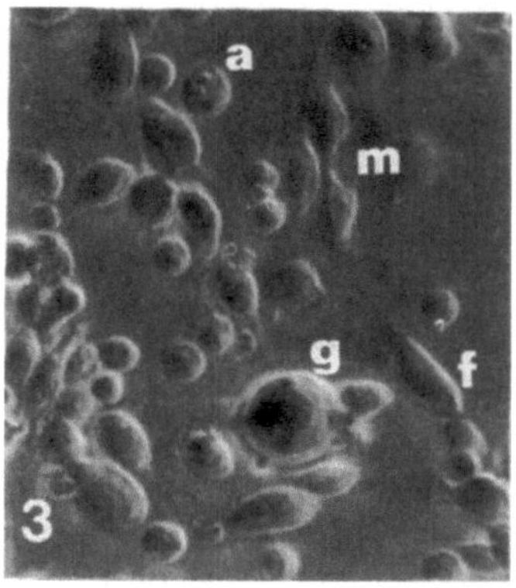

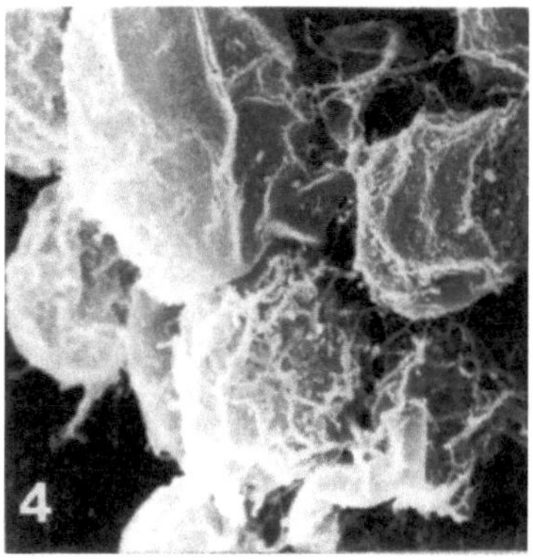

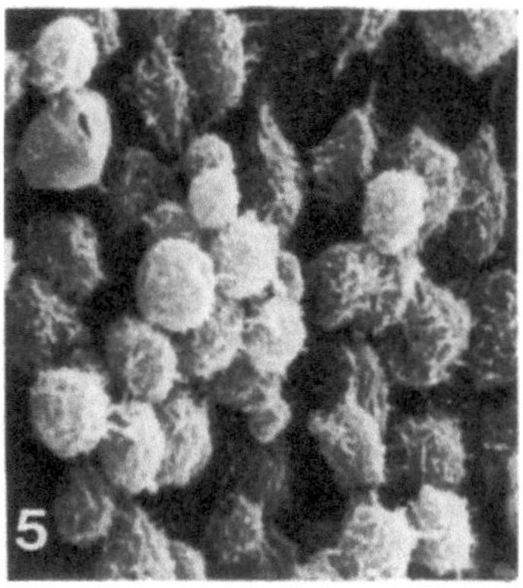

Fig.3:CF131, 4 day culture,x300 Fig.4:ATC10, 2 month culture,x3900 Fig.5:Kc-89K, 3 month culture,x1600

Design of medium Bi6030 for mass cell production

a) Culture of established cell lines in medium Bi6020.

Cell lines have been subcultured from usual media into media Bi6020 supplemented with 2% of FCS. Their in vitro evolution corroborates the former hypothesis relative to metabolic selection induced by culture media during the establishment of cell lines. The results can be listed as follows:

- non-reversible breakdown of hemocyte lines derived from cockroach (HPa, OPa), Drosophila (B3) and Lymantria (IZD 1307). Other Lepidoptera cell lines of unidentified tissue origin (IPLB 1075, NIAS Px64, OL12) die rapidly —— drastic cell selection of heterogeneous cell lines. From the first subculture, important cytolysis occurs, but cell clones develop. Cell transferring becomes successful during the second month. Then, cells grow regularly and can be maintained in media Bi6020 with only 2% of serum. This is true for cockroach APa lines, Drosophila 770M3 and Schneider's lines, Aedes ATC10 (Fig. 4) and NIAS AeA12, and for Lepidoptera lines (Sf21 Ae11, CLS 85, Tni 368, NIAS Mb19 and 32, IPRI Cf1 (see Fig. 3), NIAS LeSe 17, NIAS Px58, OL 13) —— apparent immediate growth of cell clones.

Cell proliferation of Aedes C636 line gradually increases during the first two weeks. Then, cultures can be transferred weekly. Immediate growth occurs for Drosophila cell clones isolated from Kc lines by G. Echalier. (Tables 2 and 3).

Table 2 : Factorial design of nondefined supplements. (inoculum:$5x10^6$ 89K cells,10ml medium BI6020, 150ml Erlenmeyer flask,Orbital shaker:50 rpm)

culture media (number)	LH (g/l)	YE (g/l)	NCS (%)	5day cell population(10^6) immobilized culture	stirred culture
1	2	2	2	11	116
2	8	2	2	13	22
3	2	8	2	0	0
4	8	8	2	0	0
5	2	2	8	35	ND
6	8	2	8	40	108
7	2	8	8	0	0
8	8	8	8	0	0
9	5	5	5	22	3
10	5	5	5	23	5

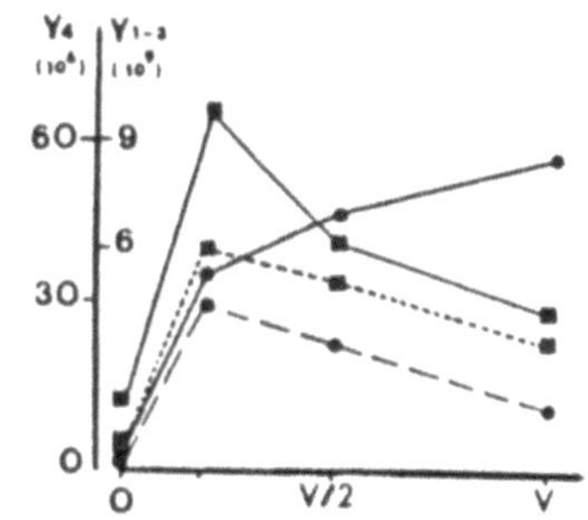

Fig. 6:Effect of vitamin concentration on 89K(o)and 23(o) cells. Numeration of first three(--)and 4th (--) cultures.

Stirring cell cultures favours cell growth (Table 2). Adding lactalbumin hydrolysate (LH) has limited effects, whereas increasing Yeast Extract (YE) negatively interacts with serum supplementation.

Table 3 : Growth of Kc-89K cells in culture media D22 and Bi6020. (25 cm^2 plastic flasks. Y_0=inoculum of $3x10^6$ cells. Y_5=5day population)

Serum supplementation	Medium D22 5% FCS	Medium Bi6020 2% FCS	2% NCS
Ratio Y_5/Y_0	3.8	9.4	13.4

Nevertheless, cell growth of 89K cells under different experimental conditions appears quite modified during serial subcultures (Fig. 6). As sub-clone 23 always gives constant and reproducible responses, it was chosen for further experiments.

b) Optimization of culture media.

Factorial analysis offers the possibility of testing essentiality of nutriments, establishing interactions and fixing optimal concentrations (13). Relation between cell growth(Y) and variables can be statistically analyzed according to equation:

$$Y = m + b_i x_i + b_{ij} x_i x_j + b_{ii} x_i^2 + e$$

where b_i, b_{ij} represent Linear, interactive and quadratic coefficients.

For example, the resolution of factorial designs of Table 2 gives :

$Y_{immob.}$ =12.3 + 0.8LH <u>-12.3YE</u> + <u>6.4NCS</u> -0.8LHxYE + 0.4LHxNCS -6.4YExNCS

$Y_{stir.}$ =40.7 - 8.2LH -40.7YE + 6.2NCS +8.2LHxYE +15.2LHxNCS - 6.2YExNCS

Significant values have been underlined with p=94.2% and 40.8%,respectively (F test)

c) Results

Experiments on 23 clone confirm the necessity of modifying the high redox potential and organic content of insect culture media, enlightening promoting interactions between amino acids and vitamins. The progressive optimization of medium Bi6020 opens on serum and protein free (SPF) medium Bi6030 (Fig. 7).

Such statements have far-reaching implications. This new technology applies to other non-hemocyte lines. Preference must be given to the Na/K ratio of homologous hemolymphs,except when the diet involves ionic excess <u>in vivo.</u> Serum supplementation can be suppressed for maintenance of Diptera and Lepidoptera cell lines over one year. Culture media Bi6030 are safer, cheaper and easier to use. Since cell adhesion increases in SPF media, cell transfers by scraping slash up to 60% of cells. Adding lipid precursors to these media improves resistance

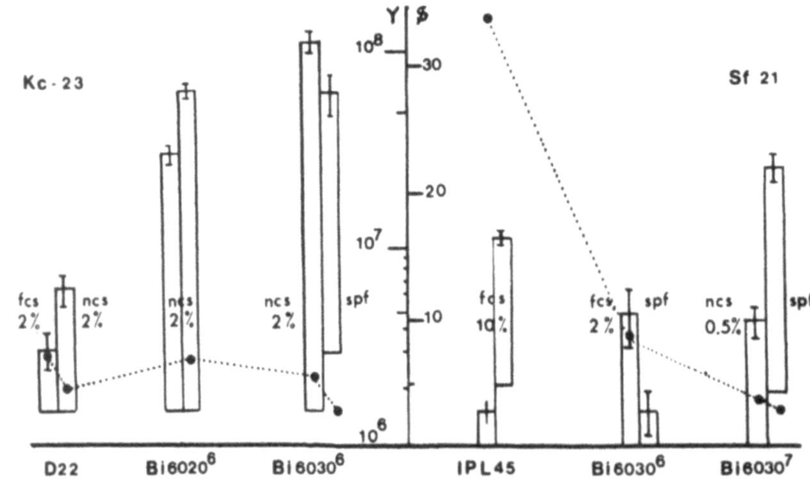

Fig. 7: Efficiency of the new technology. Diagrams represent Y0 and Y7day cell numerations. Cost prices of organic supplementation, Sigma catalog 1987(...o...).

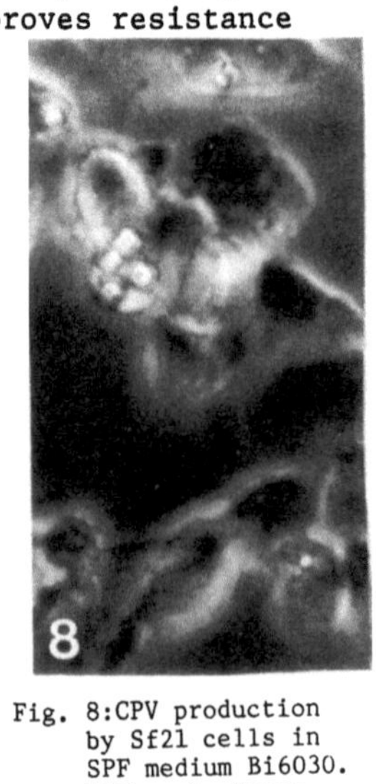

Fig. 8:CPV production by Sf21 cells in SPF medium Bi6030. 15th subculture. x 750.

of cell membranes and still increases cell growth and viability.

It can be concluded that cell and virus (Fig. 8) production in serum and protein free media should be more efficient from immobilized cultures.

References

1. Marks, E. P., Tissue Culture Methods and Applications (Kruse, P. F. and Petterson, M. K. eds.), pp.153-156, Academic Press 1973.
2. Landureau, J. C., Invertebrate Tissue Culture (Kurstak, E. and Maramorosch, K. eds.), pp.101-130, Academic Press 1976.
3. Becker, J. and Landureau, J. C., In Vitro, 17:471-479, 1981.
4. Landureau, J. C. and Toulmond, A., Experientia, 36:966-967, 1982.
5. Mitsuhashi, J., Invertebrate Cell Culture Applieactions (Maramorosch, K. and Mitsuhashi, J. eds.), pp.9-51, Academic Press 1982.
6. Quennedey, A., Quennedey, B., Delbecque, J. P. and Delachambre, J., Cell Tissue Res., 232:493-511. 1983.
7. Szopa, T. M., Lenoir-Rousseauc, J. J., Yuncker, C. and Happ. C. M., Development Biol., 107:325-336, 1985.
8. Quennedey, A. and Lenoir-Rousseaux, J. J., In Proc. 7th Int. Colloq. Invertebrate Tissue Culture, Ohito, 1987.
9. Blais, C. and Lafont, R., Wilhelm Roux's Archives, 188:27-36, 1980.
10. Landureau, J. C. and Philippe, C. In Insect Endocytobiosis (Schwemmler, W., ed.) C. R. C. Press, in press.
11. Biosys, S. A., 21 quai du clos des roses, 60200 Compiegne (France). Commercial Patent ANVAR 86.5104.00.
12. Philippe, C., These de Doctorat, Université P. et M. Curie, Paris, Oct. 1986.
13. Latge, J. P. and Sanglier, J. J., Can. J. Bot., 63:68-85, 1985.
14. Lenoire-Rousseaux, J. J., Gautron, G. and Happ, G. M. Arch. Anat. Micr., 75 : 206, 1986.

ADAPTATION OF AEDES ALBOPICTUS CLONE C6/36 CELLS TO SERUM FREE GROWTH MEDIUM

A. Igarashi

Department of Virology, Institute of Tropical Medicine, Nagasaki University, Sakamoto-machi, Nagasaki 852, Japan

High virus-yielding clone C6/36 of Aedes albopictus cells was adapted to grow with serum-free medium or with medium containing less fetal calf serum (FCS), by gradual replacement of the ordinary medium (10% FCS in Eagle's medium with 0.2 mM each of nonessential medium) by (1) 0.5% lactalbumin hydrolysate in Eagle's medium, or by (2) GIT medium from Daigo Eiyo Co. Ltd. The replacement with GIT medium could be achieved in the relatively short period of a month, compared with several months for 0.5% lactalbumin hydrolysate medium. The latter replacement could not be achieved completely and the cells required the presence of 1 % FCS even in the presence of lactalbumin hydrolysate.

Infection with Japanese encephalitis (JE) and dengue viruses has been prevalent in the Asia monsoon area, resulting in severe manifestations of encephalitis (in the case of JE) or dengue haemorrhagic fever (DHF) in the case of dengue viruses. We have been studying these viruses and virus infections using Aedes albopictus clone of C6/36 cells which was isolated from Singh's Aedes albopictus cell line for its characteristics of high yield of these viruses as well as chikungunya virus. The cells had been adapted to grow in Eagle's medium in Earle's saline supplemented with 0.2 mM each of nonessential amino acids and 10% heat-inactivated fetal calf serum. However, FCS is sometimes rather expensive for wide-scale use of the cell line, especially in the developing countries in Asia. Therefore, attempts were made to adapt the cell line to serum-free medium or to a medium requiring less amount of FCS. The adaptation was carried out by gradually replacing the complete growth medium by (1) GIT medium from Daigo Eiyo Co. Ltd. or by (2) Eagle's medium supplemented with 0.2 mM each of nonessential amino acids and 0.5% of lactalbumin hydrolysate. The adaptation to the GIT medium was achieved in the relatively short period of a month, in contrast to the latter attempt, which required several months and still required the presence of 1 % FCS even in the presence of lactalbumin hydrolysate.

The adapted cell lines as well as the unadapted one were compared for their capacity to support the growth of JE virus and the results indicated that the C6/36 cell line did not lose its high virus yielding character even after the adaptation to GIT medium or the medium with 1% FCS and 0.5% lactalbumin hydrolysate.

Further attempts will be required to adapt the cell line to less expensive medium for more wide-scale use of the cell line in the future.

GROWTH FACTORS IN THE LACTALBUMIN HYDROLYSATE TO THE CELLS OF THE FLESH FLY, SARCOPHAGA PEREGRINA

H. Nagasawa[1)], J. Mitsuhashi[2)], and A. Suzuki[1)]

1) Department of Agricultural Chemistry, Faculty of Agriculture, The University of Tokyo, Bunkyo-ku, Tokyo 113, Japan
2) Forestry and Forest Products Research Institute, Inashiki-gun, Ibaraki 305, Japan

The commercially available lactalbumin hydrolysate had many growth factors to the cells, NIH-SaPe-4, derived from the embryonic tissues of the flesh fly, *Sarcophaga peregrina*. They were water-soluble and heat-stable, but inactivated by acid hydrolysis. They were classified into two groups, Fractions I and II, according to their molecular size. From the behavior of the growth factors on ion-exchange chromatographies, the growth factors in Fraction I were basic, while Fraction II contained basic and acidic growth factors.

Various culture media for insect cells have been developed. Some of them contain chemically undefined components essential for growth such as fetal bovine serum, insect hemolymph or protein hydrolysate. The insect cells, NIH-SaPe-4, derived from the embryonic tissues of the flesh fly, *Sarcophaga peregrina*, can grow in the medium, MTCM-1103 (1). This medium contains commercially available lactalbumin hydrolysate (LAH) as the only, chemically undefined component. To establish a completely chemically defined medium for the cells, we have tried to characterize and purify growth factors from LAH.

Materials and Methods

Cells and Medium The cells used in this experiment were NIH-SaPe-4, originating from the embryonic tissues of *S. peregrina*. This continuous cell line was established by Takahashi et al. (2). It was initially cultured in Mitsuhashi-Maramorsch's (MM) medium, then adapted to the serum free medium (MM-SF) and further to a MTCM-1103 medium (1). The cells were cultured at 25°C and passed every 7-10 days. The doubling time of this cell line in the MTCM-1103 medium was 3-4 days. This cell line consists of at least two types of cells, spherical cells and spindle-shaped cells. The MTCM-1103 medium consists of inorganic salts, glucose, inosine, vitamins and lactalbumin hydrolysate (LAH) (Table 1). The LAH used was an enzymatic digest purchased from Difco.

Bioassay method for growth factor Test samples were lyophilized and dissolved in a test medium, which was modified from MTCM-1103 medium by substituting LAH with a mixture of amino acids (Table 2). The modified medium contained the same amounts of free amino acids as those in MTCM-1103 medium. Then they were sterilized by filtering through a cellulose acetate membrane (0.22 mμ, Corning). The NIH-SaPe-4 cells could not proliferate in this modified medium. A series of 2-5 fold dilutions of the test solution (0.6 ml each) were put separately into wells of 24-well micro test plate (Corning), to each of which 0.075 ml of a cell suspension (ca. 1 x 10^5 cells) was added. Most of the cells were soon attached to the bottom of the plastic

Table 1. Composition of Culture Medium, MTCM-1103.

Component	mg/100 ml
NaCl	700
NaH_2PO_4	20
$NaHCO_3$	12
KCl	20
$CaCl_2 \cdot 2H_2O$	20
$MgCl_2 \cdot 6H_2O$	10
Glucose	500
Inosine	20
Thiamine hydrochloride	0.016
Rivoflavin	0.016
Folic acid	0.016
Biotin	0.008
Calcium pantothenate	0.016
Pyridoxine hydrochloride	0.016
p-Aminobenzoic acid	0.016
Niacin	0.016
i-Inositol	0.016
Choline chloride	0.16
Lactalbumin hydrolysate	1500
pH (with KOH)	6.3

Table 2. Composition of Free Amino Acids in Lactalbumin Hydrolysate (LAH).

Amino acid	mg/1500 mg
Asp	10
Glu	37
Gln	25
Ser	20
Thr	15
Pro	6
Cys_2	2
Gly	3
Ala	30
Tyr	1
Met	15
Val	40
Phe	40
Ile	35
Leu	130
Trp	1
His·HCl	8
Lys·HCl	120
Arg·HCl	35

well. The well was kept at 25°C for 5-10 days, and then the growth stimulating activity was judged by observing the cells under a microscope (x 200). If the sample had a growth stimulating activity, the cells proliferated to form clumps. The activity was expressed by three grades, ++, + and -; ++ means that the sample has nearly the same activity as MTCM-1103 medium, + means that the sample is less active than MTCM-1103 medium, and - means inactive.

Results and Discussion

Proliferation activity of some known compounds Table 3 shows the activity of some known compounds in our bioassay system. MTCM-1103 medium contains as much as 1500 mg of LAH in 100 ml. When it was decreased to one-fourth (375 mg), the cell growth did not change. But when it was decreased to one-eighth (188 mg), the growth became deteriorated, and at the concentration below one-sixteenth the cells failed to proliferate. Casein hydrolysate (Sigma) was also active at nearly the same range of concentration as LAH. However, all the other compounds tested so far were inactive. α-Lactalbumin (Sigma), not digested, was inactive. Bovine insulin (Sigma) which is known to serve as a growth factor to some mammalian cells and 4K-PTTH now called bombyxin which is a silkworm 4 kilodalton prothoracicotropic hormone were also inactive. The latter has the structure resembling insulin and stimulates prothoracic glands to produce ecdysone (3). 20-Hydroxyecdysone did not show any stimulatory effect at the concentration of 0.05-50 mg/100 ml. However, relatively high concentrations of ecdysone (0.3-5 mg/100 ml) seemed to increase the number of spindle-shaped cells. Considering the growth rate of the cells, the spindle-shaped cells seemed to be a transconfiguration of the spherical cells by 20-hydroxy-ecdysone. Thus, only two samples were active, both of which were derived from cow's milk, suggesting that both samples contained certain inherent, possibly common, substance(s) with growth stimulating activity.

Chemical properties of growth factor in LAH The chemical properties of the growth factor in LAH were examined concerning solubility and stability. When LAH was dissolved in water and extracted with ethyl acetate, the activity was recovered

Table 3. Growth Stimulating Activity of Some Known Compounds.

Compound	Activity (mg/100 ml)
Lactalbumin hydrolysate	++ (375-1500); + (188); - (47-94)
Casein hydrolysate	++ (375-1500); + (188); - (47-94)
α-Lactalbumin	- (0.0002-19)
Bovine insulin	- (0.0004-0.4)
Bovine serum albumin	- (19)
Bovine plasma albumin	- (19)
Ovomucoid	- (19)
4K-PTTH*	- (1.3×10^2-1.3×10^5)**
Cholesterol	- (0.2)
20-Hydroxyecdysone	- (0.05-50)

*Partially purified preparation of 4 kilodalton prothoracicotropic hormone from adult heads of Bombyx mori.
**Samia units, corresponding to 0.013-13 μg of the pure hormone.

only in the aqueous layer. Then, this layer was successively extracted with n-butanol. The activity was observed again in the aqueous layer. These results indicated that the growth factor was water-soluble. The growth stimulating activity was maintained even after heating in a boiling water bath for 5 min or after autoclaving at 115°C for 15 min. Any visible precipitate was not formed by these treatments. But the activity was completely lost by acid hydrolysis (6N HCl) at 110°C for 20 hr.

Purification of growth factor   On the basis of the above-mentioned results, we tried to purify the growth factor from LAH. To get information about the molecular weight of the growth factor, LAH was first applied to a gel-filtration column of Sephadex G-50 (Fig. 1). The elution profile monitored by an absorbance at 280 nm indicated that most of the materials in LAH were low molecular weight compounds. However, the growth stimulating activity was observed in two separate fractions,

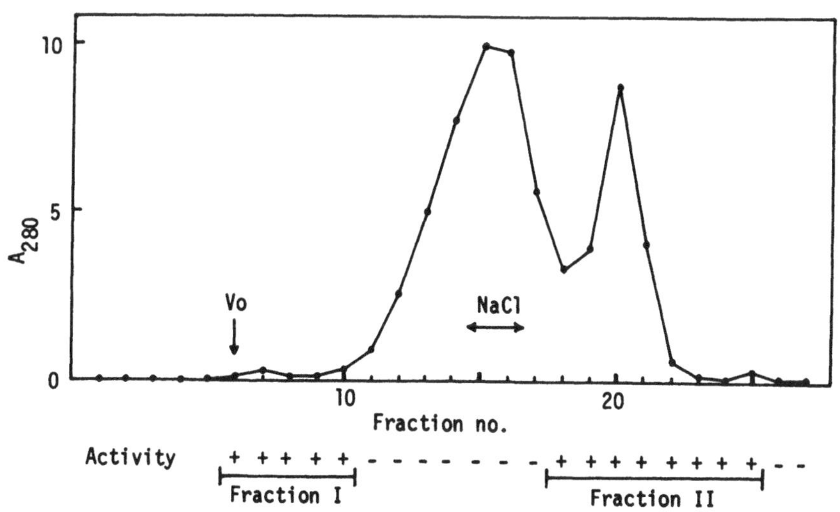

Fig. 1. Gel-filtration of LAH on Sephadex G-50. LAH (1.0 g) was applied to a column of Sephadex G-50 (superfine) (1.2 x 97.0 cm). Elution was done with 0.2 M ammonium acetate containing 2 % n-butanol. Fractions of 200 drops were collected. Vo (No. 6) means the void volume of this column and sodium chloride was eluted at Nos. 15 and 16. Each fraction was bioassayed.

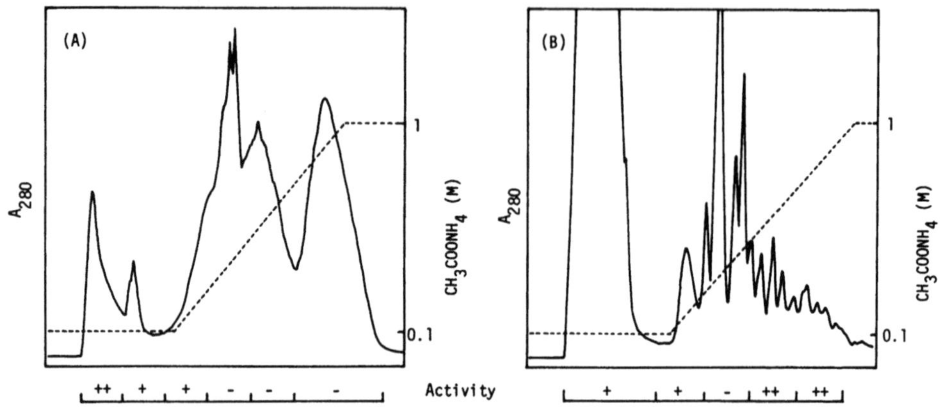

Fig. 2. Anion-exchange high performance liquid chromatography of Fraction I (A) and Fraction II (B) on a TSKgel DEAE-5PW column (7.5 x 75 mm). Dotted lines show the concentration of ammonium acetate (pH 7.0). Several fractions were collected in each HPLC and bioassayed.

Nos. 6-10 and 18-25, which were named Fraction I and Fraction II, respectively. Since both fractions had no clear peak of activity and were extended to a relatively wide range, they were considered to have multiple growth factors. The molecular weight of the compounds in Fraction I was estimated to be 4000-40,000. α-Lactalbumin was eluted mainly at No. 9 from the same column. Considering the molecular weight (14,400) and a lack of activity of α-lactalbumin, the growth factors in Fraction I are not likely to be α-lactalbumin or its enzymatic digests. The presence of active compounds with higher molecular weight than α-lactalbumin suggested that the preparation of lactalbumin before enzymatic digestion was not sufficiently pure. This seems to be one of the reasons why the growth stimulating activity of LAH changes from lot to lot. The molecular weight of the growth factors in Fraction II may be small and they must have some affinity to the Sephadex gel, because they were eluted later than sodium chloride.

Then, various methods for further purification of the growth factors in Fractions I and II were examined: hydrophobic chromatography, reverse phase high performance liquid chromatography (HPLC), gel permeation HPLC and anion- and cation-exchange HPLC. Among them, only ion-exchange HPLC's gave good results in terms of separation and recovery. Fig. 2 shows the elution patterns of Fractions I and II by anion-exchange HPLC on TSKgel DEAE-5PW. Materials were eluted with a linear gradient of 0.1-1.0 M ammonium acetate, because this salt could be removed by lyophilization. The growth factors in Fraction I were not adsorbed to this column, while they were completely adsorbed to TSKgel SP-5PW and eluted with 0.5-0.8 M ammonium acetate (pH 5.0) (data not shown), indicating that they have a basic nature. On the other hand, the growth factors in Fraction II were eluted in several fractions. Similar scattering of activity was also observed in cation-exchange HPLC. These results suggested that Fraction II had many growth factors with different ionic nature. Further purification of these active fractions is now in progress.

References

1. Mitsuhashi, J. Appl. Ent. Zool., 17:575-581, 1982.
2. Takahashi, M., Mitsuhashi, J. and Ohtaki, T. Develop. Growth and Differ., 22:11-19, 1980.
3. Nagasawa, H., Kataoka, H., Isogai, A., Tamura, S., Suzuki, A., Mizoguchi, A., Fujiwara, Y., Suzuki, A., Takahashi, S. and Ishizaki, H. Proc. Natl. Acad. Sci. USA, 83:5840-5843, 1986.

Part 2

PHYSIOLOGY AND ENDOCRINOLOGY OF CELLS *IN VITRO*

JH SECRETING ACTIVITY OF CORPORA ALLATA FROM FOURTH AND FIFTH INSTAR LARVAE OF BOMBYX MORI

M. Nariki, M. Kimura, and T. Ohtaki.

Department of Biology, Faculty of Science, Kanazawa University, 1-1 Marunouchi, Kanazawa 920, Japan

Abstract---The secretion of juvenile hormone (JH) *in vitro* by isolated corpora allata (CA) from 4th and 5th instar of *Bombyx* larvae was determined by JH radioimmunoassay (RIA). JH-I conjugated with human serum albumen was injected into rabbits repeatedly. The antibody obtained possessed a high reactivity to JH-I, but lower reactivity to JH-II and little reactivity to JH-III. To select the favorable medium, CA were cultured in 8 different culture media and yield of JH was determined in each medium. Among them MGM-450 was found to be the best for CA culture.

By these systems, it was shown that a pair of CA isolated soon after 3rd and 4th molting produced about 400 pg and 500 pg of JH-I for 8 h culture, respectively. In the 5th instar, the activity of CA gradually decreased from day 2, but a small surge of JH was observed at day 6, at which secretion of ecdysone for gut purge was started.

Insect metamorphosis is strictly controlled by two hormones: ecdysone and juvenile hormone (JH). In *Bombyx mori*, pupal commitment of the cells is evoked by the absence of JH at the particular period of the last larval instar, although JH-lacking period required for commitment change of imaginal tissues is different among them (1). Therefore, determination of JH-secretory activity of corpora allata (CA) and of JH level in haemolymph at various stages of the development is important to understand physiological mechanisms which are involved in larval-pupal or pupal-adult transformation. Here, we studied change of JH secretion by CA from 3rd larval molting to pupal development of the silkworm using a combination of *in vitro* culture of isolated CA and radioimmunoassay of JH.

Materials and Methods

Larvae of a racial hybrid (J124×C124) of *Bombyx mori* were reared on either fresh mulberry leaves or artificial diet (Nosanko Ltd, Yokohama) at 25+1°C under a 12 h light: 12 h dark photo regime through larval stage. Larvae were staged at each larval molt and at "gut-purge" to synchronize their developmental stage.

Anti-JH antiserum was prepared according to Baehr et al (2) with a minor modification (3). JH-I was hydrolyzed with sodium hydroxide in methanol. The resulting JH acid was then esterized with N'-hydroxysuccinimide in the presence of N-ethyl-N'(3-dimethylaminopropyl)-carbodiimide hydrochloride in dioxane. This ester was coupled with human serum albumin in a mixture of tetrahydrofran and water to produce an immunogen for immunization. Anti-JH antiserum was obtained after 3 challenges by the antigen to a rabbit at 4 week intervals. Fig. 1 shows the titration curve of this antiserum. Ligand was tritiated JH I with a specific activity of 16 Ci/m mol (New England Nuclear, Boston). Each tube contained approximately 10 000 cpm and 50 µl of antisera solution. The antiserum was diluted with 0.1 M phosphate buffer containing 0.1 % BSA and 0.1 % rabbit IgG. Forty percent binding was obtained by dilution of the antiserum with the buffer to 150 volumes. Therefore, in all experiments mentioned below, this dilution of the antiserum was used for radioimmunoassay.

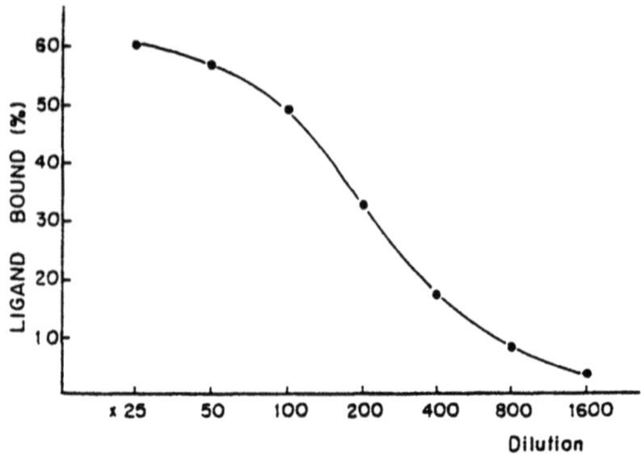

Fig. 1. Per cent binding of ^{3}H-JH to diluted anti-JH antisera

Results and Discussion

First, the specificity of this antiserum to JH I, JH II and JH III was studied. As shown in Fig. 2, the antiserum could recognize JH I very well. The detection limit for JH I was about 50 pg. The cross reactivity to JH II and JH III was 20.4 and 3.2 % of JH I, respectively. Major JH in the haemolymph of Bombyx silkworm was shown to be JH I (3). JH II was a minor component of JH and no JH III was detected in the haemolymph.

In vitro studies on the incorporation of 14C-methyl group of methionine to JH indicated that CA of early 5th instar larvae produced mainly JH I, a small amount of JH II and no JH III (3). Therefore, this antiserum can be satisfactorily used for determination of the JH secretory activity of CA *in vitro*. In the experiments of methionine incorporation to JH, Grace's medium was used for *in vitro* culture of CA. However, JH was poorly produced by the cultured CA in this medium.

Table 1. Amount of JH produced by cultured CA in 10 different media (pmol JH I equivalent/pair of CA)

Grace's medium	---
Grace's medium + 10% F.B.S.*	0.50
Grace's medium + 0.1% B.S.A.**	0.65
MM serum free	0.36
MM + 3% F.B.S.	0.34
MM + 10% F.B.S.	0.51
MGM 448 + 10% F.B.S.	0.91
MGM 450 + 10% F.B.S.	1.43
MTCM 1103 serum free	0.42
Insect Ringer	---

*Fetal bovine serum,**Bovine serum albumin

To examine the medium suitable for CA culture, we tried 10 different media including Grace's medium and insect Ringer's solution. As shown in Table 1, CA did not produce detectable amount of JH in Grace's medium without F.B.S. or insect Ringer's solution. Among the media tested, MGM 450 (4) supplemented with 10% F.B.S. gave the best yield of JH by cultured CA. In this medium, a pair of CA just after the 4th moult produced 1.43 p mole in 8 h incubation. MGM 450 supplemented with 10 % F.B.S., therefore, was used in subsequent experiments.

To study the dynamics of secretory activity, 8 pairs of CA were dissected from newly moulted 5th instar larvae and incubated in 150 µl of the medium. Every 2 h, 100 µl of the medium was taken for JH radioimmunoassay, and 100 ul of fresh medium was added into the culture immediately. Cumulative amounts of JH secretion were calculated from the results of 2 h cultures and plotted as shown in Fig. 3. It is clear that CA produced JH in a linear manner for over 10 h.

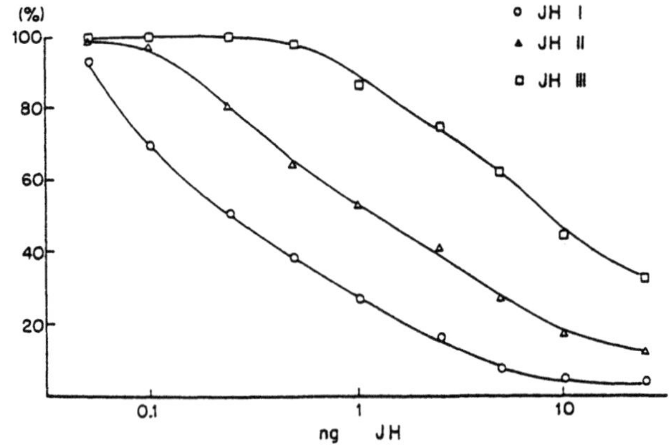

Fig. 2 Cross reactivities of JH I, JH II and JH III against ^{3}H-JH I, expressed as % recovery of radioactivity

Using the same culture conditions, changes in JH-secretory activity of CA were studied through 4th instar to pupation. The activity of JH seretion was high at early phase of 4th instar. Then, the activity gradually decreased and became about half of the initial level on day 3. Soon after the 4th larval moult, CA again recovered a high activity and reached a maximum at day 1 of 5th instar. Then, the activity gradually decreased toward day 4. At the gut-purge stage a small surge of JH from the cultured CA was observed. At the cocooning stage CA produced JH at a minimum rate (Fig. 4).

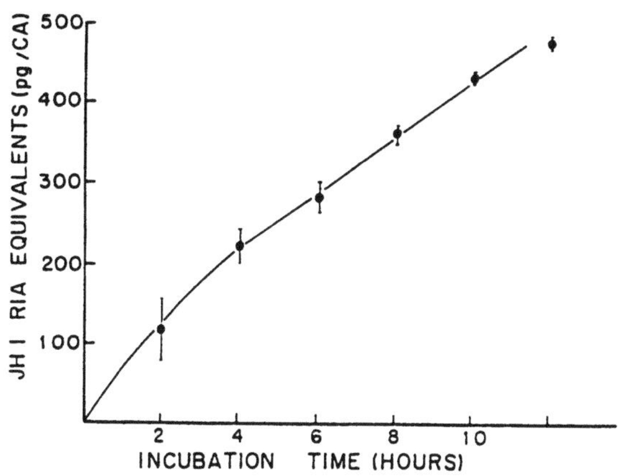

Fig. 3. In vitro production of JH by CA cultured for 2, 4, 6, 8 and 10 h.

During this experimental period, the body weight of the silkworm increases remarkably, starting at ca. 200 mg just after the 3rd moult, 800 mg in newly moulted 5th instar and to more than 4.5 g at day 6 which is the lifetime maximum body weight. In considering the change of body weight, JH secretion per g body weight from the beginning of the 4th instar to the time of pupation would be highest on day 0 of the 4th instar and decrease gradually toward pupation, making a small peak on day 1 and a very small surge on day 6 of the 5th instar.

This estimation is consistent with the change in JH titer in haemolymph determined by bioassay (5). In the bioassay data, however, JH titer in haemolymph decreased to an undetectable level on day 3 of the 5th instar and no increase was observed thereafter. Implantation of the CA dissected from 5th instar larvae on day 3 into allatectomized 4th instar larvae failed to induce larval molt of the host animals (6). Recently, a French group reported JH levels in haemolymph of the silkworm from the end of 3rd instar to pupation determined by another system of radioimmunoassay (7).

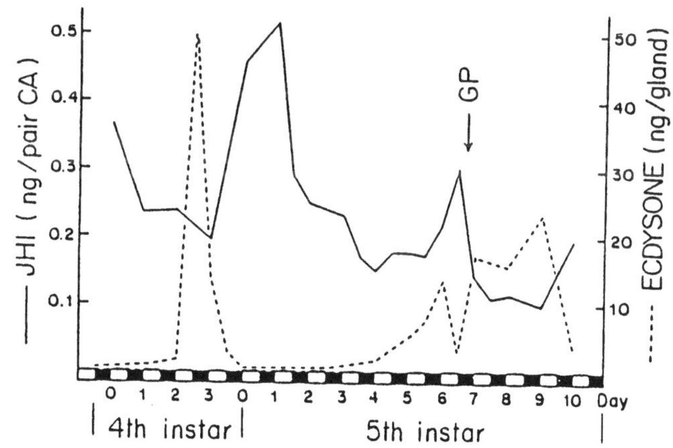

Fig. 4. JH secretion by CA during 4th and 5th instars of B. mori. For comparison, ecdysone secretion by prothoracic gland is shown (11). GP: gut purge

In this report, however, relatively high titer of JH was found at the late 3rd and 4th instars, and there was no clear peak of JH titer in early stage of the 5th instar. This is in contrast to the data of JH-bioassay (5, 8) and also the data of the present experiment. The small peak of JH-secreting activity in gut purge stage is observed in a wide variety of insect species. In Manduca sexta, cultured brain-CC-CA complex produced immunoreactive JHs, determined by antisera which was prepared according to Baehr's method, in the early stage of the 5th instar and the later period from day 5 to 7 with a large peak on day 6 (9). If JH was determined by radiochemical assay using 14C-methyl methionine, or by bioassay, this 2nd peak of JH production could either not be detected or detected as a very small surge. Anti-JH antisera prepared by Baehr's method completely cross-react with JH-acids. Therefore, both JH and JH-acid were measured in the present studies as well as in the report described by Granger et al (9). It was recently reported that the activity of JH esterase in CA became gradually higher toward the pupation in *Galleria mellonella* (10). Therefore, it is highly possible that the second peak of JH-secretion by cultured CA is composed mainly of JH-acid but not JH itself. Further studies to determine JH and JH-acid separately on both culture medium of CA and haemolymph are indispensable to elucidate physiological mechanisms of JH in the course of metamorphosis.

Acknowledgements

The authors are very indebted to Dr. J. Mitsuhashi for the kind supply of the culture media and for his helpful discussions and to Dr. Y. Fujimoto for preparing the antigen.

References

1. Ohtaki, T., F. Yamanaka and S. Sakurai (1986) J. Insect Physiol., 32, 635.
2. Baehr, J. C., J. P. Caruelle, P. Porcheron and P. Cassier (1981) In Juvenile Hormone Biochemistry (eds. Pratt, G. E. and Brooks, G. T.) pp. 47-57. Elsevier North-Holland.
3. Kimura, M., S. Sakurai, T. Nakamachi, Y. Fujimoto, M. Nariki, S. Niimi, N. Ikekawa and T. Ohtaki (1987) in preparation.
4. Mitsuhashi, J. and H. Inoue (1987) Appl. Ent. Zool., (in press).
5. Sakurai, S. (1984) J. Insect Physiol., 30, 657.
6. Fukuda, S. (1944) J. Fac. Sci. Imp. Univ. Tokyo, Sec. IV, 6, 477.
7. Plantevin, G., G. Bosquet, B. Calvez and C. Nardon (1987) Comp. Biochem. Physiol., 86B, 501.
8. Sakurai, S. Personnal communication.
9. Granger, N. A., S. M. Niemiec, L. I. Gilbert and W. E. Bollenbacher (1982) Mol. Cell. Endocrinol., 28, 587.
10. Wisniewski, J. R., M. Muszynska-Pytel, K. Grzelak and M. Kochman (1987) Insect Biochem., 17, 249.
11. Okuda, M., S. Sakurai and T. Ohtaki (1985) J. Insect Physiol., 31, 455.

FEEDBACK-REGULATION OF ECDYSONE C_{20}-HYDROXYLATION IN PRIMARY CELL CULTURES FROM *DROSOPHILA* EMBRYOS

A. Dübendorfer

Institute of Zoology, University of Zürich,
Winterthurerstrasse 190, CH-8057 Zürich, Switzerland

Cells of dissociated post-gastrulation embryos of *Drosophila melanogaster* differentiate *in vitro* into the various larval cell types and, upon treatment with ecdysone, eventually metamorphose to form a variety of adult organs. The application of radiolabelled ecdysone revealed that the primary cell cultures convert ecdysone into 20-hydroxyecdysone, an enzymatic process that is stage-dependent and feedback-regulated in this system. In ecdysone-treated cultures this internal control mechanism guarantees a slow increase in the titre of the active ecdysteroid, 20-hydroxyecdysone, and thus generates optimal conditions for metamorphosis *in vitro*.

When cultivated in appropriate culture media, cells from dissociated early *Drosophila* embryos can complete embryonic and larval differentiation *in vitro* (1, 2). Among the various larval cell types, such as neurons, muscles, haemocytes, fat body cells, chitin secreting cells and tracheal cells there are also populations of undifferentiated, imaginal cells (3) which, upon transplantation into metamorphosing larvae or treatment with ecdysteroid hormones, produce metamorphosed adult organs (4, 5). The differentiation *in vitro* of adult cuticle, furnished with trichomes and bristle organs (4), of adult tubular and fibrillar muscle (6), and of adult fat body (7) have been described in detail.

Metamorphic differentiation in primary cell cultures can be induced by supplementing the culture medium with either ecdysone (1 µg/ml) or 20-hydroxyecdysone (0.1 µg/ml). Even though 20-hydroxyecdysone is known to be biologically more active (8) and actually represents the main ecdysteroid of the *Drosophila* larva (9), differentiated adult fibrillar muscles and bristle-bearing vesicles of adult epidermis were significantly more frequent when ecdysone was used to initiate metamorphosis (6). Best results were obtained when cultures were left to develop for at least 15 days without hormone, then were given ecdysone (2 µg/ml), followed 4 days later by a 50% medium change with hormone-free medium and, after 4 more days, a second application of ecdysone, bringing the hormone concentration back to 2 µg/ml. We also observed that cell viability was markedly better with ecdysone, whereas 20-hydroxyecdysone caused much cell detachement and necrosis. This supposedly paradoxical situation may find its explanation in that the cells, to complete metamorphosis, require a gradually increasing titer of 20-hydroxyecdysone, rather than a sudden shock. After the application of the immediate precursor (ecdysone) such a slow release of 20-hydroxyecdysone would in fact be expected, given the cell cultures provided an active ecdysone-20-monooxygenase system for the controlled conversion of ecdysone into 20-hydroxyecdysone. We tested this hypothesis using primary cell cultures from dissociated *Drosophila* embryos and incubating them with radiolabelled ecdysone under various experimental conditions.

Primary cell cultures were set up from wildtype embryos of *Drosophila melanogaster* by mechanical dissociation after 6 to 8 hours of development at 25°C, as described earlier (5). Each culture started off with about 1.6 x 10^5 cells in a 50 µl column drop of Shields and Sang's *Drosophila* culture medium (10). At given stages of development *in vitro* 30 µl of the culture medium was replaced by medium containing 0.05 µCi of 23,24- [^{3}H]-ecdysone (specific activity 63,5 Ci/mM), either with or without inhibiting agents. After various incubation periods, the ecdysteroids were extracted from the medium and analysed by thin layer chromatography (TLC) and high performance liquid chromatography (HPLC) as previously described (11). Finally, the relative amounts of labelled ecdysone and 20-hydroxyecdysone were determined by liquid scintillation counting.

Ecdysteroid Conversion in Mature Primary Cultures

In 20-day-old primary embryonic cultures the larval cell types are in a state of differentiation that corresponded to the last larval instar, as judged from histochemical criteria (9). When such cultures received a very low amount of tritiated ecdysone (0.05 µCi per culture = 9.0 ng/ml = 2 x 10^{-8} M), most of this material was gradually converted to tritiated 20-hydroxyecdysone, such that after 5 days 80-90% of the radioactivity collected from the ecdysone and 20-hydroxyecdysone fractions was 20-hydroxyecdysone, as shown by the open circles in Figure 1. When related to the total radioactivity recovered from a culture, the 20-hydroxecdysone portion was lower (about 60%) since some other ecdysone metabolites of higher and lower polarity were also produced (7). An incubation period of 25-30 h was necessary for the cultures to complete 50% of their maximal conversion activity.

Feedback Inhibition of Conversion by 20-Hydroxyecdysone

When various doses of non-labelled, exogenous 20-hydroxyecdysone were added to 20-day-old primary embryonic cultures simultaneously with the labelled ecdysone, the C_{20}-hydroxylation of this material was affected in a dose-dependent manner. Figure 1 shows that a concentration of 3,3 x 10^{-8} M 20-hydroxyecdysone (0,015 µg/ml) did not reduce conversion activity whereas a ten-fold higher dose (3,3 x 10^{-7} M = 0.15 µg/ml) clearly did. At about 2 x 10^{-6} M (= 0.75 µg/ml) 20-hydroxyecdysone blocked off most C_{20} - hydroxylation of ecdysone which shows that the process is endproduct-inhibited.

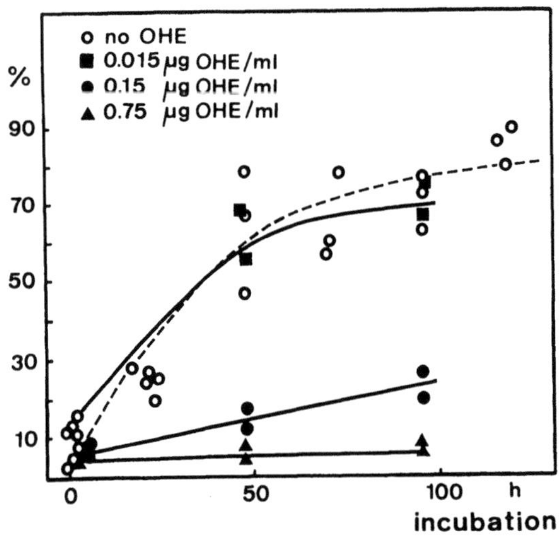

Figure 1. Relative amounts of radiolabelled 20-hydroxyecdysone (ordinate) extractable from 20-day-old embryonic primary cultures after various incubation periods (abscissa) with 0.05 µCi [^{3}H] -ecdysone (open circles, broken line). The same experiment carried out with various doses of unlabelled 20-hydroxyedysone (OHE) present in the incubation medium (dark symbols, solid lines) shows the effect of 20-hydroxyecdysone on ecdysone-C_{20}-hydroxylation in the cell cultures. Each point in the graph represents one culture.

Specific Inhibition of the Steroid Hydroxylase System

A number of inhibitors of cytochrome P-450 dependent steroid hydroxylase systems in vertebrates have been shown to inhibit ecdysone-C_{20}-hydroxylation in insects as well (12, 13). We have used one of these inhibitors, metyrapone (a gift from Dr. W.E. Bollenbacher of the University of North Carolina, Chapel Hill), which prevented any metamorphic development in our system of embryonic cells when applied at a concentration of 10^{-4} M and above. When various doses of metyrapone were added to cultures containing [^{3}H]-ecdysone we found that conversion to [^{3}H]-20-hydroxyecdysone was effectively inhibited by the drug. A concentration of 10^{-4} M was sufficient for 50% inhibition of the hydroxylation activity (Figure 2). This result suggests that an active ecdysone-20-monooxygenase system is a prerequisite for cultured embryonic cells to metamorphose when ecdysone is introduced into the system. Furthermore, it shows that metamorphic events are triggered by 20-hydroxyecdysone rather than ecdysone.

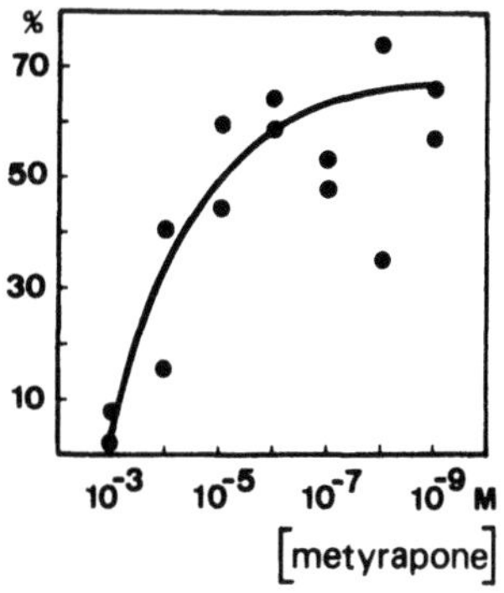

Figure 2. Relative amounts of radiolabelled 20-hydroxyecdysone (ordinate) extractable from 20-day-old embryonic primary cultures incubated with 0.05 µCi [^{3}H]-ecdysone for 72 hours in the presence of various concentrations of metyrapone (abcissa). Each point in the graph represents one culture.

Dose Requirement of 20-Hydroxyecdysone for Metamorphosis in vitro

Primary cultures of embryonic cells that are competent to react to ecdysteroids require a dose of 1-2 µg ecdysone per ml medium for successful induction of metamorphosis (4, 6). Since it is not ecdysone, but its 20-hydroxylated derivative that induces metamorphosis, we wondered how much of the originally added ecdysone was in fact converted to 20-hydroxecdysone in the metamorphosing cultures. We therefore diluted the radiolabelled ecdysone with non-labelled ecdysone, in various ratios, such that each culture received 0.1 µCi of [^{3}H]-ecdysone and a varying surplus of nonlabelled ecdysone. The relative amount of 20-hydroxyecdysone found in a culture after the incubation period of 72 hours was then not just representative for the converted [H^3]-ecdysone, but rather indicated which percentage of the total added ecdysone (labelled and non-labelled) could be expected to be C_{20}-hydroxylated. This value, reduced by a factor of 0.25 to account for ecdysone metabolites other than 20-hydroxyecdysone, could then provide an approximate actual concentration of 20-hydroxyecdysone in metamorphosing cultures.

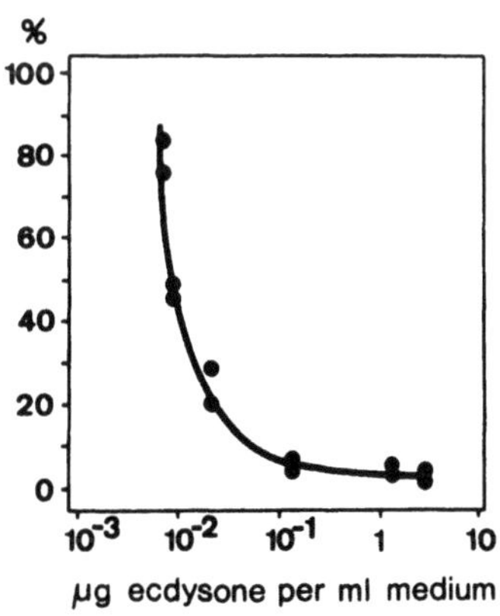

Figure 3. Relative amounts of radiolabelled 20-hydroxyecdysone (ordinate) extractable from 20-day-old embryonic cultures incubated with 0.05 µCi [^{3}H]-ecdysone plus various amounts of non-labelled ecdysone for 96 h. The abscissa represents the total concentration of ecdysone (labelled and non-labelled) at the beginning of the incubation period. Each point in the graph represents one culture.

Figure 3 shows the interdependence of ecdysone concentration and C_{20}-hydroxylation. A clear dosage relation can be seen between 10^{-1} and 10^{-2} µg/ml ecdysone, indicating that in each experiment the titre of 20-hydroxyecdysone did not exceed an estimated 0.005 µg/ml (= 10^{-8} M). This seems to indicate that the actual dose of 20-hydroxyecdysone required for inhibiting further ecdysone-20-monooxygenase activity is about 30 times lower than estimated from the inhibition experiments with exogenous 20-hydroxyecdysone. It also means that the dose of 20-hydroxyecdysone necessary for the induction of metamorphic development at the cellular level is considerably lower than generally concluded from experiments that used exogenous 20-hydroxyecdysone for the induction of metamorphosis in insect tissues *in vitro*.

Conclusions

The results have shown that the cells in primary cultures from dissociated *Drosophila* embryos are equipped with a functional ecdysone-20-monooxygenase system and thus can convert ecdysone into 20-hydroxyecdysone. 20-hydroxyecdysone inhibits enzyme activity and thus operates as a feedback-regulator. The characteristics of enzyme activity and its inhibition by 20-hydroxyecdysone and other inhibitors are very similar to what has been found in wandering stage larvae of *Drosophila* (14) and in other insects (cf. 15, 16, 17). The acquisition of competence to convert ecdysone shows a close temporal correlation with the differentiation of fat body cells in primary cell cultures (7) which suggests that much of the ecdysone-20-monooxygenase activity comes from these cells. Fat body from last instar larvae has indeed been found to contain high monooxygenase activity (7, 14). The conversion data from the experiments with increasing ecdysone concentration (Figure 3) suggest that on a cellular level feedback- inhibition operates at lower concentrations of 20-hydroxyecdysone than judged from inhibition experiments with exogenous 20-hydroxyecdysone, where some of the activity is always lost by degradation and conjugation (11, 18).

We conclude that the cell cultures grown from dissociated *Drosophila* embryos produce their own 20-hydroxyecdysone from exogenously supplied ecdysone, and that this results in a slow release of 20-hydroxyecdysone rather than a sudden hormone pulse. By feedback regulation the actual titre of 20-hydroxyecdysone is not only kept at a low level of about 10^{-8} M, but it is also maintained at this level for several days. These conditions appear to be optimal for metamorphic development.

Acknowledgements

This work was supported by the Swiss National Science Foundation, grant no. 3.640-0.84. I wish to thank Johanna Nägeli for technical assistance.

References

1. Echalier, G. and Ohanessian, A. In Vitro, 6: 162-172, 1970.
2. Shields, G. and Sang, J.H. J. Embryol. exp. Morph., 23: 53-69, 1970.
3. Shields, G., Dübendorfer, A. and Sang, J.H. J. Embryol. exp. Morph., 33: 159-175, 1975.
4. Dübendorfer, A., Shields, G. and Sang, J.H. J. Embryol. exp. Morph., 33: 487-498, 1975.
5. Dübendorfer, A. and Eichenberger-Glinz, S. Invertebrate Systems In Vitro (Kurstak, E., Maramorosch, K. and Dübendorfer, A., eds.), pp. 169-185, Elsevier Biomedical, 1980.
6. Dübendorfer, A., Blumer, A. and Deak, I.I. Wilhelm Roux' Archives, 184: 233-249, 1978.
7. Dübendorfer, A. and Eichenberger, S. Metamorphosis (M. Balls and M. Bownes, eds.), pp. 145-161, Oxford University Press, 1985.
8. Richards, G. J. Insect Physiol., 24: 329-335, 1978.
9. Borst, D.W., Bollenbacher, W.E., O'Connor, J.D., King, D.S. and Fristrom, J.W. Develop. Biol., 39: 308-316, 1974.
10. Shields, G. and Sang, J.H. Dros. Info. Service, 52: 161, 1977.
11. Dübendorfer, A. Insect Biochem., 16: 645-651, 1986.
12. Smith, S.L., Bollenbacher, W.E., Cooper, D.Y., Schleyer, H., Wielgus, J.J. and Gilbert, L.I. Molec. Cell. Endocr., 15: 111-133, 1979.
13. Smith, S.L. and Mitchell, M.J. Insect Biochem., 16: 49-55, 1986.
14. Mitchell, M.J. and Smith, S.L. Insect Biochem., 16: 525-537, 1986.
15. Svoboda, J.A., Kaplanis, J.N., Robbins, W.E. and Thompson, M.J. Ann. Rev. Entomol., 20: 205-220, 1975.
16. Koolman, J. Insect Biochem., 12: 225-250, 1982.
17. Weirich, G.F., Svoboda, J.A. and Thompson, M.J. Biosynthesis, Metabolism and Mode of Action of Invertebrate Hormone (Hoffmann, J. and Porchet, M., eds.), pp. 227-233, Springer, 1984.
18. Dübendorfer, A. and Maroy, P. Insect Biochem., 16: 109-113, 1986.

1731, A NEW RETROTRANSPOSON WITH HORMONE MODULATED EXPRESSION

F. Peronnet[1], J.L. Becker[1], J. Becker[1], A. Lescault[1], L. D'Auriol[2], F. Galibert[2], and M. Best-Belpomme[1]

1) U.A. CNRS 1135, Université Pierre et Marie Curie
7, quai Saint-Bernard 75005 PARIS, France

2) Centre G. Hayem, Hôpital Saint-Louis
2, place du Docteur Fournier 75010 PARIS, France

Permanent cultured cells of Drosophila melanogaster have been shown to be responsive to a steroid hormone, the 20-hydroxyecdysone. Screening for hormone modulated genes, we isolated and sequenced a new transposon named 1731. This element is a 4.6 kb sequence framed by direct long terminal repeats. Its characteristics lead to classify 1731 in the retrotransposon families and to propose a description of its intracellular life cycle.

I. Drosophila Cells are Ecdysone Target Cell : A Survey

Drosophila melanogaster cells cultured in vitro were shown to be 20-hydroxyecdysone (ecdysterone) responsive cells. This property to respond to a steroid hormone which is known to play a key role in the cellular differentiation and the development of insects was demonstrated as follows. When physiological concentrations of ecdysterone were added to the growth medium of these cells, a wide variety of responses was observed.

Indeed, the steroid induces morphological modifications of the cells, the arrest of their divisions, the variation of the rate of synthesis of several polypeptides, the increase of a restricted number of enzymatic activities, i.e., acetylcholinesterase, β-galactosidase, protein kinase, catalase and superoxide dismutase. All these responses to the hormonal stimulus were accompanied by the enhancement of the oxygen consumption of the cells.

Concerning the mechanism of action of ecdysterone, it was observed that the cells did not metabolize the added hormone. This strongly suggests that the active steroid into the cells was actually the ecdysterone itself. Moreover, it was demonstrated that inside these cells exists an ecdysterone receptor which binds the hormone. The presence of this receptor seems to be required to get hormonal responses since it is detectable in the responsive cells whereas it is not detectable in the non responsive cells. Finally, a series of arguments indicates that the level of action of the ecdysterone receptor-complex is the chromosomes where it likely modulates the rate of the transcription of a restricted number of genes which consequently can be referred to as ecdysterone regulated genes.

Since it was clear that Drosophila melanogaster cells were ecdysterone responsive (i.e. target) cells as summarized above (for reviews see references 1-5), it could be concluded that these cells were an adequate biological material for searching for ecdysterone regulated genes. For that, poly A^{+}RNAs were extracted from control cells or from ecdysterone treated cells and after labelling used for probing a library of genomic DNA of Drosophila melanogaster embryos. Two categories of genes were isolated (6). The first category contained the genes referred to as "inducible" genes since their transcripts increased after the addition of the hormone. The second category gathered genes referred to as "repressed" genes since their transcripts decreased after the ecdysterone stimulation. Among the genes of the second category, we isolated, characterized and sequenced a new retrotransposon which we called 1731.

II. 1731, A New Retrotransposon with Hormone Modulated Expression

According to the terminology of Boecke et al. (7) a retrotransposon is a genetic transposable element (transposon) which shares many characteristics with the retroviruses. In other terms, the retrotransposons are "crossroads" where the interests of transposons and of retroviruses meet. Taking 1731 as an example, we will observe the similarities between this transposon and the retroviruses.

1. 1731 is a Transposon.

Using the Southern blots analysis, we found that the localizations of 1731 varied even if the comparisons were made between two isogenic strains of fly. This observation indicates that 1731 can changes its loci, in other words, can transpose (8).

2. Nucleotidic Sequence of 1731.

1731 is 4654 bp long. It is surrounded by two direct repeats of 5 bp likely due to a short duplication of the genome which accompanies its insertion. 1731 contains and is limited by two long terminal repeats (L.T.Rs) of 336 bp (8). Each LTR begins by TG and ends by CA (which are parts of two short inverted repeats) and contains a "TATA box", a signal of polyadenylation and a polyadenylation site. Thus, it can be divided in three portions, U3-R-U5, recalling the LTR architecture of the retroviruses. It is of interest to note that the U3 part of the 1731 LTR contains an octonucleotide identical to the one known to play a fundamental role in the SV40 enhancer. This might imply that the 1731 LTRs can act on the rate of the transcription of flanking sequences even if they are not near. This effect at a distance could have important consequences if the influenced flanking gene(s) play(s) a role, for instance, in the differentiation or in the rate of the mitosis of the cells.

The 1731 LTRs contain also one hexanucleotide identical and three hexanucleotides practically identical (8) to the hexanucleotide part of the regions of the Mouse Mammary Tumor Virus LTR which binds the glucocorticoid receptor. Our hypothesis is that these sequences can bind the ecdysone receptor. Just downstream of the 5' LTR we found a sequence beginning by TGG which could bind a tRNA primer and just upstream the 3' LTR a purine rich region (8).

Inside 1731 we located two open reading frames (ORF). The ORF1 downstream of the 5' LTR should be for a gag-like protein. Indeed, it presents homologies with, for instance, a gag protein of HIV/ARV2. The ORF2 (upstream of the 3' LTR) should code for a polyprotein. This polyprotein can be divided in three non overlapping domains corresponding to an endonuclease, to a RNAse-H and to a reverse transcriptase. A third ORF evoking the env gene of the retroviruses does not exist.

3. The Transcription of 1731.

The transcription of 1731 generates a major poly A^+RNA of 4.6 Kb in length. The poly A^-RNAs are not detected. The transcription proceeds in only one direction, from the 5' LTR to the 3' LTR. The Northern blots show that the level of the 1731 transcripts decreases in the ecdysterone treated cells.

4. The 1731 Virus-like Particles.

It was demonstrated that a *Drosophila melanogaster* cell line contains intracellular Virus-like Particles (VLPs) associated with a "reverse transcriptase-like activity" and with the transcripts of copia (9). We confirmed the existence of these VLPs in our cells. Moreover, we showed that the reverse transcriptase activity associated with these particles had all the enzymatic characteristics of the retroviral reverse transcriptase (as an example, the activity was inhibited by a low concentration of HPA 23). Using this "true" reverse transcriptase attached to the VLPs we obtained a labelled DNA complementary (cDNA) to the inner RNAs, which hybridized with cloned copia and 1731. For this reason, among others, we concluded that some VLPs contain the transcripts of 1731. Analyzing these VLPs in term of enzymatic activities we observed that three enzymes were attached to them : not only a reverse transcriptase activity but also an endonuclease and an RNAse-H like one. This result fits well with the assignment of the three enzymatic domains of the polyprotein coded by the 1731 ORF2.

5. Extrachromosomic 1731.

The genomic repeated sequences of 1731 mentioned above are, of course, integrated ones. But we characterized also non integrated extrachromosomic forms of 1731. These are double stranded circular DNA of 4.3 or 4.6 Kb in length after linearization. Knowing that each LTR is 0.3 Kb long we conclude that the circular extrachromosomic forms of 1731 contained likely one or two LTRs.

6. The Intracellular Life cycle of 1731.

Two arguments indicate that 1731 remains in the cells : reverse transcriptase activity is not detectable in the growth medium and 1731 has no env-like gene. Its intracellular life cycle could thus be the following one. The poly A^{+}RNAs are translated into a gag like protein and a polyprotein containing the three enzymatic domains defined before. These proteins and the other 1731 transcripts are used to build the 1731 VLPs. The reverse transcriptase activity copies the RNAs into a linear double stranded DNA. The circularization of this last form generates the circular extrachromosomic forms which can be integrated into the genome by using an ("integrase") endonuclease activity. If 1731 frequently followed this life cycle, the probability of mutations by its insertion in a structural gene or by disturbing the rate of transcription of flanking sequences would exponentially increase. Obviously, that is not the case. We are thus led to postulate a very strict control which could be, for instance, an inhibitor of the reverse transcriptase activity. Reciprocally we are also led to speculate on the possibility that this control is outflanked in rare circumstances. Another possibility which has never been observed, perhaps because it is a very rare and practically undetectable event, is that retrotransposons carry an onc gene after "illegitimate recombination" as the retroviruses do (10).

References

1. Best-Belpomme, M., Courgeon, A.M. and Echalier, G. Progress in Ecdysone Research (Hoffmann, J.A., ed.), pp. 379-392, Elsevier/North-Holland Biomedical Press. 1980.
2. Couderc, J.L., Sobrier, M.L., Giraud, G., Micard, D., Dastague, B., Ropp, M., Becker, J.L., Maisonhaute, C., Peronnet, F., Courgeon, A.M. and Best-Belpomme, M. Biosynthesis, Metabolism and Mode of Action of Invertebrate Hormones (Hoffmann, J. and Porchet, M., eds), pp. 293-304, Springer-Verlag. 1984.
3. Cherbas, L., Fristrom, J.W. and O'Connor, J.D. Biosynthesis, Metabolism and Mode of Action of Invertebrate Hormones (Hoffmann, J. and Porchet, M., eds.), pp. 305-322, Springer-Verlag. 1984.
4. Morganelli, C.M. and Berger, E.M. Ecdysone from Metabolism to Regulation of Gene Expression (Bownes, M., ed.), pp. 233-240, Pergamon Press. 1986.
5. Peronnet, F., Ropp, M., Rollet, E., Becker, J.L., Becker, J., Maisonhaute, C., Pernodet, J.L., Vuillaume, M., Courgeon, A.M., Echalier, G. and Best-Belpomme, M. Techniques in In Vitro Invertebrate Hormones and Genes (Kurstak, E., ed.), pp. C206 1-20, Elsevier Scientific Publishers Ireland Ltd. 1986.
6. Couderc, J.L., Becker, J.L., Sobrier, M.L., Dastugue, B., Best-Belpomme, M., Lepesant, J.A. and Pardue, M.L. Chromosoma (Berlin), 89:338-342, 1984.
7. Boecke, J.D., Garfinkel, D.J., Styles, A. and Fink, G.R. Cell, 40:491-500, 1985.
8. Peronnet, F., Becker, J.L., d'Auriol, L., Galibert, F. and Best-Belpomme, M. Nucl. Acids Res., 14:9017-9033, 1986.
9. Shiba, T. and Saigo, K. Nature, 302:119-124, 1983.
10. Duesberg, P.H. Proc. Natl.Acad. Sci. USA, 84:2117-2124, 1987.

NMR SPECTROSCOPIC STUDY OF NEMATODE: CARBOHYDRATE METABOLISM IN CULTURED EGGS OF A. CANTONENSIS

M. Takahashi, M. Nishina[1)], E. Hori[1)], K. Matsushita[2)], K. Kato[3)], and A. Ohsaka[4)]

Department of Medical Entomology, [3]Department of Enteroviruses and [4]Department of Applied Immunology, National Institute of Health, Kamiosaki, Shinagawa-ku, Tokyo 141 [1]Department of Parasitology, Saitama Medical School, Moroyama, Iruma-gun, Saitama 350 and [2]NMR Application Laboratory, JEOL, Ltd., Akishima, Tokyo 196, Japan

^{1}H-NMR and ^{13}C-nuclear magnetic resonance (NMR) spectroscopy was used to study the aerobic glucose metabolism in eggs of *Angiostrongylus cantonensis* in a NCTC-109 medium supplemented with calf serum. Without performing any pretreatment of the spent medium, we were able to identify and quantitate, by means of NMR, the endproducts of glucose metabolism in eggs after cultivation for 2, 4, and 8 days. We demonstrated that *A. cantonensis* eggs took up glucose rapidly; among the major endproducts were found lactic acid, acetic acid and alanine.

This nematode is known as a neurotropic parasite that requires a period of development in the central nervous system of the final host, and eventually causes eosinophilic meningoencephalitis in man. Little information is available, however, on the nutritional requirements, morphology or development of this egg (1). We have investigated the development of this embryo and the changes in concentration of amino acids and glucose in the in vitro cultivation(2) and have applied ^{1}H- and ^{13}C-NMR spectroscopy to a variety of pathogens and have demonstrated that the metabolites present in suspension of intact pathogens can be identified and quantitated in a noninvasive fashion. In this presentation we report our results, demonstrating that *A. cantonensis* eggs took up glucose rapidly through the eggshell from the surrounding medium to produce various metabolites, that is, lactate, acetate, succinate and alanine.

Materials and Methods

Life cycle: The adult males and females live in the pulmonary arteries and the right side of heart of rodent hosts. Various species of molluscs serve as intermediate hosts. The first-stage larvae infect the molluscan intermediate host either by direct penetration or ingestion. The rodent final hosts acquire infection through the oral route, the larvae enter the circulation and the majority of the infective third-stage larvae reach the brain in one or two days.

Collection of adult worms and eggs: *A. cantonensis* was maintained in aquatic snails, *Biomphalaria glabrate* and albino rats. For infection,rats were given third-stage larvae orally and the adult worms were recovered from the rats that had discharged the first-stage larvae in their feces 45 days post-infection. The arteries of the rats were placed in petri dishes of 35 mm diameter and washed twice with physiological saline.

Cultivation procedure: The eggs in the uterus of the five female worms were released with forceps into the physiological saline in a petri dish. The size of eggs was 43 μm in width and 76 μm in length. The eggs were washed three times with culture medium. About 40,000 eggs in a petri dish were given 1 ml of NCTC-109 culture medium supplemented with 50% fetal calf serum. These were maintained at 37°C in a humidified atmosphere of 5% CO_2 in air for 8 days without medium change. Cultivation experiments were carried out in triplicate and control cultures without eggs were run in parallel with test cultures. Under our culture conditions the hatchability of the eggs was approximately 10%. This result is quite similar to the results of Uga and Matsumura (1).

NMR spectroscopy: ^{1}H(270 MHz) and ^{13}C(67.7 MHz)-NMR spectra of the culture media with and without eggs were obtained from a JEOL JNM-GX 270 spectrometer operated in the Fourier transform mode (sample tube: 5 mmϕ). To suppress the proton signal of water (H_2O), the water protons were irradiated in the gated homonuclear decoupling mode. 45°-pulse (length: 7.8 μsec) was applied 64 times with a recycle of 7 sec, and 33 K data points were used for the spectrum width of 5.4 KH_2. Four-hundred microliters of the sample and 40 μl of D_2O were placed in a 5 mmϕ NMR tube, D_2O being used as an internal lock of the magnetic field. As an internal standard for chemical shift and signal intensity. TSP[3-(trimethylsilyl)propionic acid] was added to the sample at a final concentration of 1.0 mM.

Results and Discussion

Figure 1 (a-d) shows 270 MHz ^{1}H-NMR spectra of the medium (a) before cultivation, and (b) after cultivation with *A. cantonensis* eggs at 37°C for 2 days, (c) 4 days and (d) 8 days. The spectra were obtained by irradiating water protons (around 5 ppm) in a gated homonuclear decoupling mode (3,4). The region between 4 and 6 ppm from the signal of a reference material, TSP, was distorted by the irradiation and, therefore, was omitted from the figures (3,4). The spectra shown in Fig. 1 indicate the presence of various species of molecules. The relatively broad signals having appeared at around 0.9-1.0 and 2.0 ppm were attributed to the methyl protons of proteins and lipids. The two doublets (δ:1.32 and δ:1.47) were assigned to the methyl protons of lactic acid and alanine, respectively (3,4). The singlet observed at δ:1.91 was due to the methyl protons of acetic acid (3,4). A small amount of succinic acid was also observable at δ:2.34 (3,4). Table 1 summarizes the changes in the major metabolites produced by the eggs in the culture medium during the whole cultivation period. The concentration of each metabolite was

Table 1 Change in the contents of metabolites in the medium during cultivation with A.cantonensis eggs

	Days after cultivation			
	0	2	4	8
Lactic acid	22.4*±4.1	46.6±8.1[a]	59.9±12.2[a]	68.5±2.1[a]
Alanine	2.4±0.2	4.9±0.8[a]	6.7±0.5[a]	9.2±2.3[a]
Acetic acid	0.9±0.2	4.4±0.3[a]	5.1±2.2[a]	7.5±0.5[a]
Succinic acid	1.3±0.3	1.1±0.2	1.8±1.3	2.0±0.1[a]

*:Concentration of metabolite in mM(Mean ± SD).

a:Significantly different at P=0.1(df:3)from the value at day 0.

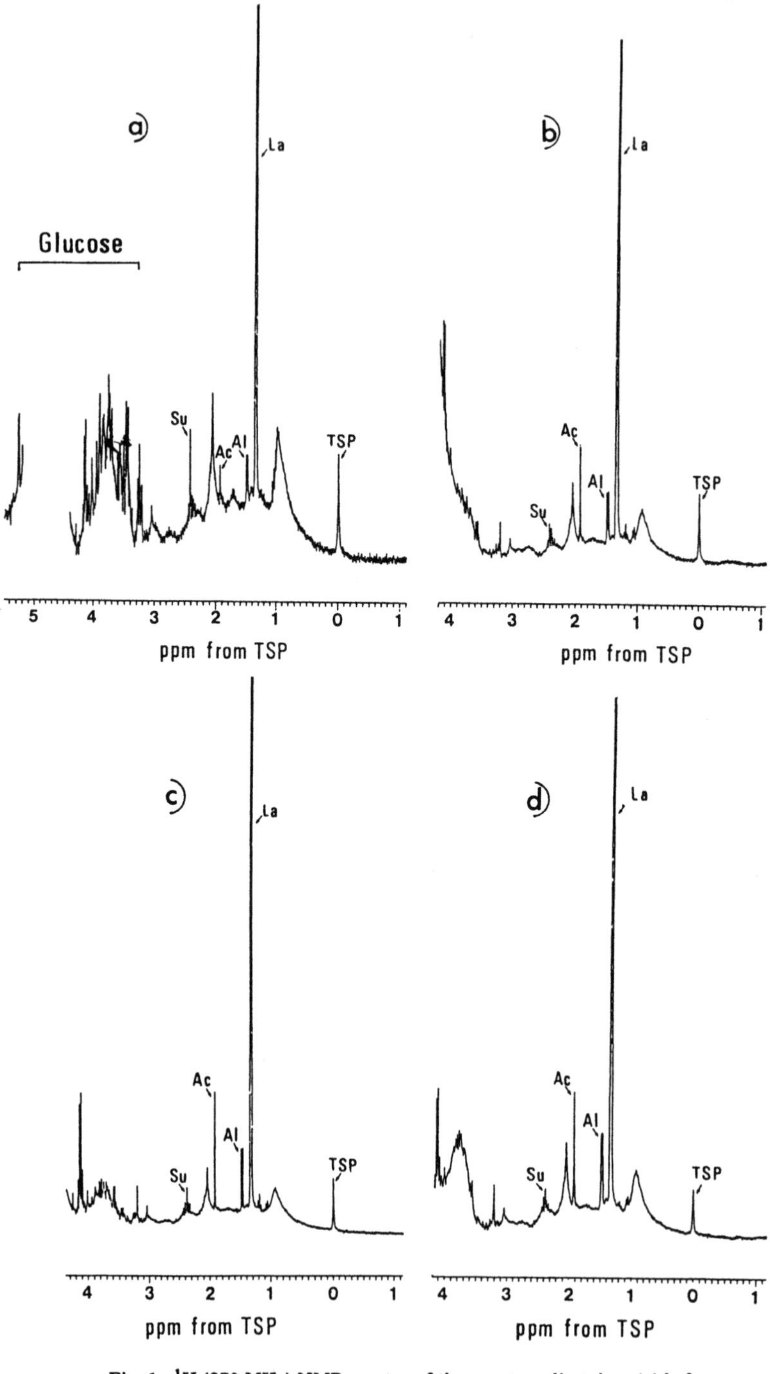

Fig. 1. ^{1}H (270 MHz)-NMR spectra of the spent media taken (a) before cultivation of *A. cantonensis* eggs, and (b) after cultivation at 37℃ for 2 days, (c) 4 days and (d) 8 days. The spectra were taken in 5-mmϕ sample tubes at room temperature. Chemical shift is shown in ppm from internal TSP [3-(trimethylsilyl)propionic acid]. La: lactic acid, Al: alanine, Ac: acetic acid, Su: succinic acid.

determined from the signal intensity by comparing it with a known amount of TSP added to the culture medium as an internal standard for signal intensity. A marked increase in the intensity of the methyl signal of lactic acid was observed after cultivation. It is also noted that the amount of acetic acid and alanine increased after cultivation, while succinic acid concentration increased only slightly. The proton signals of glucose disappeared from the culture medium after cultivation for 2 days (see Figs. 1a and 1b). This was confirmed by ^{13}C-NMR (data not shown).

In summary, we have applied an NMR technology for the study of carbohydrate metabolism of *A. cantonensis* (6). The eggs of this worm were found to take up glucose through eggshell from the surrounding medium. The metabolites were found lactate, acetate and alanine. Greater change in the content of these metabolites and that of amino acid was observed in the first four days than last four days of cultivation period. These changes should be responsible for early development of these embryos. From all these results and discussions, it is apparent that the eggs of this nematode are parasitic in a sense that the energy metabolism in them is mainly dependent upon the energy source present in outer medium. Systematic application of a similar technique to parasitic worms are now in progress in our laboratory.

References

1. Uga, S. and Matsumura, T.: Japan. J. Parasitol., 31:59-66, 1982.
2. Takahashi, M., Nishina, M. and Hori, E.: J. Saitama Med. Sch., 12:261-265 (text in Japanese), 1985.
3. Uga, S. and Matsumura, T.: Japan. J. Parasitol., 33:257-264, 1984.
4. Ohsaka, A., Yoshikawa, K., Yamaya, S. and Sugawara, T.: Physiol. Chem. Phys., 13:159-164, 1981.
5. Matsushita, K., Yoshikawa, K. and Ohsaka, A.: JEOL New Analytical Instrumentation, 18A:54-56, 1982.
6. Takahashi, M., Nishina, M., Hori, E., Matsushita, K., Kato, K. and Ohsaka, A.: Japan. J. Med. Sci. Biol., 40:27-33, 1987.

RIBOSOME METABOLISM IN MOSQUITO CELLS

A. M. Fallon

Department of Molecular Genetics and Microbiology, University
of Medicine and Dentistry of New Jersey,
School of Osteopathic Medicine,
Piscataway, NJ 08854-5635, U. S. A.

Our research interests focus on the somatic cell genetics of insect cells, and on the potential advantages of using cultured cells to work out molecular approaches that can be applied to the intact insect. We have initiated a multidisciplinary approach to studying ribosome metabolism in mosquito cells, and in mosquito fat body after a blood meal. To facilitate these studies, we have obtained cDNA clones corresponding to mosquito ribosomal proteins L8, L14 and L31, and we have also established conditions for assaying the expression of cloned DNA in cultured mosquito cells.

The contribution of insect systems to developmental biology and genetics is best exemplified by the vast amount of information that has been accumulated using the fruit fly, *Drosophila melanogaster*. Biological features of this organism that have facilitated laboratory investigation include its short life cycle, simple dietary requirements, polytene salivary gland chromosomes that are readily amenable to cytological procedures, and a small genome size. These properties have contributed to the isolation and characterization of a vast repertoire of mutant flies, initially using classical genetic approaches. More recently, the development of techniques for transforming *Drosophila* cells in culture, and for transforming *Drosophila* embryos using vectors based on the transposable P element, has provided powerful new approaches to carrying out fundamental molecular investigations using this insect. Clearly, the application of the techniques and approaches that have been successful with *Drosophila* to insect species of medical or agricultural importance constitutes an important challenge to insect physiologists and biochemists.

Our own work focuses on the mosquito, a disease vector whose impact on world health has increased significantly in recent years due to the appearance of insecticide-resistant strains (1). Our approach has been to use cultured cells as an important adjunct to physiological studies at the level of the intact organism. A brief survey of recent developments in our laboratory is described below; more comprehensive reviews of genetic and biochemical studies that have been carried out using mosquito cells in culture include those of Kurtti and Munderloh (2) and Fallon and Stollar (3).

The *Aedes albopictus* cells used in our laboratory were derived from minced neonate larvae (4), and were initially maintained in the insect tissue culture medium described by Mitsuhashi and Maramorosch (5). Sarver and Stollar (6) have more recently adapted these cells to Eagle's (7) vertebrate tissue culture medium. For routine use, we maintain *Aedes albopictus* cells in Eagle's minimal medium supplemented with nonessential amino acids, glutamine, penicillin, streptomycin, and heat-inactivated fetal calf serum at a final concentration of 5%. Cells are maintained at 28°C in a 5% CO_2 atmosphere. Under these conditions, standard ("wild type") cells have a doubling time of approximately 18 h (3).

Stollar and coworkers were the first to apply systematically the techniques and approaches of somatic cell genetics to mosquito cells in culture (8). Since the initial description of mutants resistant to bromodeoxyuridine (thymidine kinase deficient), a number of mutants affecting various intracellular processes has been accumulated, including cells resistant to α-amanitin (8), ouabain (8), methotrexate (9), cycloheximide (10), and puromycin (11). In the case of the antibiotics cycloheximide and puromycin, which act at the level of the ribosome, it has not yet been possible to obtain from vertebrate cells in culture mutants with comparable levels of resistance. Resistance to these drugs can be demonstrated in cell free lysates prepared from the mutant mosquito cells, indicating that the alteration in these mutants affects an intracellular component of the protein synthetic machinery, and not the permeability properties of the plasma membrane (10,11).

A long term goal of our research is the isolation and characterization of mosquito ribosomal protein genes. Cloned genes can be used as probes to investigate physiological processes in the intact organism. Specifically, in the mosquito, the blood meal triggers a cascade of hormonal events that culminate in the synthesis of egg yolk protein (vitellogenin) by the fat body (12). Vitellogenin is released from the fat body into the hemolymph and is selectively sequestered by the developing oocytes. Early work showed that this reproductive process included a cycle of accumulation and degradation of total RNA in the fat body of blood-fed mosquitoes (13). More recently, we have shown that the amount of total fat body RNA represents synthesis and degradation of ribosomes (14), which presumably enable the fat body to synthesize massive amounts of vitellogenin. Recent analyses of ultrastructural changes in mosquito fat body during the vitellogenic cycle were consistent with these biochemical data (15).

One approach to cloning ribosomal protein genes has been to examine antibiotic-resistant mutants, such as those resistant to cycloheximide or to puromycin, for a ribosomal protein with altered electrophoretic mobility. Using the two-dimensional polyacrylamide gel system described by Sanders et al. (16), we have identified 28-31 ribosomal proteins from the small subunit, and 36-39 proteins from the large ribosomal subunit isolated from Aedes albopictus cells (17). The mosquito ribosomal proteins range in mass from 10-53 kDa. In cultured cells, the largest protein on the small ribosomal subunit, S1, is the major phosphoribosomal protein. Aedes S1 appears to share homology at the level of peptide mapping with S6, the major phosphoribosomal protein in Drosophila melanogaster cells (17). The ribosomal proteins from Aedes aegypti fat body are virtually identical in electrophoretic mobility to those from Aedes albopictus cells in culture. However, in fat body, two proteins (S10 and A) in addition to S1 are phosphorylated (14).

The two-dimensional gel system has also been used to compare the ribosomal proteins from cycloheximide-resistant and sensitive mosquito cells. Although we have not been able to identify an altered protein from resistant cells, our most recent studies have shown that the cycloheximide-resistant cells are temperature-sensitive for growth at 34.5°C, and have increased sensitivity to G-418, another antibiotic that acts at the level of the ribosome (18). Work is in progress to obtain additional clones of Aedes albopictus cells with increased resistance to cycloheximide. In similar studies with emetine-resistant Chinese hamster cells, second-level mutants with increased resistance to the selective agent (relative to the initial mutants) facilitated identification of the altered ribosomal protein (19).

Our second approach to identifying ribosomal protein genes has been to clone into a plasmid vector, double stranded cDNA copies of size-selected mRNA from Aedes cells. Using the technique of hybrid-selection, followed by in vitro translation of the selected mRNA in the wheat germ cell-free translation system, we have recently identified cDNA sequences corresponding to ribosomal proteins L8, L14 and L31 (Durbin, Swerdel, and Fallon, submitted). Using these cDNA probes, ribosomal protein gene expression will be investigated in fat body. In addition, the cDNAs will be used to identify the corresponding gene from a bacteriophage λ library.

To develop the technology for studying expression of cloned genes in *Aedes* cells, we have attempted to introduce into these cells various recombinant vectors using the calcium phosphate coprecipitate method, which has been widely used with vertebrate cells. We found that transfection of mosquito cells was more effective when the polycation polybrene (1,5-dimethyl-1,5-diazaundecamethylene polymethobromide) rather than calcium phosphate was used as the chemical mediator (20,21). The DNA construct used to establish our transfection protocol contained a heat inducible promoter from the *Drosophila* heat shock protein 70 gene, fused in frame with the chloramphenicol acetyltransferase gene from *E. coli* (22). Since the heat shock proteins show considerable conservation during evolution, it was not surprising that the *Drosophila* heat shock protein promoter functions in mosquito cells. Interestingly, however, we have not yet detected expression in transfected mosquito cells of genes under the control of the promoter from the *Drosophila* transposable element *copia*.

Development of species-specific vectors, and establishment of techniques for stable integration of transfected DNA into insect cells other than those of *Drosophila*, requires not only the establishment and optimization of transfection protocols, but also the identification of effective promoters to drive expression of introduced genes. It is our hope that our efforts towards identifying regulatory sequences from mosquito ribosomal protein genes will ultimately contribute to the development of new genetic tools for investigation of mosquito biology.

References

1. Mouchès, C., Pasteur, N., Bergé, J.B., Hyrien, O., Raymond, M., de St. Vincent, B.R., de Silvestri, M., and Georghiou, G.P. Science 233:778-780, 1986.
2. Kurtti, T.J. and Munderloh, U.G. Advances in Cell Culture 3 (Maramorosch, K., ed.), pp. 259-302, Academic Press, New York, 1984.
3. Fallon, A.M. and Stollar, V. Advances in Cell Culture 5 (Maramorosch, K., ed.), pp. 97-137, Academic Press, New York, 1987.
4. Singh, K.R.P. Curr. Sci. 36:506-508, 1967.
5. Mitsuhashi, J. and Maramorosch, K. Contrib. Boyce Thompson Inst. 22:435-460, 1964.
6. Sarver, N. and Stollar, V. Virology 80:390-400, 1977.
7. Eagle, H. Science 130:432-437, 1959.
8. Mento, S.J. and Stollar, V. Somat. Cell Genet. 4:179-191, 1978.
9. Fallon, A.M. Insect Biochem. 14:697-704, 1984.
10. Fallon, A.M. and Stollar, V. Mutation Res. 96:201-212, 1982.
11. Fallon, A.M. and Stollar, V. Somat. Cell Genet. 8:521-532, 1982.
12. Racioppi, J.V., Gemmill, R.M., Kogan, P.H., Calvo, J.M., and Hagedorn, H.H. Insect Biochem. 16:255-262, 1986.
13. Hagedorn, H.H., Fallon, A.M., and Laufer, H. Develop. Biol. 31:281-294, 1973.
14. Hotchkin, P.G. and Fallon, A.M. Biochim. Biophys. Acta, 924:352-359, 1987.
15. Raikhel, A.S. and Lea, A.O. Tissue & Cell 15:281-300, 1983.
16. Sanders, M.M., Groppi, V.E. Jr., and Browning, E.T. Anal. Biochem. 103:157-165, 1980.
17. Johnston, A.M. and Fallon, A.M. Eur. J. Biochem. 150:507-515, 1985.
18. Nouri, N. and Fallon, A.M. *In Vitro* Cellular and Developmental Biology 23: 175-180, 1987.
19. Madjar, J.-J, Frahm, M., McGill, S., and Roufa, D. Mol. Cell. Biol. 3:190-197, 1983.
20. Durbin, J.E. and Fallon, A.M. Gene 36:173-178, 1985.
21. Fallon, A.M. Experimental Cell Res. 166:535-542, 1986.
22. Di Nocera, P.P. and Dawid, I.B. Proc. Natl. Acad. Sci. USA 80:7095-7098, 1983.

ADP-RIBOSYL TRANSFERASE IN CERATITIS CAPITATA WIED. CELLS AND EMBRYOS

R. Cavalloro[1], A. I. Scovassi[2], R. Izzo[2], and U. Bertazzoni[2]

1) Commission of the European Communities, Joint Research Centre, Ispra, Italy
2) Istituto di Genetica Biochimica ed Evoluzionistica, C.N.R., Pavia, Italy

We have studied the enzyme ADP-ribosyl transferase, involved in the DNA repair process, in Ceratitis capitata Wied. embryos at different times of development and in cultured cells treated with γ-rays. Our results indicate that the active band of the enzyme is already present at 30 min after egg deposition and remains constant for about 3 hours. Cultured cells treated with γ-rays show an increase in enzyme activity with high doses of radiation. We have also studied the survival of cells in the presence of the specific inhibitor 3-aminobenzamide. Insect cells show resistance to the treatment with this drug. In short, ADPRT is present in both Med-fly cells and embryos and plays an important role in fundamental physiological process.

The enzyme Adenosine DiPhospho-Ribosyl Transferase (ADPRT) is localized in the nucleus of animal cells where it catalyzes the transfer of ADP-ribose from Nicotinamide Adenine Dinucleotide (NAD) to several acceptor proteins (1). The ADP-ribosylation of nuclear proteins is involved in different processes of DNA metabolism and several lines of evidence indicate that the enzyme plays a significant role in DNA repair (2).

We have recently developed in Pavia an activity gel technique allowing the "in situ" detection of active polypeptides of ADPRT after Na dodecyl sulphate polyacrylamide gel electrophoresis (3, 4). Using this method, we have studied the distribution of the enzyme in different classes of organisms and traced its persistence from mammals to gastropods, establishing also its presence in Ceratitis capitata Wied. cells and embryos (5).

We have analyzed the activity of ADPRT during the development of Med-fly embryos and report in this work the results we have obtained.

We have also studied the effect of γ-irradiation on the activity of the enzyme in Ceratitis capitata cultured cells developed at the Joint Research Centre, Ispra (6, 7).

In addition, we report here the results we have obtained by treating the Med-fly cells with 3-aminobenzamide (3AB), a specific inhibitor of ADPRT, and studying its effect on cell survival.

Materials and Methods

Cell culture and embryos

Ceratitis capitata cells were obtained from the "in vitro" stabilized line CEC CC 130 and grown in monolayer in a modified Schield and Sang medium (6).

Med-fly embryos were recovered at different times of development from a permanent rearing system established at the Ispra Joint Research Centre (8).

γ-Irradiation

Irradiation gamma was performed on a cell population 48 h after transplantation, essentially as described by Cavalloro et al. (9). The material was placed in a vertical position at the centre of a Gammacell-220 irradiation chamber and exposed to a 2480 Ci of ^{60}Co source with a dose rate of 1000 rad/min.

The doses used were 600 and 2400 rad, and cells were recovered 30 min, 1 h or 2 h after irradiation.

3-Aminobenzamide treatment

Med-fly cells were grown in the presence of different concentrations (5, 10, 20, 40 mM) of the ADPRT inhibitor 3-aminobenzamide.

At different times of incubation the cell growth rate was estimated (10).

Preparation of extracts

Samples of 5 x 10^6 cells were washed with Phosphate Buffer Saline and pellets kept at -80°C until use. Embryos (0.2 g) were dechorionated with sodium hypochloride (6), washed and kept at -80°C.

For the preparation of total extracts, cells or embryos were resuspended in 100 - 200 μl of 0.6 M NaCl, 50 mM Tris HCl pH 7.5, 0.5 mM ditiothreitol, 1 mM Etilen Diamino Tetracidic Acid, 1 mM p-methyl sulfonyl fluoride, 1 μM pepstatin, 10 mM Na bisulfite, sonicated two times for 20 sec each, and centrifuged 10 min at 12000 rpm at 4°C. The supernatant was used as extract.

Activity gel procedure

Extracts prepared from Med-fly cells and embryos were analyzed by the activity gel procedure for ADPRT, as described by Scovassi et al. (3). Visualization of active bands was directly obtained by autoradiography.

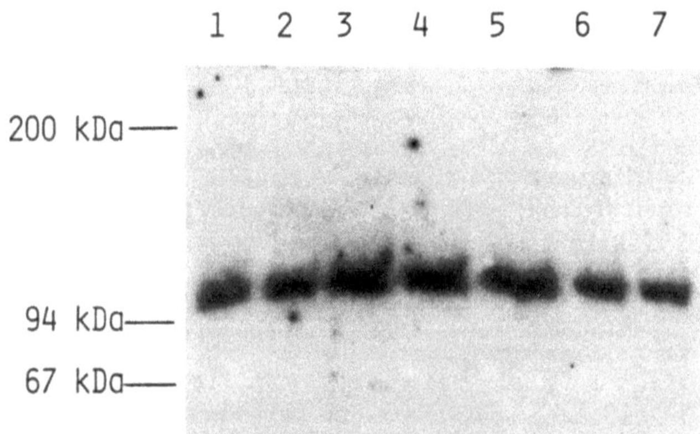

Fig. 1. Activity gel analysis of ADPRT from extracts of Ceratitis capitata Wied. embryos at different times of development.
Extracts were prepared as described in Materials and Methods.
Lanes: 1-7, embryos collected 30 min, 3 h, 6 h, 9 h, 12 h, 24 h, 48 h after egg deposition, respectively.
M_r markers were: myosin (200 kDa), phosphorylase b (94 kDa) and bovine serum albumin (67 kDa).

Results

ADPRT activity in developing Med-fly embryos

We have analyzed by the activity gel procedure the activity and the structure of ADPRT in Med-fly embryos, obtained at different times during embryogenesis.

The results of a typical experiment are reported in Fig. 1, where the dark bands represent the active polypeptide of the enzyme.

It appears that the enzyme activity is already elevated at 30 min after egg deposition (lane 1) and remains constant for about 48 h (lanes 2-7). The active bands have a molecular mass of 120000 indicating that the structure of ADPRT in Med-fly embryos is very close to that observed for the vertebrate organisms (5).

Effect of γ-irradiation on ADPRT activity in Med-fly cells

Ceratitis capitata cells were irradiated for 30 min, 1 h and 2 h at two different doses of γ-rays: 600 and 2400 rad.

In Fig. 2, the activity gel of ADPRT from control and irradiated Med-fly cells is reported.

Control cells (lane 1) exhibit an active band of 120 kDa, the same as that found in embryos (see Fig. 1). The cell treatment with 600 rad for 30 min (lane 2) causes a slight increase in the intensity of the activity band and the enhancing effect is more pronounced after 1 h and 2 h of irradiation (lanes 3 and 4). When treating the cells with 2400 rad of γ-rays, the activation of the enzyme is more evident (compare lane 5 with control lane 1) although it appears to remain constant after delivering the same dose for longer times (lanes 6 and 7).

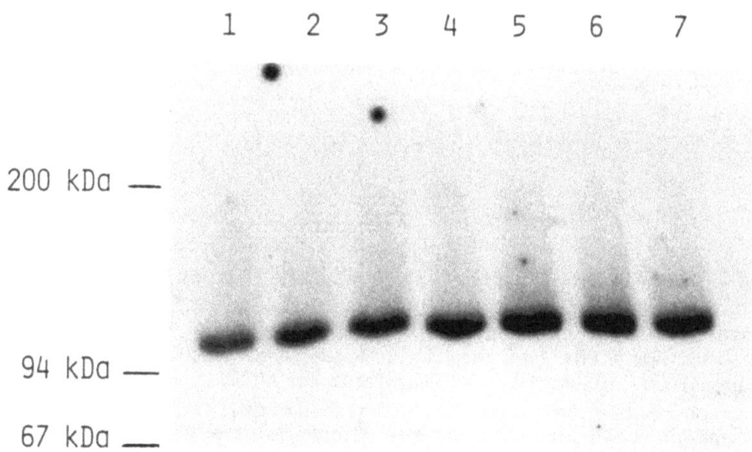

Fig. 2. Activity gel analysis of ADPRT in γ-irradiated *Ceratitis capitata* Wied. cells.
Irradiation conditions and preparation of cell extracts were as described in Materials and Methods.
Lanes: 1, control cells; 2-4, cells irradiated with 600 rad and collected 30 min, 1 h, 2 h after treatment, respectively; 5-7, cells irradiated with 2400 rad and collected after 30 min, 1 h, 2 h after treatment, respectively.
M_r markers were as in the legend of Fig. 1.

Treatment of Med-fly cells with 3-aminobenzamide

The extent of survival of cultured cells was determined after treating them with increasing concentrations of 3-aminobenzamide (5, 10, 20, 40 mM).

The results of a typical experiment are reported in Table 1.

The growth rate of control cells ranged from 1.69 at the first day to 6.52 after 3 days of culture. When the inhibitor was added to the cultured cells at the concentration of 5 mM, no effect on the growth rate was observed. It should be noted that at this dose the ADPRT activity of mammalian cells is completely inhibited and cell doubling is affected (2).

From the data reported in Table 1, it is evident that a significant decrease in all survival is obtained at 20 mM 3AB and that very high concentration of the drug (40 mM) completely blocks cell division.

These data suggest that insect cells are much more resistant to the treatment with 3AB than mammalian cells.

Treatment of Med-fly cells with 3AB was performed as described in Materials and Methods, and extent of growth rate measured by counting the total number of cells observed at a given time versus the initial number of cells in the same areas (10).

Table 1. Effect of 3-aminobenzamide (3AB) on cell growth of Med-fly cells

	1st day	2nd day	3rd day
Control	1.69	3.72	6.52
3AB 5 mM	1.64	3.49	5.74
3AB 10 mM	1.37	2.85	3.86
3AB 20 mM	0.38	1.26	1.84
3AB 40 mM	0.002	0.001	0.08

Discussion

The ADP-ribosylation reaction is considered as an important post-translational modification of nuclear proteins since it appears to induce profound changes in the protein charge and function and is involved in several processes of DNA metabolism, including DNA replication, repair and recombination.

The mechanism of action of the enzyme ADPRT, which is responsible for the transfer of ADP-ribose units to acceptor proteins, is not yet understood, although its physiological role appears to be linked to the modification of the chromatin structure (1, 2). The enzyme has been mainly studied in mammalian cells and very few reports concern its presence and function in invertebrate cells (2).

We have investigated this problem in the Med-fly system where both embryonic and cultured cells are available. By using the activity gel technique we have analyzed the levels of ADPRT during embryogenesis and we have found that the enzyme is already present after 30 min from egg deposition and that its activity remains elevated for about 48 h.

The molecular mass of the enzyme is 120000, quite similar to that observed in vertebrate cells (5). This suggests that the same overall ADPRT activity is present also in insect cells and that the enzyme plays an important role during embryo development being expressed at maximum levels at very early times after egg formation.

Since the mammalian enzyme is activated in response to cell treatment with DNA-damaging agents (2, 3), we have investigated this effect in Med-fly cells in culture after γ-irradiation. Our results indicate that high doses of γ-rays cause a significant increase in the intensity of the activity band of the enzyme, suggesting that ADPRT could be involved in the repair of radiation damage in insect cells.

We have also studied the extent of survival of Med-fly cultured cells after treatment with 3-aminobenzamide, a specific inhibitor of ADPRT (2). A significant decrease in growth rate was obtained only at very high concentrations of the drug, suggesting that the Med-fly enzyme is more resistant to 3AB inhibition than the mammalian one.

In conclusion, it appears that the ADP-ribosylation system in insect cells is carried out by an enzyme which is very similar to the vertebrate ADPRT for what it concerns the molecular mass and response to irradiation.

However the insect cell behaviour toward inhibition with 3AB appears to differ notably from that of mammalian cells. This suggests that the enzyme present in insect cells could represent a variant molecular form more resistant to 3AB and/or that a reduced permeability of Med-fly cells to the 3AB could interfere with the extent of inhibition.

References

1. Hayaishi, O. and Ueda, K. ADP-ribosylation Reactions, pp. 3-16, Academic Press Inc., 1982.
2. Shall, S.C. Adv. Radiation Biol. 11: 1-69, 1984.
3. Scovassi, A.I., Stafanini, M. and Bertazzoni, U. J. Biol. Chem. 259: 10973-10977, 1984.
4. Bertazzoni, U., Scovassi, A.I., Mezzina, M., Sarasin, A., Franchi, F. and Izzo, R. Trends Genet. 2: 67-72, 1986.
5. Scovassi, A.I., Izzo, R., Franchi, E. and Bertazzoni, U. Eur. J. Biochem. 159: 77-84, 1986.
6. Cavalloro, R. Rev. Can. Biol. 40: 181-186, 1981.
7. Cavalloro, R., Scovassi, A.I. and Bertazzoni, U. Fruit Flies of Economic Importance (Cavalloro R. ed.), pp. 212-218, A.A. Balkema, 1983.
8. Cavalloro, R. and Girolami, V. Redia LI: 315-327, 1969.
9. Cavalloro, R., Pozzi, G. and Coppola, M. Rev. Can. Biol. Exptl. 42: 109-115, 1983.
10. Cavalloro, R., Bertazzoni, U., Pozzi, G. and Scovassi, I. Redia LXVIII: 229-237, 1985.

EFFECTS OF DELTAMETHRIN ON A *DROSOPHILA MELANOGASTER* CELL LINE *IN VITRO* : CYTOTOXICITY, ACCUMULATION AND METABOLISM

A. Baeze-Squiban[1)], M. Best-Belpomme[2)], and F. Marano[1)]

1) Laboratoire de Cytophysiologie et Toxicologie cellulaire, UA CNRS 567, UNIVERSITE PARIS VII, 2, Place Jussieu, 75005 PARIS, France
2) Laboratoire de Fifférenciation moléculaire et cellulaire du Développement, UA CNRS 1135, UNIVERSITE PARIS VI, 7-9, quai Saint Bernard, 75005 PARIS, France

The cytotoxicity of deltamethrin, one of the most potent insecticides in current use, has been tested on a *Drosophila melanogaster* cell line. Deltamethrin reduces cellular proliferation, but never leads to total mortality of cellular population. Kinetics of uptake of ^{3}H-deltamethrin shows that it accumulates in cells and is metabolized. Some products, results from the hydrolysis of the molecule and other polar compounds, are presumed to be conjugated.

Recent synthetic pyrethroids such as deltamethrin (Roussel-Uclaf) are among the most potent and largely used insecticides. They are characterised by a high safety factor (insect / mammalian ratio), rapid metabolism and excretion in mammals and limited soil persistence (1). Electrophysiological studies on axons of vertebrate and insects have implicated the sodium channel as a principal site of pyrethroid action but the existence of other sites must not be excluded (2).

Concerning the effects of deltamethrin on non-excitable cells, few papers are found in the literature. Moreover, whereas the metabolic fate of deltamethrin has been widely studied in mammals, fewer studies have been done on insects *in vivo*, in insect-derived enzyme preparation *in vitro* (3) and nothing on cells *in vitro*. Considering this lack of information, we have investigated the effects of deltamethrin on *Drosophila melanogaster* cells *in vitro* (4) treated or not by ecdysterone concerning its cytotoxicity, its accumulation and its metabolism.

Cytotoxicity and Bioaccumulation

Deltamethrin reduces cellular proliferation from 10^{-5} M but never leads to total mortality of a cellular population, even at high concentrations (10^{-4} M). However, kinetics of uptake of ^{3}H-deltamethrin have shown that deltamethrin quickly accumulated in cells and in greater quantity (more than 2 fold) in cells treated by β-ecdysone. A part of radioactivity has been linked to the TCA precipitable fraction but, in contrast, was the same in cells treated or not by the hormone. After treatment with organic solvents, the remaining radioactivity in TCA-precipitable fraction was very low : only 1 % of the initial radioactivity. Electrophoresis of whole cellular proteins (15 % polyacrylamide gels - SDS) have shown that the labelling was not located on a specific polypeptide. So, these results suggest that deltamethrin would not have covalently binding sites.

Metabolism in Cells

Autoradiography of cellular extracts has shown that, after 48 hrs of incubation, many products have appeared. The main compound had the same Rf as control deltamethrin and only a part (about 10 %) of the molecules had been metabolized. Deltamethrin has been metabolized in the same way in insect cells treated or not by β-ecdysterone:

hydrolysis of the ester band (metabolites have different Rf according to the labelling of the molecule on benzyl or dimethyl groups).

hydroxylation of the intact molecule (metabolites have the same Rf whatever the labelling is).

These polar metabolites appear to be largely conjugated since, after Helix promatia enzyme treatment, most of the radioactivity was displaced from the origin and migrated to an Rf of a less polar compound. So, in insect cells, deltamethrin gave conjugated metabolites retaining the ester linkage. This pathway of metabolism has been shown for another pyrethroid, permethrin, in different insects *in vivo*. By contrast, in mammals, hydroxylation occurs on the intact ester deltamethrin but does not lead to conjugate.

In the supernatant of culture, we have found the same metabolites as were obtained in cellular extracts. We have shown that the most important part of these metabolites appeared in relation with the cellular activity. They may therfore be released from the cells in the culture medium or the cells could excrete some enzymes which induce an extracellular metabolisation. Further experiments would permit us to make a conclusion on this.

Our study has shown that deltamethrin is metabolised at the cellular level in the same way as has been demonstrated at the level of the insect *in vivo*. So, *Drosophila melanogaster* cells in culture permit us to quickly and easily evaluate the cytotoxicity, bioaccumulation and metabolism of deltamethrin. These "alternative methods" would be of great interest for the screening of new insecticides.

References

1. Elliot, M. Environ. Health. Perspect., 14:3-13, 1976.
2. Lund, A. E. and Narahashi, T. Pest. Biochem. Physiol., 20:203-216, 1983.
3. Soderlund, D. M., Sanborn, J. R. and Lee P. W. Progress in Pesticide Biochemistry and Toxicology, 3:401-435, 1983.
4. Echalier, G. and Ohanessian, A. In Vitro, 6:162-172, 1970.

IN VIVO AND IN VITRO INHIBITION OF PROTHORACIC GLAND ACTIVITY BY 1-BENZYL-5-[(E)-2,6-DIMETHYL-1,5-HEPTADIENYL]IMIDAZOLE (KK-42) IN BOMBYX MORI

M. Kiuchi and H. Akai

The Sericultural Experiment Station (MAFF), Tsukuba, Ibaraki 305, Japan

1-Benzyl-5-[(E)-2,6-dimethyl-1,5-heptadienyl]imidazole(KK-42), one of a series of imidazole derivatives which were synthesized as agents exhibiting an "anti-JH" activity, has been found to be a very potent inducer of precocious metamorphosis in the silkworm, *Bombyx mori*.

By the topical application of KK-42 to the spinning *Bombyx mori* larvae, pupal ecdysis was inhibited, and the haemolymph ecdysteroid titre, as measured by a radioimmunoassay, was lowered. The secretion of ecdysone by the prothoracic glands *in vitro* was also inhibited by as little as 10^{-8} M KK-42 in the culture medium. Thus it is suggested that the compound suppresses the prothoracic gland activity directly, though the effect on the corpora allata is yet unknown.

Since the discovery of precocenes by Bowers et al.(1), many chemical compounds were found to exhibit an "anti-JH" activity on insects though the effects were limited(2,3). Recently, a series of terpenoid imidazole derivatives were reported to be potent inducers of precocious metamorphosis in *Bombyx mori*(4,5,6,7,8), and we also reported on the induction of precocious metamorphosis by the administration of a similar imidazole compound to *Bombyx mori* at various larval stages (9,10).

Although these compounds are assumed to be anti-juvenile hormones because they induce precocious metamorphosis and the effect is antagonized in some cases by the application of a juvenile hormone analogue(4,11), the mechanism of action has not yet been elucidated. It was observed that the topical application or dietary administration of these compounds was always followed by the prolongation of the duration of the treated stage, delay or prevention of next ecdysis, when precocious metamorphosis was not induced (7,8,9,10,11,12,13,14). Thus it was suggested that the compounds exerted some effects on ecdysone which controls ecdysis.

Materials and Methods

The silkworm larvae(F_1: N140 x C145) were reared on an artificial diet at 26 °C under a 12D:12L photo regime. Allatectomy was performed from 16 to 20hr after the 3rd ecdysis. KK-42(1-benzyl-5-[(E)2,6-dimethyl-1,5-heptadienyl]imidazole) was kindly supplied by Dr.Eiichi Kuwano(Kyushu University). KK-42 was dissolved in acetone and topically applied to spinning larvae with a microsyringe. *In vitro* study of the prothoracic gland activity was carried out according to the method described by Okuda et al.(15) with slight modifications. KK-42 was mixed in Grace's insect culture medium by brief sonication followed by stirring. The haemolymph ecdysteroid titre and the amount of ecdysone secreted into the culture medium were measured by a radioimmunoassay(15,16).

Results and Discussion

When KK-42 was topically applied to larvae at the beginning of spinning, pupa-

Table 1. Effects of the topical application of KK-42 on pupation

Time of application[1)]	KK-42 (μg)	No. of larvae used	Number of responses A	B	C	D
0	0	20	20(3.7)	0	0	0
	12.5	20	20(4.6)	0	0	0
	25	20	14(5.4)	5(6.8)	1(7.0)	0
	50	20	0	14(6.8)	2(9.0)	4
	100	20	0	0	1(8.5)	19
1	12.5	20	18(4.5)	1(6.5)	1(8.5)	0
	25	20	13(4.4)	5(5.7)	1(8.5)	0
	50	20	5(4.5)	11(5.4)	3(7.2)	1
	100	20	0	8(5.3)	5(5.7)	7
2	12.5	20	15(4.6)	4(4.6)	1(5.5)	0
	25	20	9(4.5)	10(4.7)	1(5.5)	0
	50	20	4(4.5)	11(4.8)	5(5.4)	0
	100	20	2(4.6)	12(4.7)	5(6.1)	1

1)Days after the beginning of spinning.
Mean time(days) of the response from the beginning of spinning is indicated in parenthesis. A - D: See Fig. 1.

Fig. 1 Inhibition of pupation by topical application of KK-42

A: Normal pupation, B: Failure of normal ecdysis, C: Partial formation of pupal skin, D: Unchanged larval form

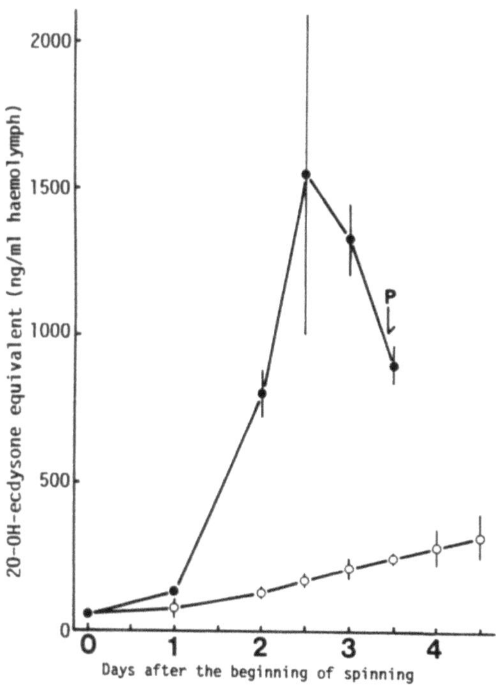

Fig. 2 Haemolymph ecdysteroid level of larvae after topical application of KK-42

-●-: Control, -○-: Application of 100μg of KK-42 immediately after the beginning of spinning, P: Pupation

Table 2. Effects of the topical application of KK-42 on the pupation of allatectomized mature larvae

KK-42 (μg)	No. of larvae used	Number of responses			
		A	B	C	D
0	10	10(2.5)	0	0	0
10	10	8(3.7)	2(4.5)	0	0
25	10	2(4.0)	7(4.4)	1(6.0)	0
50	10	1(4.0)	5(4.4)	3(5.5)	1
100	10	0	2(4.5)	6(6.4)	2

CA were excised at the early 4th instar. KK-42 was applied topically immediately after the beginning of spinning. Mean time(days after the application) of the responses is indicated in parenthesis. A - D: See Fig. 1.

Table 3. Ecdysone secretion *in vitro* by the prothoracic glands after topical application of KK-42

Time after treatment (minutes)	Amount of ecdysone secreted (ng/gland 2 hours)
Control	13.5 ± 2.5
15	6.3 ± 2.8
30	3.3 ± 0.9
60	2.0 ± 0.7
120	2.7 ± 0.8

100μg of KK-42 was topically applied two days after the beginning of spinning. Values indicate mean ± SD of 6 glands.

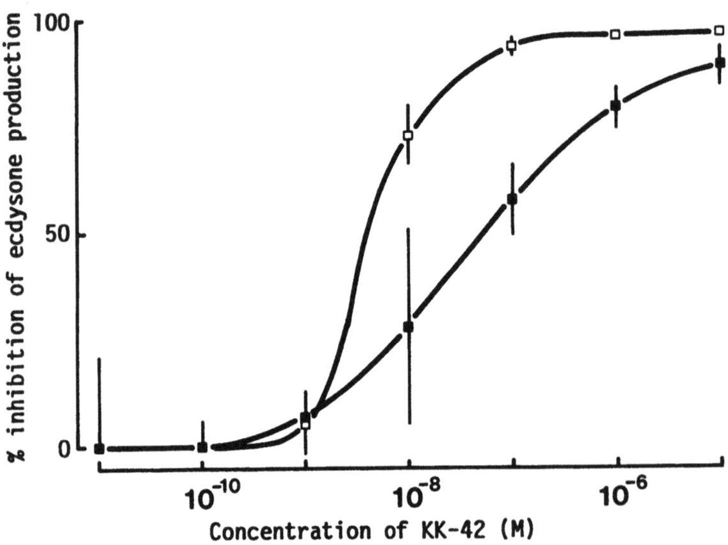

Fig. 3 *In vitro* inhibition of ecdysone secreting activity of prothoracic gland by KK-42

Prothoracic glands were dissected out from the 5th instar larvae 42-46 hours after the beginning of spinning. One prothoracic gland out of a pair was incubated in Grace's insect culture medium alone(control gland) and the other was incubated with KK-42.

tion was delayed or inhibited depending on the dose(Table 1). Lower doses caused only a delay of normal pupation. As the dose increased the number of individuals failing to undergo normal ecdysis(Fig. 1 B) and showing a partial formation of pupal skin increased(Fig. 1 C). At the dose of 100μg in almost all the larvae metamorphosis did not occur and larvae remained unchanged for more than two weeks followed by death(Fig. 1 D). Inhibitory effects on pupation were stronger when the compound was applied at earlier stages, and less pronounced when the application time was delayed. Application of KK-42 to the larvae which precociously matured by allatectomy at the early 4th instar also gave rise to a similar phenomenon(Table 2).

These results strongly suggest that KK-42 affected the prothoracic gland which secretes ecdysone responsible for ecdysis. Then the titre of ecdysteroid in the haemolymph was measured after the application of KK-42 at the beginning of spinning. As shown in Fig. 2, the haemolymph ecdysteroid level of the treated larvae remained very low compared to that of the control.

In order to determine whether KK-42 directly affects the prothoracic gland in the inhibition of ecdysone secretion, the prothoracic glands dissected from the larvae 42 - 46hr after the beginning of spinning were incubated in Grace's medium containing KK-42. As shown in Fig. 3, KK-42 inhibited ecdysone production at a concentration as low as 10^{-8} M during the first 2 hours of incubation, and during the next 4 hours, the inhibition reached 75% at 10^{-8} M and nearly 100% above 10^{-7} M.

Since KK-42 has a JH-like terpene chain and exerts its effects by topical application, the compound is thought to pass through the cuticle readily. In fact, the secreting activity of the prothoracic glands dissected out 15 minutes after application was as low as 50% of that in the untreated control(Table 3).

Thus KK-42 was found to affect the prothoracic gland directly, by inhibiting ecdysone secretion. Although it remains to be determined whether the compound directly inhibits the corpora allata activity, the mechanism for the induction of precocious metamorphosis may be ascribed to the decrease of the ecdysteroid level alone, because precocious metamorphosis was completely antagonized by the dietary administration of a small amount of 20-hydroxyecdysone(unpublished data).

Some natural and synthetic compounds are known to inhibit ecdysone synthesis, but their activity is rather low(17). Since KK-42 is very active both *in vivo* and *in vitro*, the compound may be useful in studies on the hormonal regulation of insect development, as there are no other methods of suppression of the prothoracic gland activity.

References

1. Bowers, W. S., Ohta, T., Cleere, J. S. and Marsella, P. A. Science, 193:542-547, 1976.
2. Staal, G. B. Ent. Exp. Appl., 31:15-23, 1982.
3. Staal, G. B. Ann. Rev. Entomol., 31:391-429, 1986.
4. Kuwano, E., Takeya, R. and Eto, M. Agric. Biol. Chem., 47:921-923, 1983.
5. Kuwano, E., Takeya, R. and Eto, M. Agric. Biol. Chem., 48:3115-3119, 1984.
6. Kuwano, E., Takeya, R. and Eto, M. Agric. Biol. Chem., 49:483-486, 1985.
7. Asano, S., Kuwano, E. and Eto, M. J. Pesticide Sci., 9:503-509, 1984.
8. Asano, S., Kuwano, E. and Eto, M. Appl. Ent. Zool., 19:212-220, 1984.
9. Akai, H., Kimura, K., Kiuchi, M. and Shibukawa, A. J. Seric. Sci. Jpn., 53:545-546, 1984.
10. Kiuchi, M., Kimura, K. and Akai, H. J. Seric. Sci. Jpn., 54:77-81, 1985.
11. Asano, S., Kuwano, E. and Eto, M. Appl. Ent. Zool., 21:63-69, 1986.
12. Kiuchi, M. and Akai, H. J. Seric. Sci. Jpn., 54:527-528, 1985.
13. Asano, S., Kuwano, E. and Eto, M. Ent. Zool., 21:305-312, 1986.
14. Kuwano, E. and Eto, M. Agric. Biol. Chem., 50:2919-2920, 1986.
15. Okuda, M., Sakurai, S. and Ohtaki, T. J. Insect Physiol., 31:455-461, 1985.
16. Takeda, S., Kiuchi, M. and Ueda, S. Bull. Seric. Exp. Sta., 30:361-374, 1986.
17. Retnakaran, A, Granett, J. and Ennis, T. Comprehensive Insect Physiology, Biochemistry and Pharmacology(Kerkut, G. A. and Gilbert, L. I., eds.), vol. 12, pp. 529-601, Pergamon. 1985.

WHOLE EMBRYONIC CULTURE OF SPONTANEOUS PARTHENOGENETIC EMBRYOS IN THE SILKWORM, *BOMBYX MORI*

A. Murakami[1], Y. Ohtsuki[2], and T. Kitazawa[2]

1) National Institute of Genetics, Mishima, Japan
2) Sericultural Experiment Station, Tsukuba, Japan

Four to twelve week-old spontaneous parthenogenetic diapausing eggs, after the removal of the chorion, were cultured in Grace medium. After two weeks of culture at 25°C, the extent of embryonic development was examined. The growth of these parthenogenetic embryos *in vitro* varied with the age of the eggs after oviposition. Most of the four week-old parthenogenetic diapausing eggs grew up to the late stage of morphogenesis but as partial embryos while the majority of the twelve week-old eggs developed completely, at least, up to that stage,as did normally fertilised diapausing eggs. The present findings and others clearly indicated that spontaneous parthenogenetic diapausing eggs can be partly rescued from genetic death by whole embryonic culture.

In the *Bombyx* silkworm, spontaneous parthenogenesis commonly occurs in nature (1). However, spontaneous parthenogenetic eggs usually die around the stage immediately preceding the resumption of post-diapause embryonic growth. Only a small proportion of well developed eggs survives beyond that stage. However, when normally fertilised diapausing eggs in which the chorion has been removed are explanted in Grace medium, the embryo thus cultured instantly begins to develop until the stage which immediately precedes hatching (2). In this case it is expected that spontaneous parthenogenetic eggs can be rescued due to the release of the diapause state and develop, at least, to an advanced embryonic stage.

It is well known that in the silkworm, the diapause phenomenon is dependent upon maternal traits (3,4,5,6,7,8). The silkworm egg consists of four main membrane structures, egg chorion, vitelline membrane, serosal cuticle (9) and serosal membrane. The first two structures are maternal products as in the case of spontaneous parthenogenetic and normally fertilised diapausing eggs, while the serosal cuticle (9) and serosal membrane are of zygotic origin for fertilised eggs but of maternal origin for parthenogenetic eggs. Accordingly, the functions of the membrane contributing to each phenomenon can be analysed by comparing the mortality between the normally fertilised eggs and the spontaneous parthenogenetic eggs *in vivo* and *in vitro*, and by investigating the effect of the removal of the chorion on the breaking of the diapause.

Materials and Methods

In the present experiment, eggs laid by F_1 hybrid females from a cross between *J106* female x *Cambodge* male were used, since these females showed the highest parthenogenetic capacity (32.1%) among many hybrid crosses and original stocks in the silkworm (10). When virgin female moths were placed on egg cards, they began ovipositing a few days after emergence until around the 10th day. Spontaneous parthenogenetic diapausing eggs collected in such manner were kept at 25°C to make the diapause proceed for about 4-12 weeks after oviposition until they were used for the whole embryonic culture.

All the breeding and related procedures were carried out in the laboratory at a temperature of 25°C and relative humidity ranging from *ca*. 50 to 80% depending on the stage of growth of the silkworm.

The whole embryo and yolk complex, enclosed by the serosal membranes, were removed from the egg chorion and then suspended in hanging-drops of Grace insect tissue culture medium (Grace medium (11)) at 25°C for 2 weeks according to the method of Ohtsuki and Sugiyama (2) without renewing the medium (Fig.1).

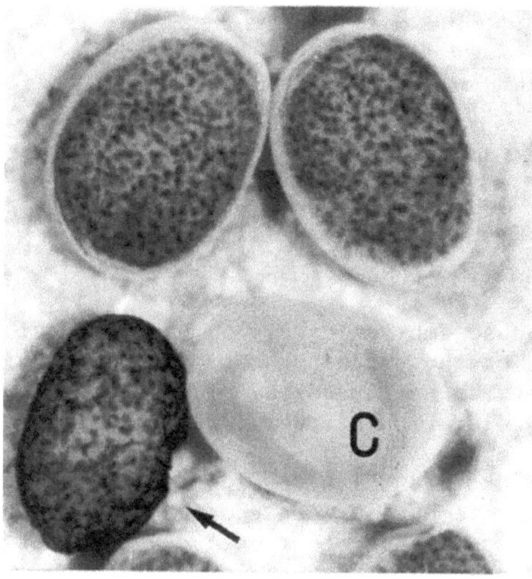

Fig.1. Dechorionated egg before the culture (arrow) and removed chorion(c).

After 2 weeks of culture, the embryos were dissected in 0.85% NaCl solution to examine the extent of embryonic development. The number of embryos which developed beyond the stage of appearance of the abdominal appendage and that of undeveloped embryos were counted. A test to determine whether embryos were alive was performed by vital staining with 0.5% trypan blue, and the embryos showing no staining were considered to have died. Developed embryos were further classified according to their developmental stages into two groups, that is, the embryos in the stage of appearance of the abdominal outgrowth to dorsal closure and the embryos in the stage of bristle formation to body pigmentation.

Results and Discussion

The results of the experiments are summarized in Tables 1 and 2, for the twelve and four week-old eggs, respectively. As shown in Table 1, all the eggs with nearly natural pigmentation which were fertilised grew up to the stage of embryos with bristle formation and pigmented body as did normally fertilised diapausing eggs. Only one embryo among them further grew to the stage of a larva ready to hatch. It is interesting to note that all the eggs with irregular pigmentation were able to develop at least to the stage of appearance of the abdominal appendage, and about half of them developed further to the stage of embryos with bristle formation but non-pigmented skin. One of these eggs reached the stage of embryo with bristle formation and body-pigmentation as observed in normally developed eggs.

Table 1 Embryonic Development of 11-12 Week-old Spontaneous Parthenogenetic Diapausing-Eggs Cultured in Grace Medium at 25°C for 2 Weeks

Pigmentation of eggs at the onset of culture	No.of eggs cultured	No.of eggs developed	Developmental stage of embryos* on the 14th day cultivation		
			No.of embryos in which abdominal appendage appeared	No.of non-pigmented embryos with bristles	No.of pigmented embryos with bristles and dark black head parts
Completely pigmented eggs	8	8	0	0	8
Irregularly pigmented eggs	7	7	3	3	1

* All the embryos showed a positive vital staining reaction.
** The embryo had already developed until the stage of bristle formation at the onset of cultivation.

As shown in Table 2, however, the majority of the four week-old eggs reached the stage of bristle formation but as partial embryos. The others remained as a mass of nondifferentiated cells. The appearance of partial embryos was ascribed to abnormal functions in an ooplasm not reactivated due to the absence of sperm.

Table 2 Embryonic Development of 3-4 Week-old Spontaneous Parthenogenetic Diapausing-Eggs Cultured in Grace Medium at 25°C for 2 Weeks

No.of eggs cultured	No.of eggs in a state of undifferentiated cells	No. of embryos* with pigmented head parts but abnormal segmentation of thoracic and abdominal parts			No.of embryos with bristles
		No segments	2-3 segments	4-5 segments	
18	5	3	6	4	0

* All the embryos showed a positive vital staining reaction.

The egg-age difference in the post diapause embryonic development is considered to reflect the differences in the ability of the parthenogenetically developing eggs used for the culture : the distinction between nearly normal pigmented eggs and others can be easily made since the external coloration of twelve week-old eggs is stable, unlike the coloration of four week-old eggs.

For this reason, it is difficult to select eggs which develop adequately in the case of culture. It is also likely that the removal of the chorion from early diapausing eggs is not suitable for a normal post-diapause embryonic development.

Under natural conditions, parthenogenetic eggs either well or poorly developed (or pigmented) die during the long diapause period of estivation and hibernation, shrinking like nonpigmented unfertilised eggs. Only a small proportion of the well developed eggs is able to survive beyond the diapause period. However, the findings obtained in this experiment clearly indicated that through the use of whole embryonic culture, spontaneous parthenogenetic diapausing eggs can be partly rescued from genetic death.

As stated already, the eggs of silkworms consist of four extra-embryonic membranes (Fig.2). The chorion and vitelline membrane are maternal products and the serosal cuticle and serosal membrane are zygotic products. Accordingly, the origin of all these membranes in parthenogenetic eggs is maternal in contrast with the situation in fertilised eggs. Since the vitelline membrane is an amorphous structure, it does not appear to play an essential role in the diapause phenomenon. In spite of such a difference, when the spontaneous parthenogenetic whole embryo enclosed by the serosal membrane was explanted in Grace medium after the removal of the chorion, embryonic growth rapidly resumed as in the case of the fertilised eggs. This finding clearly indicates that the removal of the chorion enables to end the diapause, while the serosal membrane is not involved in this activity. From the genetical viewpoint mentioned above, it is also obvious that the egg diapause trait

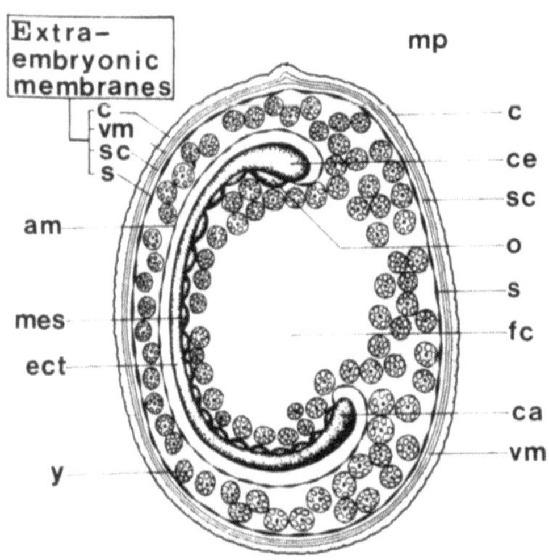

Fig.2. Diagram showing the extra-embryonic membranes of the diapausing-egg of *Bombyx mori*.

c:chorion sc:serosal cuticle
vm:vitelline membrane s:serosal membrane

in the *Bombyx* silkworm is dependent upon the physical and chemical properties of the chorion which control the permeability of oxygen and preservation of water (12,13). Various artificial treatments applied to break the diapause or developmental blockage may change the properties of the chorion so as to increase the permeability of oxygen into the whole embryo and yolk complex covered by the serosal membrane. It appears that mechanical wounding which simulates the removal of the chorion ultimately modifies the physiological conditions of the embryo-yolk complex leading to the conversion of polyalcohols, sorbitol and glycerol into glycogen used for the nutrition of the reactivated embryos (14).

As mentioned above, most of the spontaneous parthenogenetic diapausing eggs in nature were not able to survive beyond the diapause period,in contrast to the fertilised diapausing eggs. When the parthenogenetic eggs were given a shelter prior to the termination of the diapause, most of them were able to develop to a fairly advanced embryonic stage as did the fertilised eggs. In general, diapausing eggs of the silkworm are remarkably resistant to an unfavorable environment because they are highly inactive physiologically. On the other hand, parthenogenetic eggs, due to a certain defect in the serosal membrane, presumably of genetic origin, cannot initiate the preliminary steps for the termination of the diapause *in vivo*, while they are able to do so *in vitro*. Since the whole embryo and yolk complex are able to develop adequately in either liquid paraffin (14,15) or physiological saline (15), it is likely that the Grace medium may play an accessory role in providing water (or maintaining the osmotic pressure) to the embryo rather than being a source of nutrients. In brief, the high mortality of the spontaneous parthenogenetic diapausing eggs in nature may be caused by a genetical defect in the function of the serosal membrane and/or its secondary product, the serosal cuticle.

The present findings and others suggested that the *Bombyx* silkworm has a high capacity for natural spontaneous parthenogenesis.

References

1. Astaurov,B.L. Advances in Morphogenesis, 6:199-257, 1967.
2. Ohtsuki,Y. and Sugiyama,H. Bull. Sericul. Exp. Sta. Japan, 25:199-217, 1972. (in Japanese with English summary)
3. Toyama,K. Jour. Genetics, 2:351-404, 1913.
4. Watanabe,K. Bull. Sericul. Exp. Sta. Japan, 3:397-437, 1918. (in Japanese)
5. Watanabe,K. Bull. Sericul. Exp. Sta. Japan, 4:87-106, 1919. (in Japanese)
6. Watanabe,K. Bull. Sericul. Exp. Sta. Japan, 6:411-455, 1924. (in Japanese)
7. Umeya,Y. Jap. J. Genetics, 3:155-182, 1925. (in Japanese with English summary)
8. Murakami,A. Ann. Rep. Natl. Inst. of Genetics (Japan), (in press)
9. Takei, R. and Nagashima,E. J. Seric. Sci. Japan, 44:118-124, 1975. (in Japanese with English summary)
10. Murakami,A., Ohtsuki,Y. and Kitazawa,T. Ann. Rep. Natl. Inst. of Genetics (Japan), 36:46-47, 1986.
11. Grace,T.D.C. Nature, 195:788-789, 1962.
12. Beck,S.D. Insect Photoperiodism, p.288, Academic Press, New York, 1968.
13. Lees,A.D. The Physiology of Diapause in Arthropods, p.151, Cambridge Univ. Press, 1955.
14. Okada,M. Experientia, 27:658-660, 1971.
15. Ohtsuki,Y. and Kitazawa,T. J. Seric. Sci. Japan, 43:111-118, 1974. (in Japanese with English summary)

EFFECT OF GONAD-STIMULATING SUBSTANCE ON CYCLIC AMP PRODUCTION AND 1-METHYLADENINE SECRETION IN STARFISH FOLLICLE CELLS

M. Mita[1], N. Ueta[1], and Y. Nagahama[2]

1) Department of Biochemistry, Teikyo University School of Medicine, Kaga, Itabashi-ku, Tokyo 173, Japan
2) Laboratory of Reproductive Biology, National Institute for Basic Biology, Okazaki 444, Japan

Follicle cells were isolated from ovaries of the starfish, *Asterina pectinifera*, by treatment with 1-methyladenine (1-MeAde). When the isolated follicle cells were incubated with sea water containing gonad-stimulating substance (GSS), the 1-MeAde produced was observed in the media. Level of cAMP in follicle cells increased 5-6 times following the administration of GSS. Isobutylmethylxanthine which is a potent inhibitor of phosphodiesterase also induced 1-MeAde production in follicle cells. These results suggest that 1-MeAde biosynthesis in follicle cells strongly depends on the intracellular cAMP content. It appears that after binding of GSS with its receptor adenylate cyclase is activated to increase the cAMP level.

In starfish, fully grown oocytes within the ovary are arrested at the prophase of the first maturation division just before spawning. A ripe ovary contains a large number of such oocytes of almost the same size. These (oocytes, namely immature oocytes) still possess a large nucleus (germinal vesicle) and fail to undergo normal fertilization. Just before spawning, the immature oocytes resume meiotic division, as indicated by the dissolution of the nuclear membrane (germinal vesicle breakdown, GVBD), and undergo the subsequent processes of oocyte maturation until the discharge of the second polar body. Maturation division can be resumed by 1-methyladenine (1-MeAde)(1-4)(Fig. 1), which is considered to be a general maturation-inducing substance among starfishes. This substance is produced by follicle cells under the influence of a hormonal peptide, gonad-stimulating substance (GSS), released from the supporting cells of the neural system (5,6). GSS is a single peptide (7) and its reactions seem to correspond to those of a pituitary gonadotropin in vertebrates. The stimulatory effects of many peptide hormones on the functions of target cells and tissues are considered to occur as a result of increase in cyclic adenosine 3',5'-monophosphate (cAMP) content (8). Therefore, the relation between 1-MeAde production and intracellular cAMP level in starfish follicle cells was studied.

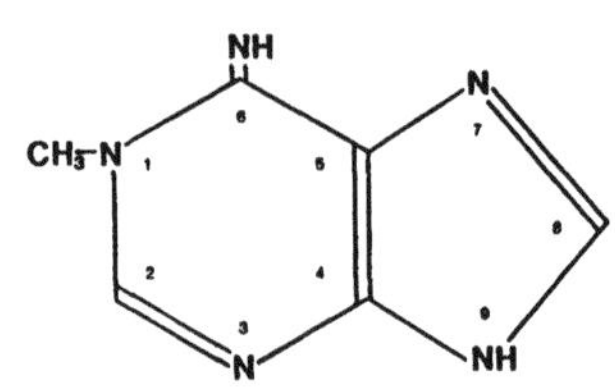

Fig. 1. 1-Methyladenine

Follicle cells were isolated from ovaries of the starfishes, *Asterina pectinifera*, as shown in Fig. 2 (9). Ovaries were washed with cold modified Van't Hoff's artificial sea water (ASW); 460 mM NaCl, 10 mM KCl, 9.2 mM $CaCl_2$, 35.9 mM $MgCl_2$ and 17.5 mM $MgSO_4$, buffered to pH 8.2 with 20 mM borate-NaOH (10). They were torn with fine forceps to isolate folliculated oocytes in an ice bath. Isolated folliculated oocytes were treated with 1-MeAde (10^{-6} M) for 40 min at room temperature, after washing several times with cold ASW. The follicle cells were separated from oocytes by allowing the latter to sediment by gravity. The supernatant containing the follicle cells was collected and filtered through nylon mesh (40 μm) to remove contaminating oocytes. The filtrate was centrifuged at 3,000 rpm for 5 min at 4°C. The sedimented follicle cells were cultured *in vitro*. The amount of follicle cells was calculated by original number of oocytes, since an oocyte is enclosed with about fifty follicle cells (personal information from Shirai).

The isolated follicle cells (5×10^5 or 10^6 cells) were suspended in 1.0 ml ASW containing GSS prepared from the lyophilized radial nerves of *A. pectinifera* (11, 12). After incubation, the cell suspension was centrifuged at 5,000 rpm for 1 min. The supernatant was assayed with respect to the amount of 1-MeAde, according to the method of Shirai (12,13). The 1-MeAde produced was found in the incubation media. The continuous presence of GSS in these media brought about an almost linear 1-MeAde production for 6 h (Fig. 3).

However, GSS-induced 1-MeAde production depended on temperature of ASW (Fig. 3). The follicle cells hardly produced 1-MeAde at 0°C. The amount of 1-MeAde increased in higher temperatures (10°C - 20°C). The maximal activity of 1-MeAde production in follicle cells was

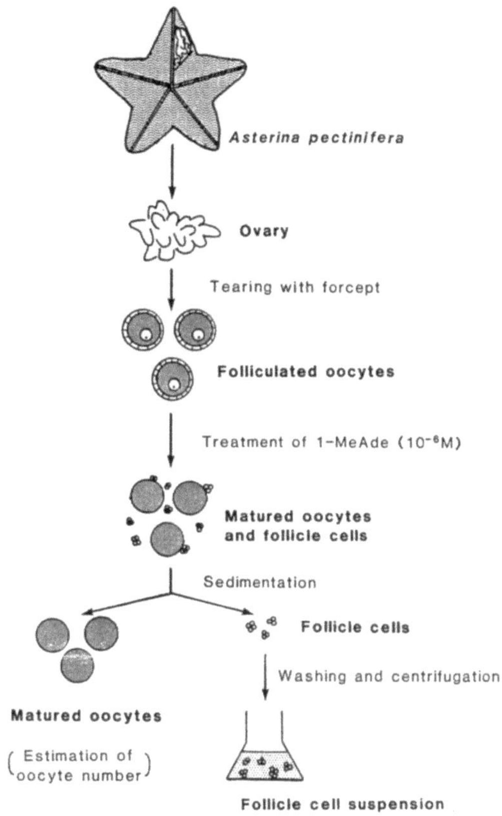

Fig. 2. Preparation of follicle cells

Table 1. Effects of actinomycin D, cycloheximide and puromycin on 1-MeAde production by follicle cells

Condition	1-MeAde production (nmol/2 h/5×10^5 follicle cells)
Control (ASW alone)	0
GSS (0.05 mg/ml)	0.32 ± 0.04
GSS + actinomycin D (0.01 mg/ml)	0.31 ± 0.04
GSS + cycloheximide (0.01 mg/ml)	0.32 ± 0.06
GSS + puromycin (0.01 mg/ml)	0.31 ± 0.04

Values represent the means ± S.E.M. obtained in three separate experiments.

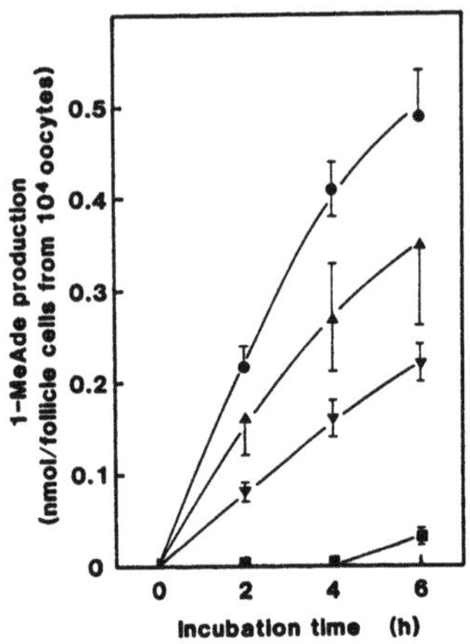

Fig. 3. Effect of temperature on 1-MeAde production in follicle cells. The follicle cells were incubated with GSS (0.05 mg/ml) at 0°C (■), 10°C (▼), 20°C (●), and 30°C (▲). Each point shows the mean ± S.E.M. of three separate experiments.

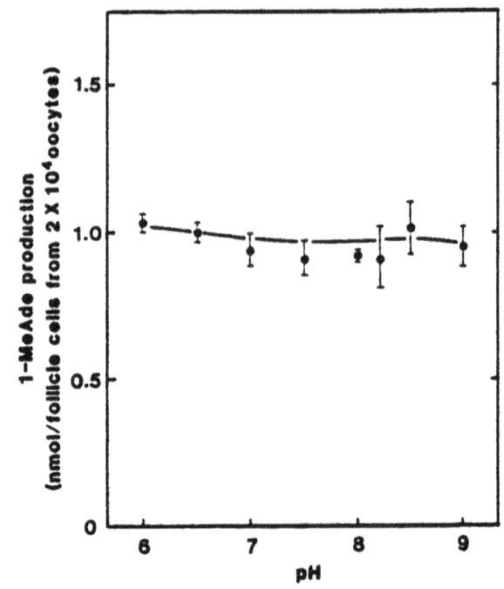

Fig. 4. Effect of pH of ASW on 1-MeAde production in follicle cells. The follicle cells were incubated with GSS (0.1 mg/ml) for 2 h at 20°C. The pH of ASW was adjusted with 20 mM borate buffer. Each point shows the mean ± S.E.M. of four separate experiments.

obtained at 20°C, but it reduced at 30°C. On the other hand, 1-MeAde biosynthesis in follicle cells was not influenced by the pH (between 6 and 9) of ASW (Fig. 4). Therefore, culture of follicle cells may be best done in ASW at pH 8.2 at 20°C.

After incubation for 2 h at 20°C, approximately 0.32 nmoles 1-MeAde was produced by 5×10^5 follicle cells (Table 1). Without GSS, the follicle cells never produced 1-MeAde. In general, hormonal actions are mostly accompanied by RNA synthesis and protein synthesis. However, neither actinomycin D which is an inhibitor of RNA synthesis nor cycloheximide and puromycin which are inhibitors the protein synthesis had any effect on 1-MeAde production induced by GSS (Table 1). This suggests that 1-MeAde biosynthesis in follicle cells does not require of synthesis of RNA and/or protein.

Table 2. Effects of GSS and IBMX on 1-MeAde production and the cAMP content in follicle cells

Condition	1-MeAde production (nmol/2 h/10^6 follicle cells)	cAMP content (pmol/10^6 follicle cells)
Zero time	——	12.3 ± 1.8
Control	0	11.2 ± 2.2
GSS (0.1 mg/ml)	0.87 ± 0.10	60.8 ± 12.0
IBMX (0.1 mM)	0.30 ± 0.03	52.1 ± 5.8
GSS + IBMX	0.92 ± 0.18	60.8 ± 7.7

Values represent the means ± S.E.M. obtained in three separate experiments.

The intracellular content of cAMP was determined with a commercial radioimmunoassay kit (Yanasa Shoyu Co., Chiba, Japan)(14). The amount of cAMP in 10^6 follicle cells was about 12 pmoles at zero time control (Table 2). The cAMP content increased 5-6 times following the administration of GSS (0.1 mg/ml). It thus appears that the effect of GSS on follicle cells results in the receptor-mediated formation of cAMP. The regulatory function of cAMP in 1-MeAde production was also found in the result that isobutylmethylxanthine (IBMX) as a potent inhibitor for cAMP phosphodiesterase induced 1-MeAde production in follicle cells (Table 2). At the same time, the cAMP content in follicle cells increased following treatment with IBMX. However, the 1-MeAde production in the presence of both GSS and IBMX was almost the same as that in the presence of GSS alone. It is likely that GSS (0.1 mg/ml) is sufficient to cause maximal activation of 1-MeAde biosynthesis.

In conclusion, cAMP was found to be a second messenger of GSS as a function of 1-MeAde production by starfish follicle cells. It is probable that cAMP activates 1-MeAde biosynthesis through the action of protein kinase, but independently of RNA synthesis and protein synthesis.

Acknowledgements

This study was carried out under the auspices of the NIBB Cooperative Research Program (86-171) and was supported in part by grants from the Ministry of Education, Science and Culture of Japan (No. 62740435) to M. M.

References

1. Kanatani, H. Exp. Cell Res., 57:333-337, 1969.
2. Kanatani, H., Shirai, H., Nakanishi, K. and Kurokawa, T. Nature (London), 221:273-274, 1969.
3. Kanatani, H. Int. J. Cytol., 35:253-298, 1973.
4. Kanatani, H. Biol. Fert., 1:119-140, 1985.
5. Hirai, S. and Kanatani, H. Exp. Cell Res., 67:224-227, 1971.
6. Hirai, S., Chida, K. and Kanatani, H. Dev. Growth Differ., 15:21-31, 1973.
7. Kanatani, H., Ikegami, S., Shirai, H., Oide, H. and Tamura, S. Dev. Growth Differ., 13:151-164, 1971.
8. Steinberg, D. and Huttunen, J. K. Adv. Cyclic Nucl. Res., 1:47-62, 1972.
9. Mita, M. Dev. Growth Differ., 27:563-572, 1985.
10. Kanatani, H. and Shirai, H. Dev. Growth Differ., 12:119-140, 1970.
11. Kanatani, H., Shirai, H. and Taguchi, S. Dev. Growth Differ., 18:25-31, 1976.
12. Shirai, H. Methods in Cell Biol., 27:73-88, 1986.
13. Shirai, H. Exp. Cell Res., 87:31-38, 1974.
14. Honnma, M., Satoh, T., Takezawa, J. and Ui, M. Biochem. Med., 18:257-273, 1977.

Part 3

PRODUCTION OF SPECIFIC PROTEINS BY CULTURED CELLS *IN VITRO*

SYNTHESIS OF SARCOPHAGA LECTIN AND SARCOTOXINS IN NIH-SAPE-4, AN EMBRYONIC CELL LINE OF SARCOPHAGA PEREGRINA

H.Komano, E.Kasama, *Y.Nakanishi, K.Matsuyama, K.Ando, Y.Nagasawa,and S.Natori.

Faculty of Pharmaceutical Sciences, University of Tokyo, Bunkyo-ku, Tokyo 113, Japan and *Faculty of Pharmaceutical Sciences, Kanazawa University, Kanazawa-shi, Kanazawa 920, Japan

An established cell line originating from a *Sarcophaga peregrina* embryo, NIH-Sape-4 was found to synthesize *Sarcophaga* lectin and sarcotoxins, but not typical fat body proteins.
From a series of experiments with *Sarcophaga* lectin and sarcotoxins, we have shown that these proteins participate in the defence mechanism of *Sarcophaga*.
Thus, these results indicate that NIH-Sape-4 synthesizes certain defence proteins constitutively and will be useful for large scale purification of these proteins.

I. *Sarcophaga* Lectin and Sarcotoxins

We have been studying a galactose specific lectin named *Sarcophaga* lectin and antibacterial proteins named sarcotoxin induced in the hemolymph of third instar larvae of *Sarcophaga peregrina* when their body wall is pricked with a hypodermic needle. These proteins are supposed to participate in the defence mechanism of this insect, since defence mechanism is expected to be enhanced under these conditions to eliminate pathogens invading the body through the damaged body wall.

1. What is the function of *Sarcophaga* lectin?

Sarcophaga lectin is a protein with molecular mass of 190,000 and consists of α and β subunits which have similar primary structure. The subunit structure is $\alpha_4\beta_2$.(1)
We found that the induction of *Sarcophaga* lectin was enhanced in response to invasion of foreign substances and that this lectin actually participates in the elimination of externally introduced sheep red blood cells into the abdominal cavity of larvae (2). These results show that *Sarcophaga* lectin certainly participates in the defence mechanism of this insect.

2. Antibacterial proteins induced in the hemolymph of *Sarcophaga* larvae.

Our results indicate that three groups of antibacterial proteins differ in mobility on polyacrylamide gel electrophoresis under non-denaturing conditions and are induced in the hemolymph when the body wall of a third instar larva is pricked with a hypodermic needle (3). Two of these groups of proteins have been purified. Sarcotoxin I is a group of proteins with a molecular mass of about

4,000 and consists of three proteins, sarcotoxin IA, IB, IC, which have almost the same primary structure (4). Sarcotoxin II, another group of proteins consists of three similar proteins, sarcotoxin IIA, IIB, and IIC, with a molecular mass of about 24,000 (5).

3. The expression of *Sarcophaga* lectin gene and sarcotoxin IA gene during normal development.

Recently, we have isolated cDNA clones for *Sarcophaga* lectin and sarcotoxin IA (6,7). Northern blot hybridization experiments using these cDNA clones as probes showed that the *Sarcophaga* lectin gene is transiently expressed in the early embryonic and early pupal stages (8), and that the sarcotoxin IA gene is also expressed in the early pupal stage, although its expression in the early embryonic stage is not clear. In fact, we actually detected *Sarcophaga* lectin in lysates of animals at these stages by radioimmunoassay (8,9). These results strongly suggest that in *Sarcophaga*, certain proteins participating in the defence mechanism are also indispensable for the ontogenic process.

II. Synthesis of *Sarcophaga* Lectin and Sarcotoxins in NIH-Sape-4, an Embryonic Cell Line of *Sarcophaga peregrina*.

NIH-Sape-4 is an established cell line originating from a *Sarcophaga peregrina* embryo (10). We found that this cell line synthesizes *Sarcophaga* lectin and sarcotoxin I and II, and secretes them into the culture medium.

1. Northern blot analysis of NIH-Sape-4 with four fat body cDNA clones.

First, using Northern blot hybridization with four cloned fat body protein cDNAs as probes, we investigated whether this cell line synthesizes mRNA for fat body proteins. Of these proteins, *Sarcophaga* lectin and sarcotoxin IA are synthesized in response to injury of the body wall, whereas the storage protein and 25-kDa protein are developmentally regulated, and are synthesized in the fat body of third instar larvae during normal development. As shown in Fig.1, the genes for both *Sarcophaga* lectin and sarcotoxin IA were expressed in the cultured cells, whereas those for the storage protein and 25-kDa protein were not expressed significantly. This result suggests that this embryonic cell line constitutively synthesizes the defence proteins that are synthesized in the larval fat body in response to body injury, but does not synthesize other fat body proteins.

2. Detection of antibacterial proteins in the culture medium of NIH-Sape-4

Since mRNA for sarcotoxin IA was detected in NIH-Sape-4 cells, we examined the antibacterial activity of the culture medium. For this, we subjected the culture medium to CM-cellulose chromatography and measured the antibacterial activity in the eluate, since the antibacterial proteins are known to be adsorbed to this resin under the conditions used. As a result, significant antibacterial activity was eluted from the resin. Then, we subjected fractions containing antibacterial activity to polyacrylamide gel electrophoresis, and located the antibacterial activity on the gel. As shown in Fig.2, three distinct main bands of material inhibiting the growth of E.coli were detected. Probably, this cell line produces various antibacterial proteins including sarcotoxin IA.

3. Purification of Sarcophaga lectin from the culture medium of NIH-Sape-4

Northern blot analysis showed that this cell line synthesizes mRNA for Sarcophaga lectin. If NIH-Sape-4 secretes this lectin, culture medium of these cells should be a better starting material than hemolymph for large-scale purification of this lectin.
We found the culture medium contained a protein that cross-reacted with antibody against Sarcophaga lectin and a substance that inhibited the hemagglutinating activity of Sarcophaga lectin. This Sarcophaga lectin could be separated from inhibitor on a column of butyl-Toyopearl, and purified to homogeneity.
The results of a typical purification are summarized in Table 1. The specific activity of the purified lectin was about the same as that of authentic Sarcophaga lectin from the hemolymph. About 50μg of purified Sarcophaga lectin could be obtained from about 400ml of the culture medium of NIH-Sape-4 cells.

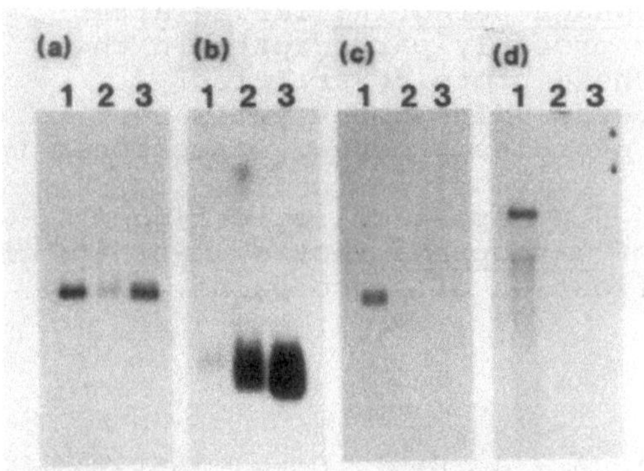

RNA was extracted from the fat body of injured larvae or NIH-Sape-4 cells and examined by Northern blot hybridization. The DNA probes used were (a) Sarcophaga lectin gene, (b) sarcotoxin IA gene, (c) 25-kDa protein gene, (d) storage protein gene. Lane 1, total RNA from the fat body; lane 2, total RNA from cultured cells; lane 3, poly(A)+RNA from cultured cells.

Fig.1 Detection of Sarcophaga lectin and sarcotoxin IA mRNA by RNA blot hybridization.

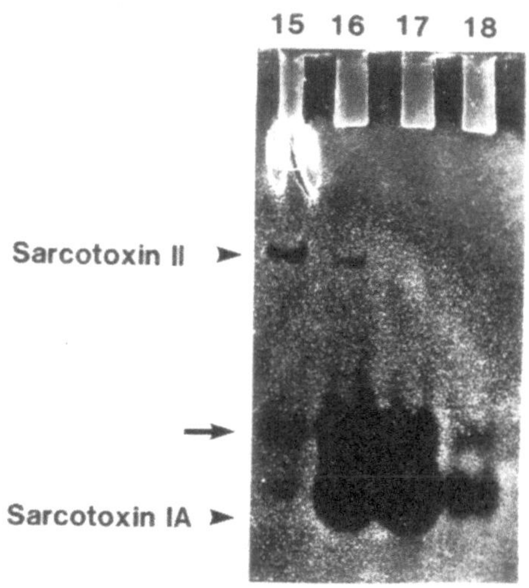

The culture medium was applied to a column of CM-cellulose and adsorbed material was eluted. The sample of each fraction was subjected to polyacrylamide gel electrophoresis, and antibacterial activity was located by examination of growth of E.coli overlaid on the gel. Lane numbers correspond to fraction numbers in the chromatogram on a column of CM-cellulose. The arrow indicates a spot of unidentified material.

Fig.2 Antibacterial activities of the culture medium.

Table 1. Summary of the purification of *Sarcophaga* lectin from the culture medium of NIH-Sape-4

	Total protein(mg)	*Sarcophaga* lectin(μg)*	Recovery (%)	Specific activity(titer/mg)
Culture medium	67	398		—
CM-cellulose	48	294	100	—
$(NH_4)_2SO_4$ precipitation	29	132	46	—
Butyl-Toyopearl	1.3	81	29	446**
Galactose-agarose	0.053	53	19	2792**

*The amount of protein was determined by radioimmunoassay.
**Hemagglutinating activity was assayed with sheep red blood cells and defined as the reciprocal of the maximum dilution of the test sample causing hemagglutination (titer)

We demonstrated that NIH-Sape-4 cells secrete *Sarcophaga* lectin and antibacterial proteins. Originally, these proteins were found to be synthesized in the fat body of *Sarcophaga peregrina* larvae after injury of their body wall, and they probably participate in the defence mechanism of this insect. These proteins are also indispensable for the ontogenic process. NIH-Sape-4 cells seem to synthesize selectively the fat body proteins produced in response to injury of the body wall.

We found that the culture medium of NIH-Sape-4 cells is a good starting material for purification of *Sarcophaga* lectin. Perhaps it will be also a good source for purification of sarcotoxins.

References

1. Komano, H., Mizuno, D. and Natori, S. J. Biol. Chem. 255:2919-2924. 1980.
2. Komano, H. and Natori, S. Devel. Comp. Immunol. 9:31-40, 1985.
3. Okada, M. and Natori, S. Biochem. J. 211:727-734, 1983.
4. Okada, M. and Natori, S. J. Biol. Chem. 260:7174-7177, 1985.
5. Andoh, K., Okada, M. and Natori, S. Biochemistry 26:226-230, 1987.
6. Takahashi, H., Komano, H., Kawaguchi, N., Kitamura, N., Nakanishi, S. and Natori, S. J. Biol. Chem. 260:12228-12233, 1985.
7. Matsumoto, N., Okada, M., Takahashi, H., Qu, X.-M., Nakajima, Y., Nakanishi, Y., Komano, H. and Natori, S. Biochem. J. 239, 717-722, 1986.
8. Takahashi, H., Komano, H. and Natori, S. J. Insect Physiol. 32:771-779, 1986.
9. Komano, H., Nozawa, R., Mizuno, D. and Natori, S. J. Biol. Chem. 258:2143-2147, 1983.
10. Takahashi, M., Mitsuhashi, J. and Ohtaki, T. Develop. Growth and Differ. 22:11-19, 1980.

ISOLATION OF A MORPHOGEN THAT INDUCES VESICLE FORMATION IN A CELL LINE DERIVED FROM IMAGINAL WING DISCS OF *TRICHOPLUSIA NI*

S. M. Ferkovich and H. Oberlander

Insect Attractants, Behavior, and Basic Biology Research Laboratory, Agricultural Research Service, U.S. Department of Agriculture, Gainesville, Florida 32604, U. S. A.

A factor that promotes vesicle formation (VPF) in an insect cell line TND1 was purified partially from larval hemolymph of the cabbage looper, *Trichoplusia ni* (Hübner). The estimated molecular mass for VPF was between 20,500 and 37,500 daltons by gel permeation, 22,500 daltons by the Ferguson plot on nondenaturing PAGE and 16,880 daltons on SDS PAGE. VPF fractions were isolated from hemolymph by gel permeation and then subjected to preparative isoelectric focusing. Two active fractions containing the polypeptide from the isoelectric separations had isoelectric points of 6.18 and 6.32 and had specific activities of 34 and 32 vesicles formed per test culture chamber per µg protein, respectively.

The study of imaginal wing discs and their response to hormones *in vitro* has led to a better understanding of how insect hormones control metamorphosis in holometabolous insects (1, 2). More recently, hormone-responsive cell lines from specific tissues such as wing discs have been established (3). One such cell line, IAL-TND1, was developed from imaginal wing discs of the cabbage looper moth, *Trichoplusia ni*. The cell line grows in aggregate (clumped) form (Fig. 1A) but the first year in culture grew as multicellular vesicles (Fig. 1B). Addition of larval hemolymph reversed the morphology from multicellular aggregates to multicellular vesicles (4). Preliminary characterization of the hemolymph factor indicated that it was of a polypeptide nature. Moreover, additional information indicated that the transformation of the vesicle to the aggregate form was accompanied by the cells acquiring the capacity to produce 20-hydroxyecdysone (5, 6).

In this chapter, we present information on the isolation and purification of the vesicle promoting factor (VPF) from larval hemolymph. We also show how this cell line, which both produces and responds to 20-hydroxyecdysone, is a favorable model system for studying the mechanism by which a hormone and morphogen interact to control cellular morphogenesis.

I. General Features of the Cell Line and Vesicle Bioassay

The cell line, designated as IAL-TND1, was established from epithelial cells of imaginal wing discs of *T. ni*. It grows as suspended multicellular vesicles and aggregates. Its kariology is near diploid with a mode of 56 and has a growth rate increase in density of 3- to 8-fold per week (7).

Vesicle promoting activity was assayed by adding the sample (15-60 µg protein) to the cell aggregates in multiwell plates and counting the number of vesicles produced under an inverted microscope. Hemolymph-free medium was used as a control (4). Also, 500 to 600 µg of protein from whole hemolymph was run in the bioassay as a standard.

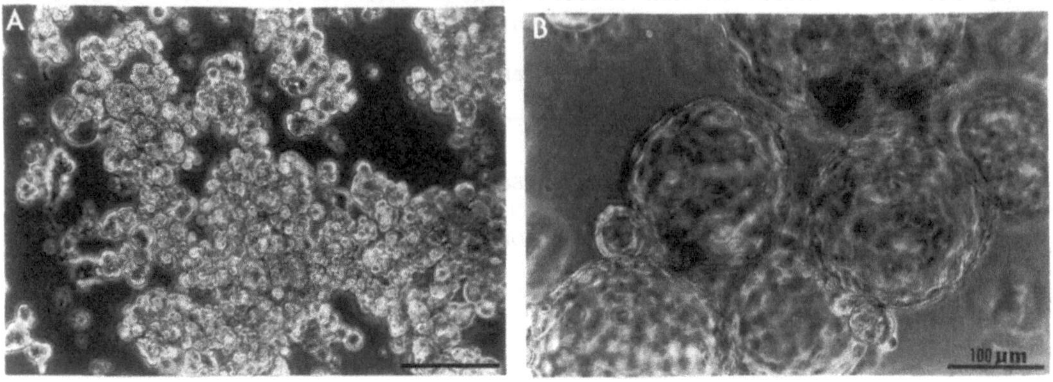

Fig. 1. Results of experiment in which TND1 cells in aggregate form (A) were exposed to 2% T. ni hemolymph that contained VPF and changed to multicellular vesicle form (B) after 6 days of exposure (5).

II. General Features of the Vesicle Promoting Factor

Several properties of the VPF that were determined earlier (4) are as follows: 1) response to hemolymph in the vesicle bioasay is concentration dependent; 2) vesicle promoting activity was present in every developmental stage tested including fourth and fifth instar (last) larvae and pupae; 3) the VPF was detected in three other species of Lepidoptera tested, Spodoptera frugiperda, Galleria mellonella and Manduca sexta, although the greatest activity occurred in T. ni hemolymph; 4) the VPF was tissue specific, with greatest activity in the fat body and the imaginal wing discs; 5) the VPF was heat labile, nondialyzable (10,000 daltons cut off membrane) and sensitive to a nonspecific protease (protease XIV from Streptomyces griseus) but not trypsin. Based on this information, we decided to look for a polypeptide in the larval hemolymph of T. ni. Previously, we reported that exclusion of VPF by dialysis membrane suggested an apparent molecular mass that was greater than 60,000 daltons. However, we know now that VPF readily forms aggregates and that the polypeptide has a molecular mass less than 60,000 daltons (8).

III. Purification of VPF

Initially, hemolymph was separated on a gel permeation column (1000-700,000 daltons fractionation range, Fracto-gel®) (Fig. 2) (8). VPF activity eluted between the 43,000 daltons ovalbumin standard and the 13,500 daltons standard of ribonuclease A. The 25,000 daltons standard of chymotrypsinogen eluted within the area of VPF activity, indicating that the factor had a molecular weight close to this standard. Next, we prepared a calibration curve from a gel with a more narrow fractionation range (500-80,000 daltons). The molecular mass of the VPF was estimated to be between 20,500 to 37,000 daltons. After the gel permeation step, the active fraction was unstable during subsequent separations and storage (8).

The active fractions from the gel permeation column then were lyophilized and separated by preparative isoelectric focusing with a pH range of 5-7. Figure 3A shows the amount of protein in the fractions and Figure 3B shows the bioassay results. Fraction 8 had the most activity, while some activity was present in fraction 9 but not in the other fractions. When we examined these fractions on a nondenaturing polyacrylamide gel, a broad band, labeled b, stained heavier in fraction 8 and lighter in fraction 9. A slower migrating band, labeled a, also was present in fractions 8 and 9, but did not have vesicle promoting activity. Fraction 7 (Fig. 3B), which was applied to lane 1 (Fig. 3C) contained band a but not b, and did not have VPF activity. When fractions 8 and 9 were run on an SDS dissociating gel, the most evident bands were a 108,000 dalton band and a 73,000 dalton band and a broad, heavy staining band with a molecular mass of 16,880 daltons. Because the gel permeation data indicated the VPF was a low molecular weight polypeptide, we suspected that the 16,880 dalton band was the active band and that it originated from band b on the nondenaturing gel in Figure 3C. When band a was cut from the nondenaturing gel and run on an SDS gel (Fig. 4), only a 73,000 dalton band was evident (lanes 2, 4 & 6). When band b was cut from the nondenaturing gel and run on an SDS gel, only a 16,880 dalton band was evident. As indicated in Figure 3B,C, fraction 7, containing band a, did not have VPF activity but fractions 8 and 9, containing band b, (16,880 daltons) had vesicle promoting activity. A band with an Rf similar to band b also was purified by chromatofocusing and shown to have VPF activity (8).

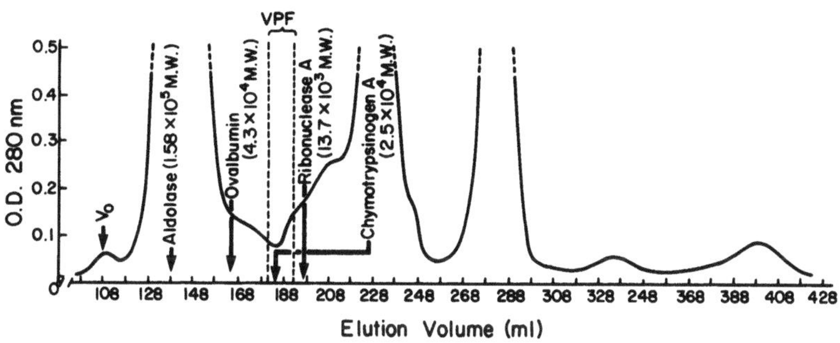

Fig. 2. Molecular-sieve chromatography of cabbage looper larval hemolymph on Fracto-gel. VPF indicates tubes that contain vesicle promoting activity in cell bioassay (8).

The molecular weight of the active fraction 8 from the isoelectric focusing separation (Fig. 3B) was estimated on nondenaturing polyacrylamide gels using the Ferguson plot. The apparent molecular weight of the broad band observed in Figure 3C (band b) had a molecular weight of 22,500 daltons (8).

In summary, the estimated molecular weight for VPF was between 20,500 and 37,500 daltons by gel permeation, 22,500 daltons by the Ferguson plot on nondenaturing PAGE and 16,880 daltons on SDS PAGE (8).

Of the total hemolymph protein, the VPF comprises 0.75% and was active in the VPF bioassay at 10^{-6} M.

IV. Inhibitors

As an initial approach to the mode of action of VPF we tested various inhibitors of cellular processes for their effect on vesicle formation (Table 1). Colchicine and vinblastin, which inhibit microtubule formation and prevent ecdysone-induced tracheole migration in imaginal wing discs (9) also inhibited vesicle formation. On the other hand, cycloheximide, a protein synthesis inhibitor, cytochalasin B, a contractile microfilament inhibitor, and tunicamycin, an inhibitor of glycosylation, did not affect vesicle formation. Actinomycin D was toxic to the cells. Additional information is needed at the ultrastructural level of the cells to interpret the effects of the inhibitors on vesicle formation. However, the inhibitory effects of colchicine and vinblastin suggested that the formation of vesicles is dependent on the integrity of the microtubules.

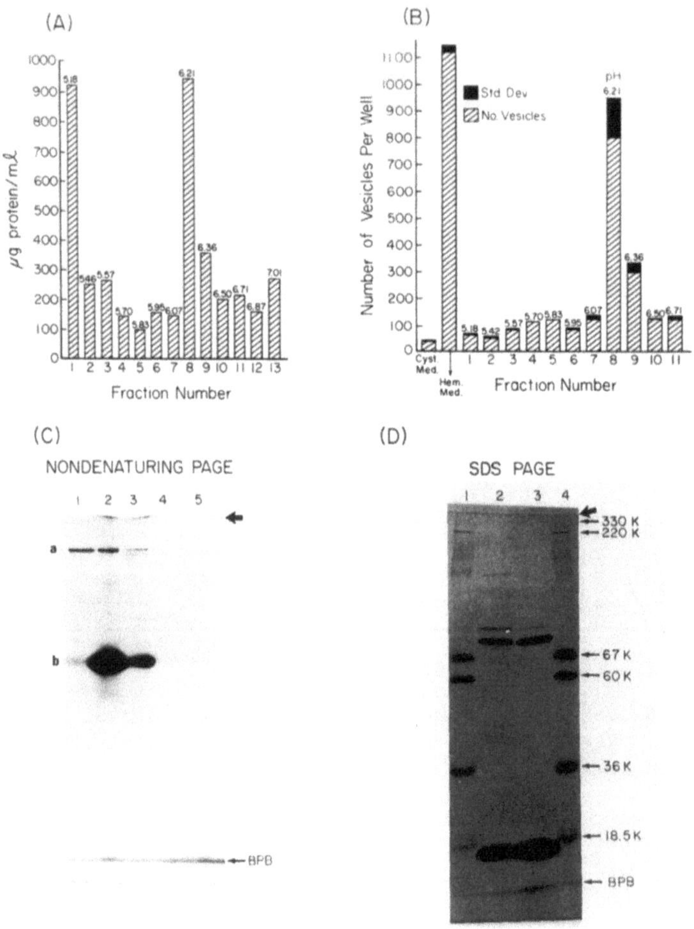

Fig. 3. Isoelectric focusing of combined fractions that contain VPF from molecular-sieve chromatography of cabbage looper larval hemolymph. (A) Protein content and pH of fractions. (B) VPF activity and pH of fractions. (C) Nondenaturing PAGE of fractions 7, 8, 9, 10 and 11 in lanes 1, 2, 3, 4, and 5, respectively. BPB, bromophenol blue marker. (D) SDS PAGE of fractions 8 and 9 in lanes 2 and 3. Molecular weight standards in lanes 1 and 2: thyroglobulin (330 Kd), ferritin (220 Kd), albumin (67 Kd), catalase (60 Kd), lactate dehydrogenase (36 Kd), ferritin subunit (18.5 Kd) and BPB, bromophenol blue marker. Heavy arrows in C and D indicate surface of separating gel. Gels (12.5 %) were stained with Coomassie blue (8).

Table 1. Effects of inhibitors on hemolymph-induced vesicle formation.

Inhibitor	Concentration (μg/ml)	% of Hemolymph control
Colchicine	0.02	33.3
	0.2	12.0
Vinblastine	0.02	13.4
	0.2	8.5
Cycloheximide	0.2	100.0
	2.0	100.0
Cytocholasin B	2.0	100.0
	20.0	100.0
Tunicamycin	0.02	100.0
	0.2	100.0
Actinomycin D*	0.02	0.0
	0.2	0.0

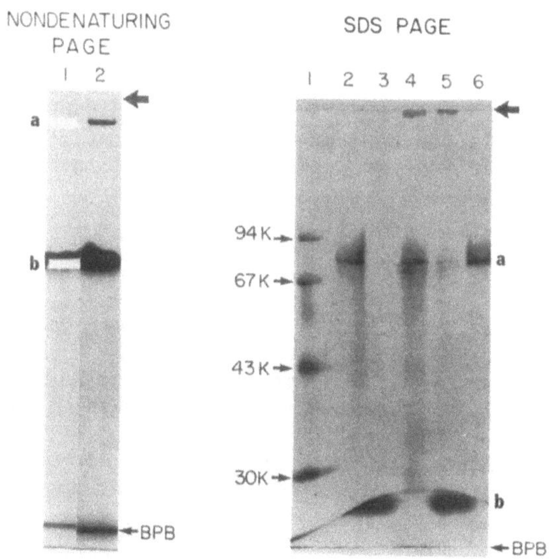

Fig. 4. Protein bands from a VPF active fraction cut out of nondenaturing polyacrylamide gel and run on SDS PAGE. The VPF fraction was isolated by preparative isoelectric focusing as described in Figure 2. Nondenaturing PAGE: lanes 1 and 2, VPF active fraction from isoelectric focusing separation, isoelectric point 6.21. The bands in lane 1 were cut out of the gel and placed on the surface of the SDS gel and then electrophoresed. SDS PAGE: lane 1, molecular weight standards, carbonic anhydrase (30 Kd); lanes 2, 4 and 6 band labeled a from lane 1, nondenaturing PAGE; and lanes 3 and 5 band labeled b from lane 1, nondenaturing PAGE. BPB, bromophenol blue marker, stained with Coomassie blue.

V. Ecdysteroids

20-Hydroxyecdysone at physiological levels of the hormone (0.1 and 1.0 µg/ml) was inhibitory to vesicle formation (6). 22-Iso-ecdysone, which is a known inactive analog of the hormone in the molting process, also was inactive in the assay. Thus, the hormone specificity and concentration of 20-hydroxyecdysone required for inhibition of the response are similar to that of a positive response to the hormone in wing discs in culture.

Additionally, the aggregate form of the cell line is now known to produce 20-hydroxyecdysone (5, 6). It was found that imaginal wing discs which were cultured in aggregate-conditioned medium changed morphologically in a manner similar to treatment with physiological levels of 20-hydroxyecdysone by undergoing evagination and peripodial membrane separation, later steps in disc development (5). Since the VPF induces the aggregate form to produce vesicles and the effect of VPF can be overridden by addition of physiological levels of 20-hydroxyecdysone, the question then becomes, "does VPF destroy the ecdysteroids produced by the cells?" This hypothesis was tested by incubating hemolymph with 20-hydroxyecdysone and then placing imaginal discs in the medium to see if they would exhibit some of the characteristic responses to ecdysteroids (5). Two percent hemolymph plus 500 ng or 10^{-6} M 20-hydroxyecdysone did not prevent action of the hormone. When the hemolymph concentration was raised to 20% and the conc. of hormone lowered 10-fold, the hormone was still effective. Therefore, VPF did not appear to act by destroying or inactivating the hormone, and must induce vesicle formation by some other means.

VI. Conclusions

The cell line and the VPF provide a promising model system for studying the control of cell morphogenesis in vitro. We have isolated and purified a polypeptide that induces vesicle formation and an inhibition of this process can be effected by addition of 20-hydroxyecdysone. The relationship of this in vitro system to development in vivo is the subject of continued investigation. Mention of a proprietary product does not constitute an endorsement or the recommendation for its use by the USDA.

References

1. Oberlander, H., and Lynn, D. E. Advances in Cell Culture Volume 2, (Maramorosch, K., ed.), pp. 237-265. 1982.
2. Doctor, J. S., and Fristrom, J. W. Comprehensive Insect Physiology, Biochemistry, and Pharmacology, Volume 2, (Kerkut, G. A. and Gilbert, L. I., eds.), pp. 201-235. 1985.
3. Lynn, D. E., and H. Oberlander. In Vitro Invertebrate Hormones and Genes (Kurstak, E. ed.), Techniques In the Life Sciences C213, pp. 1-12, Elsevier Scientific Publishers, Ireland. 1986.
4. Lynn, D. E., Oberlander, H., and Ferkovich, S. M. In Vitro Cell Dev. Biol. 21: 277-281. 1985.
5. Oberlander, H., Leach, C. E., Lanka, S., and Willis, J. Arch. Insect Biochem. & Physiol. 5: 81-89. 1987.
6. Lynn, D. E., Feldlaufer, M. F., and Lusby, W. R., Arch. Insect Biochem. Physiol. 5: 71-79. 1987.
7. Lynn, D. E., Miller, S. G., and Oberlander, H. Proc. Natl. Acad. Sci., USA, 79: 2589-2593, 1982.
8. Ferkovich, S. M., Oberlander, H., Dillard, C., and Leach, C. E. Arch. Insect Biochem. Physiol. (in press).
9. Oberlander, H. Insect Biology of the Future (Locke, M., and Smith, D. S. eds.), pp.423-438. Academic Press, N.Y. 1980

LEPIDOPTERAN CELL FUSION AND FORMATION OF INSECT-MAMMALIAN HETEROKARYONS

Chen Q., Shi X.,
Yu Z., and Xiong X.

Insect Virus Research Laboratory, Department of Biology, Huazhong Normal University, Wuhan, China

This paper is concerned with polyethylene glycol-mediated lepidopteran cell fusion and the formation of insect-mammalian heterokaryons as well as their development to synkaryons. Seven lepidopteran cell lines were fused with mouse thymic cells (MTC) or HeLa cells by treatment with PEG and DMSO. The mixture of 45% PEG and 5% DMSO gave the best fusion incidences. Pretreatment or simultaneous treatment with trypsin promoted cell fusion frequency. Some of the heterokaryons of Bombyx mori cells and HeLa cells developed to synkaryons after fusion.

Artificial cell fusion as a tool has been contributing a lot to cell biology since Okada's discovery (1) in the 1960's that Sendai virus could mediate cell fusion. But polyethylene glycol (PEG) rapidly substituted for Sendai virus for its high fusogenicity, reproductivity, and simplicity. In order to know the fusibility of insect cells and to establish insect-mammalian hybrids which may finally be used to produce biologically active peptides, somewhat like the work of monoclonal antibody hybridoma (2), or to locate genes, as conducted by Ruddle (3), attempts have been made to fuse lepidopteran cells with mammalian cells. The main results are reported here.

Materials and Methods

The seven lepidopteran cell lines, TN-CL1 of Trichoplusia ni (TN) (4), SCLd-135 of Lymantria dispar (Ld) (5), BCIRL-SO3-HNU1 of Spodoptera ornithogalle (SO3) (6), IPLB-SFP-21 of Spodoptera frugiperda (SFP) (7) and Bm-21E-HNU5 of Bombyx mori (Bm) (8), IMC-Hz-1 of Heliothis zea (IMC-1)(9), BCIRL-PX2-HNU3 of Plutella xylostella (PX) (10), were cultured in TC-199-MK medium with 10% calf serum at 28°C. When used for fusion, cells in logarithmic growth stage were collected. Fresh mouse thymic cells (MTC) were asceptically separated from mouse thymi. HeLa cells were cultured conventionally.

Methods for fusion mainly followed Davidson's (11). For selection of heterokaryons, the method that has been advanced by Wright (12) was adopted. Briefly, the insect cells were treated with 2.0 mM iodoacetamide (IAM), which blocks the activities of mercapto-molecules, before they fused with MTC, while HeLa cells were treated with 3.5 mM IAM before they fused with Bm cells.

Results and Discussion

Unlike mammalian cells, most of the seven species of insect cells were very sensitive to the toxicity of PEG. Only SFP and IMC-1 could tolerate the treatment of 50% PEG. The others lost their viabilities progressively with increasing concentrations of PEG and had no viabilities at all when PEG reached 50%. But cell fusion frequency plummets when PEG concentration is slightly lower than 50%, as concluded by Davidson (11). Therefore, a sharp contradiction exists between cell viability and cell fusion frequency.

Synergic effect of dimethyl sulfoxide (DMSO). Low concentration of PEG may effectively induce cell fusion when it is combined with DMSO for DMSO itself could mediate cell fusion (13). Under microscope, the three combined solutions, 35%PEG+15% DMSO, 40%PEG+10%DMSO and 45%PEG+5%DMSO, not only caused cell aggregation but also were toxically permissible to cells, suggesting they were usable. Table 1 shows that

Table 1. Fusogenicity of fusogenic solution[a]

Solvent	Fusogenic solutions		CFF	Fusion incidences of		
	C of PEG	C of DMSO		DK	TK	PK
TC-199-MK	50%		7.24	4.51	0.55	2.18
	45%	5%	9.89	7.63	1.41	0.85
	40%	10%	10.45	5.89	1.70	2.86
	35%	15%	10.47	6.39	1.27	2.82

[a]SFP cells were used.
CFF : cell fusion frequency (%); C : concentration;
DK : dikaryon; TK : trikaryon PK : polykaryon.

the solution of 45% PEG + 5% DMSO is the best one because it induced the highest incidence of dikaryons (as much as 77% of the total fused cells). It also reveals that DMSO did promote the fusogenicity of PEG. And above all, the decrease of PEG reduced its toxicity. So, Bm and TN cells grew well after treatment with 45% PEG + 5% DMSO solution.

Promotion of trypsin in cell fusion. When used in cell fusion, 0.025% and 0.01% trypsin was adopted. The result (Table 2) showed that either pretreatment or

Table 2. Promotion of trypsin in cell fusion[a]

Treatment	no T	45%PEG+5%DMSO	25%PEG+5%DMSO+0.01%T	
		0.025%T pretreated for 5 min	Simultaneously treated for 2 min	for 5 min
CFF (%)	9.89	21.89	8.67	19.67

simultaneous treatment of trypsin promoted cell fusion frequency, even if PEG concentration was as low as 25%. Postponed treatment of trypsin could increase cell fusion frequency, too.

Trypsin digests extrinsic proteins and the outside portions of intrinsic proteins because arginine and lysine residues, whose carboxyl group-formed peptide bonds are broken by trypsin, are distributed on the surface of membrane for their hydrophil. As results, exclusion volume of membrane glycoproteins (14) dwindles and membrane fluidity (15) increases, which make cells more easily induced together and the intramembranous proteins more readily aggregated by PEG. The above also explains that protracting the treatment of trypsin increases cell fusion frequency.

Influence of pH values. Different solvents affected the fusogenicity of fusogenic solutions. When Puck's balanced salt solution (PBSS) and TC-199-MK were used as solvents to prepare 45% PEG + 5%DMSO solution, respectively, 11.89% and 9.89% SFP cells were fused. But when their pH values were adjusted to 6.7, the fusion frequencies they mediated were nearly the same, 11.03% of PBSS and 11.19% of TC-199-Mk, suggesting that it was the pH values of the solvents that affected the fusogenicity. The relationship between pH values and cell fusion frequencies is shown in Figure 1.

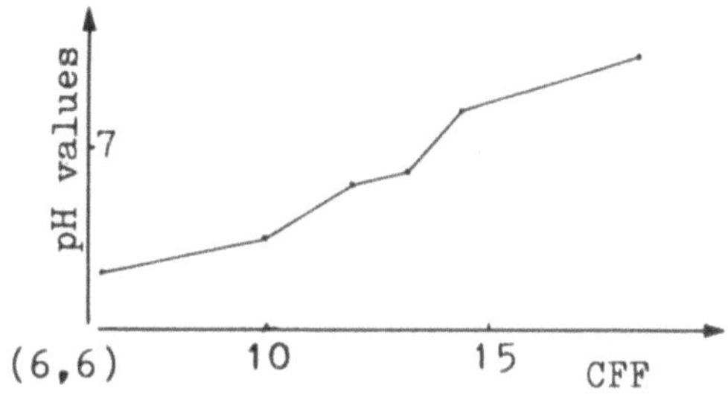

Fig. 1. The relationship between pH values and cell fusion frequencies (SFP cells were used.)

Fusibility of lepidopteran cells and mammalian cells. Lepidopteran cells were not so fusible as mammalian cells. Their fusion frequencies were much lower than those of mammalian cells under the same fusion condition (Table 3).

Formation of interphylum heterokaryons. Mouse thymic cells and their nuclei are much smaller and much more deeply stained with Giemsa's than lepidopteran cells (Figure 2). And every nucleus of HeLa cells has

Table 3. Fusibility of lepidopteran cells and mammalian cells[a]

	Lepidopteran cells			Mammalian cells	
	Bm	SFP	IMC-1	MTC	HeLa
CFF (%)	16.9	9.89	3.49	41.67	35.29

[a] Fusogenic solution was 45% PEG + 5% DMSO + TC-199-MK (pH 6.5).

two to three nucleoli while that of Bm cells does not when rapidly stained with Giemsa's in culture (Figure 3). These morphological differences helped a lot to discriminate heterokaryons. About 3.2% heterokaryons of SFP-MTC, Bm-MTC or Bm-HeLa were induced by the solution of 45% PEG + 5% DMSO + TC-199-MK (pH 6.5).

Twenty-four hours after fusion Bm- or SFP-MTC heterokaryons lost their viability, maybe owing to the undivided MTC-formed portion of the heterokaryon cannot supply the heterokaryon with the molecular activities that had been inhibited by IAM in the contribution of Bm or SFP cell. In the fusion of HeLa cells with Bm cells, almost all of the unfused HeLa cells and their homokaryons died and became detached after IAM treatment. They were aspired away 24 hours later, along with Bm cells and Bm homokaryons because Bm cells are floating culture. On the bottom of the glass flask, only the viable heterokaryons and a few unkilled HeLa cells remaine attached.

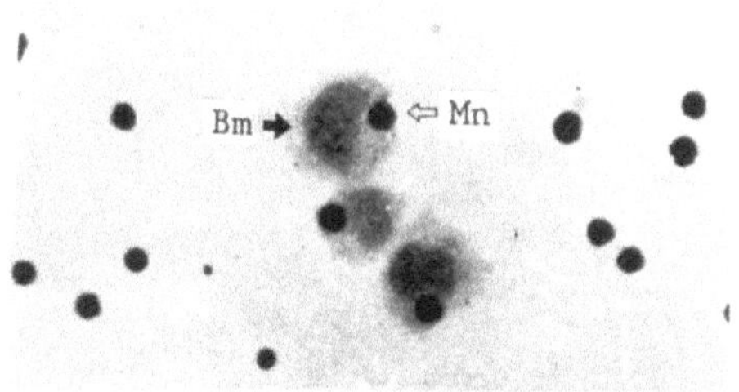

Fig. 2. Bm-MTC heterokaryons
Bm: nucleus of Bm cell
Mn: nucleus of MTC.

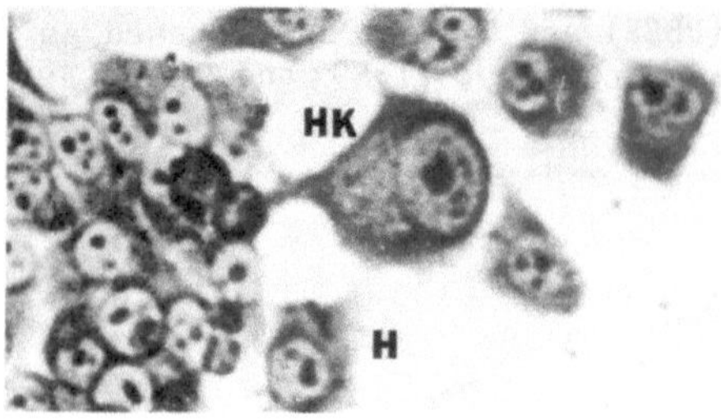

Fig. 3 Bm-HeLa heterokaryons
HK: heterokaryons:
H: HeLa cell, having two or three nucleoli.

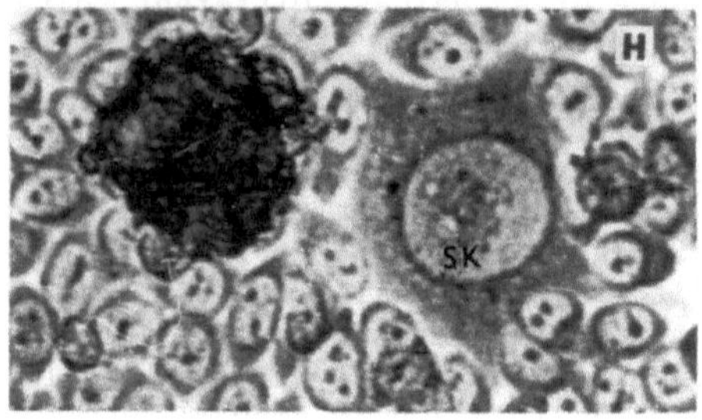

Fig. 4 Bm-HeLa synkaryon
SK: synkaryon;
H: HeLa cell.

Some of the heterokaryons began to develop to synkaryons forty-eight hours after fusion (Figure 3) and synkaryons could be visualized ten days later (Figure 4).

References

1. Okada, Y. Exp. Cell Res., 26:98-107, 1962.
2. Köhler, G. and Milstein, C. Nature, 256:495-497, 1975.
3. Ruddle, F. H. Nature, 294:115-120, 1981.
4. McIntosh, A. H., and Rechtoris, C. In Vitro, 10:1-5. 1974.
5. Hink, W. F. (unpublished).
6. Chen Q. J. Central China Teacher's College (Natural Sci. Ed.), No. 3:101-108, 1984.
7. McIntosh, A. H. and Ignoffo, C. M. J. Invertebr. Pathol., 38:395-403, 1981.
8. Chen, Q., et al. (in press).
9. Hink, W. F. and Ignoffo, C. M. Exp. Cell Res., 60:307-309. 1970.
10. Chen Q. J. Central China Teacher's College (Natural Sci. Ed.), No. 3:99-103, 1983.
11. Davidson, R. L. and Gerald, P. S. Methods in Cell Biology (E. M. Prescot, Ed.) Vol. 15:325-327, Academic Press, New York, 1977.
12. Wright, W. E. Somat. Cell Genet., 7:769-775, 1981.
13. Klebe, R. J. and Mancuso, M. G. Somat. Cell Genet., 7:473-478, 1981.
14. Knutton, S. J. Cell Biol., 36:64-72, 1979.
15. Majundar, S. and Baker, R. F. Exp. Cell Res., 126:175-182, 1980.

LIPOPROTEIN (VLDL & HDL) SYNTHESIS BY EEL-HEPATOCYTES CULTURED IN SERUM-FREE MEDIUM

S. Hayashi, N. Nishi, and Z. Ooshiro

Faculty of Fisheries, Kagoshima University,
Shimoarata 4-50-20, Kagoshima 890, Japan

Hepatocytes of eel, *Anguilla japonica*, were cultured in completely defined medium without serum in a plastic culture-dish precoated with fibronectin. The medium consisted of Williams' medium E, insulin, glucagon, EGF, prolactin, growth hormone, and H_2SeO_3. Insulin, glucagon and prolactin were necessary for the maintenance of normal morphology of the cells. Many kinds of proteins were synthesized and secreted into the medium by the cells. Most of the secreted proteins were the proteins which were present in the eel serum. The main proteins in the eel serum were lipoproteins, which were 35 to 45% of all the proteins. Very low density lipoprotein (VLDL) and high density lipoprotein (HDL) were secreted by the cells after an 8 h lag period.

We have used freshly isolated hepatocytes(1) or cultured hepatocytes(2,3) of eel to investigate metabolism in fish liver. We have found that the cultured hepatocytes have a higher rate of protein and glycogen synthesis than the freshly isolated hepatocytes(2). Similar results have been obtained from comparative studies on various activities of freshly isolated hepatocytes of adult rats and the various activities of similar cells that have been cultured for a few days(4).

Recently we have done the primary culture of eel hepatocytes in serum-free medium (3). It is obvious that a serum-free medium is more suitable for investigating the effect of hormones on cultured cells and the synthesis of secreted proteins by cultured cells than a medium containing serum.

This report describes the morphological effect of hormones on the eel hepatocytes cultured in serum-free medium and the synthesis of serum lipoproteins by the cultured hepatocytes.

Materials and Methods

Eel. Japanese cultured eels, *Anguilla japonica*, weighing 200 to 250 g were used.

Materials. Insulin from bovine pancreas was purchased from Fluka, AG.; glucagon from sheep pituitary from Sigma Chemical Co.; EGF from mouse submaxillary glands from Bioproducts Co.; and plastic dishes for cell culture from Corning Co. and Becton and Dickinson Co. L-(U-^{14}C)leucine and (^{14}C)methylated proteins were obtained from Radiochemical Centre, Amersham. Collagenase(EC3.4.24.3) from *Cl.histoliticum* was purchased from Wako Pure Chemical Industries.

Primary culture of eel hepatocytes. Isolated eel hepatocytes were prepared by perfusing liver with Ringer solution containing collagenase(5). Plastic dishes precoated with fibronectin at the density of 2.5 to 5.0 μg/cm^2 were used. Fibronectin was isolated from horse serum by gelatin-agarose column(2). The culture medium reported originally by Enat et al.(6) was modified slightly and the modified medium consisted of Williams' medium E, insulin (10 μg/ml), glucagon (1 μg/ml), EGF (50 ng/ml), prolactin (20 mU/ml), growth hormone (10 μU/ml), 3 nM H_2SeO_3, and 23 mM $NaHCO_3$. Isolated eel hepatocytes were cultured in the medium at 30 °C in a humidified 5% CO_2 atmosphere.

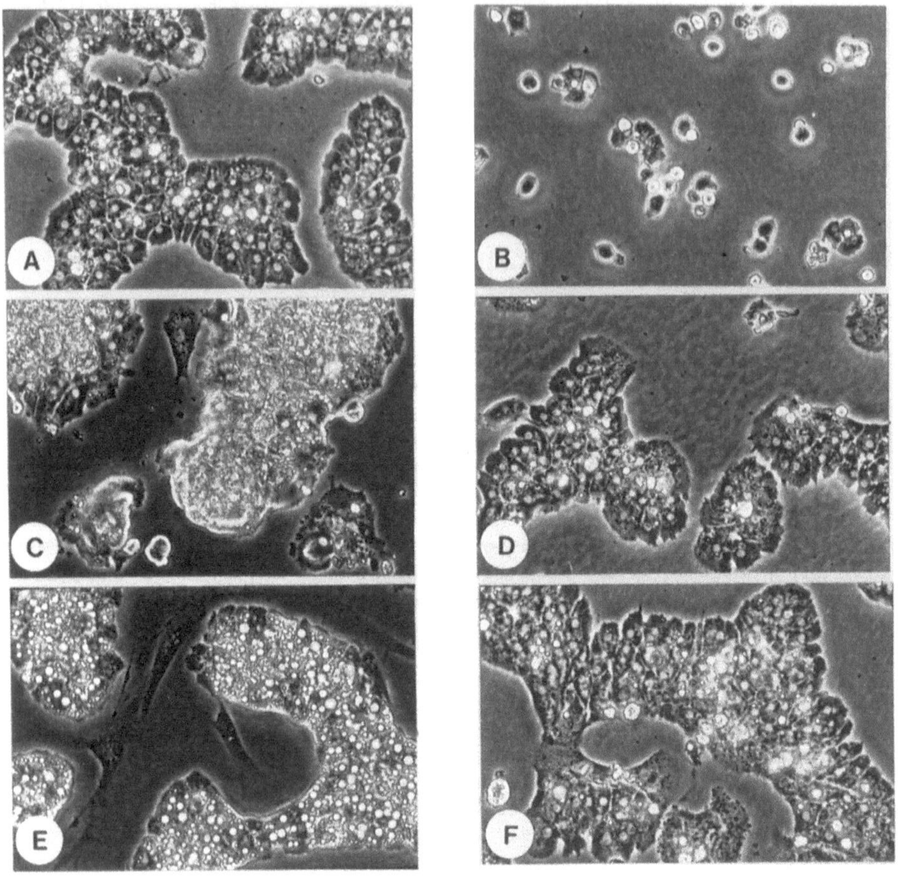

Fig. 1. Effect of the medium lacking one component on the hepatocytes cultured for 11 days. Phase-contrast micrographs were made and all magnifications are x150. Hepatocytes were cultured in the complete medium (A), and insulin (B), glucagon (C), EGF (D), prolactin (E), and growth hormone deficient medium (F), respectively.

Isolation of lipoprotein. Potassium bromide was added to eel serum to attain a density of 1.21 and the serum was centrifuged at 40,000 rpm for 40 h at 15 °C. Then lipoprotein fraction was pooled from the upper layer in a centrifugal tube and applied on a Bio-Gel A15m column, which was equilibrated with 0.2 M phosphate buffer (pH 7.4) -0.1 M NaCl-0.01% EDTA. Molecular weight of apolipoprotein was determined by the method of Laemmli(7) and its isoelectric point was determined by electrofocusing with immobiline(LKB Product Co.).

Analysis of lipid. Lipid was extracted from lipoproteins with chloroform:methanol (1 : 2) and analyzed by a thin layer chromatographic apparatus with a flame ionization detector(Iatroscan TLC-FID system) and gas chromatography.

Protein synthesis. After the incubation of cultured hepatocytes with ^{14}C-leucine, 0.1 mM leupeptin and pepstatin were added to the medium. Then the medium was applied on a Sephadex G-25 column, the protein fraction was pooled and lyophilized.

Lyophilized proteins were dissolved in a small amount of 0.05 M Tris-Cl (pH 6.8) -1.67% SDS-0.56 M 2-mercaptoethanol and used for the determination of radioactivity as extracellular proteins and for fluorography. On the other hand, to the cells on the dish trichloroacetic acid (TCA) was added and TCA insoluble fraction was precipitated by centrifugation. The precipitate was dissolved by 0.01 N NaOH and the radioactivity was measured as intracellular proteins.

Results and Discussion

Effect of hormones and EGF on the morphology and cell number of cultured hepatocytes

The spreading of the hepatocytes cultured in the completely defined medium was observed from one day culture and the attachment between hepatocytes and the clear nuclei was observed from 3 days culture.

To investigate the morphological effect of each hormone and EGF in the medium on hepatocytes, hepatocytes were cultured in the medium lacking one component in the completely defined medium. The insulin deficient medium remarkably affected the hepatocytes (Fig. 1-B). The cells maintained a spherical shape during 11 days culture, almost the same morphological feature as the freshly isolated hepatocytes.

Good spreading and the attachment of the hepatocytes cultured in glucagon or prolactin deficient medium were observed within 6 days culture. However, after 7 days culture the hepatocytes accumulated vacuoles like oil droplets and the spreading was inhibited (Fig. 1-C,E). The morphology of the hepatocytes cultured in growth hormone or EGF deficient medium was almost the same as that of those cultured in the complete medium.

When the number of hepatocytes cultured in the complete medium for 11 days was 100%, those of the hepatocytes cultured in insulin, glucagon, and EGF deficient medium were 53, 58, and 59%, respectively. As described above, the EGF deficient medium had little morphological effect on the hepatocytes, but had remarkable effect on the cell number.

These results indicated that insulin was essential for the culture of eel hepatocytes.

Isolation and identification of very low density lipoprotein (VLDL) and high density lipoprotein (HDL) in eel serum

Lipoprotein fraction obtained by centrifugation of eel serum at 40,000 rpm for 40 h was separated to Fractions I and II by gel-filtration on a Bio-Gel A15m column. From the determination of protein content and molecular weights of proteins by SDS-polyacrylamide gel electrophoresis and the analysis of lipid compositions, Fractions I and II were identified as VLDL and HDL, respectively (Table 1).

The amount of proteins in VLDL and HDL was 35 to 45% of all eel serum proteins.

VLDL and HDL synthesis by cultured hepatocytes

After the addition of ^{14}C-leucine into the completely defined medium the hepatocytes were incubated for 2, 4, 8, 16, and 32 h and the incorporation of ^{14}C-leucine into intra- and extracellular proteins at each incubation time was determined. As shown in Fig. 2 the incorporation into intracellular proteins was higher than that into extracellular proteins during the indicated incubation.

Table 1. Identification of VLDL and HDL in the eel serum.

Analysis	Fraction I		Fraction II
Protein	4.7 %		35.1 %
SE	2.0		18.1
TG	56.0		2.8
FS	3.4		1.9
PL	33.9		42.1
M.W.	180,000-200,000		a) 28,000 b) 14,000
pI			a) 5.3 b) 5.4-5.5
Particle Size	Fraction I	>	Fraction II
	VLDL		HDL

SE, steryl esters; TG, triglycerides; FS, free sterols; PL, phospholipid

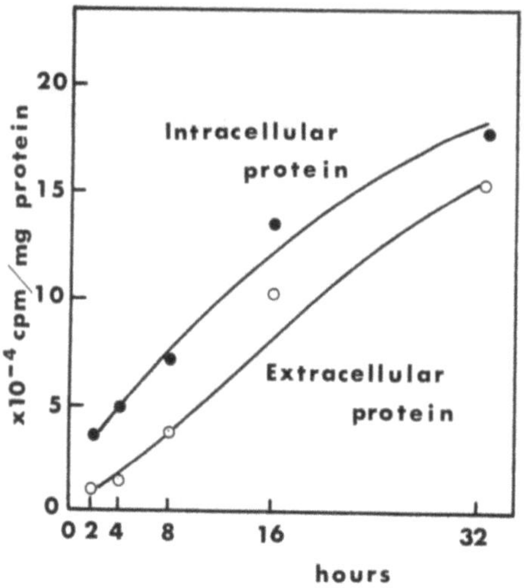

Fig. 2. Protein synthesis by cultured hepatocytes. Hepatocytes were cultured for 8 days and hours in the figure are incubation time after the addition of ^{14}C-leucine.

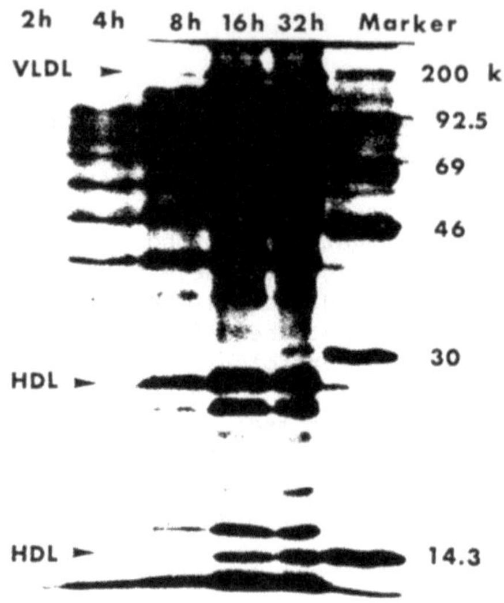

Fig. 3. Fluorogram of extracellular protein synthesized by cultured hepatocytes. Apolipoproteins of VLDL and HDL are shown by arrows.

As molecular weights of apolipoproteins of VLDL and HDL were revealed, it was found from the fluorogram of extracellular proteins that VLDL and HDL were secreted into the medium after 8 h incubation by cultured hepatocytes (Fig. 3.). Although the intense incorporation of ^{14}C-isoleucine into albumin at 3 h incubation was observed in cultured rat hepatocytes(8), only faint incorporation into albumin-like proteins by cultured eel hepatocytes was observed at 2 and 4 h incubation. This result and the fact that the main protein in eel serum is lipoprotein suggest that lipoprotein in eel serum has some functions which albumin in mammals has.

Acknowledgements

We are grateful to Yuji Koga, Dr.Shinichi Teshima, and Dr.Akio Kanazawa for help in analysis of lipid.

References

1. Hayashi,S. and Ooshiro,Z. J.Comp.Physiol., 132: 343-350, 1979.
2. Hayashi,S. and Ooshiro,Z. Bull.Japan.Soc.Sci.Fish., 51: 765-771, 1985.
3. Hayashi,S. and Ooshiro,Z. Bull.Japan.Soc.Sci.Fish., 52: 1641-1651, 1986.
4. Ichihara,A., Nakamura,T., Tanaka,K., Tomita,Y., Aoyama,K., Kato,S. and Shinno,H. Differentiation and Carcinogenesis in Liver Cell Cultures (Borek,C. and Williams, G.M., eds.), pp.77-84, The New York Academy of Sciences. 1980.
5. Hayashi,S. and Ooshiro,Z. Bull.Japan.Soc.Sci.Fish., 44: 499-503, 1978.
6. Enat,R., Jefferson,D.M., Ruiz-Opazo,N. and Gatmaitan,Z. Proc.Nat.Acad.Sci.USA, 81: 1411-1415, 1984.
7. Laemmli,U.K. Nature, 227: 680-685, 1970.
8. Tanaka,K., Sato,M., Tomita,Y. and Ichihara,A. J.Biochem., 84: 937-946, 1978.

Part 4

CELL DIFFERENTIATION AND GENE EXPRESSION

DIFFERENTIATION OF EMBRYONIC CELLS OF *DROSOPHILA* STUDIED WITH ELECTRON MICROSCOPE

Y. Kuroda and Y. Shimada

Department of Ontogenetics, National Institute of Genetics,
Mishima, Shizuoka 411, Japan

Undifferentiated cells from post-gastrula embryos of *Drosophila melanogaster* were cultured in the presence or absence of ecdysterone. In the absence of ecdysterone, cells differentiated into various embryonic or larval-type cells. Under an electron microscope, the formation of typical myofibrillar structure was observed in muscle cells. In epithelial cells, interdigitations, gap junctions and desmosomes were formed between adjacent cells. In nerve cells, neurofilament and neurotubules were found. In the presence of ecdysterone, cells differentiated into various structures which were very similar to some imaginal discs and adult tissues. In the present paper, the detailed process of the formation of trachea was described.

All cells in the blastoderm in *Drosophila* embryos are predetermined for their prospective fate in normal development (1,2). Some cells will differentiate into various types of embryonic and larval tissues or organs, and other cells will differentiate into adult tissues or organs. We established an experimental system in which undifferentiated cells from post-gastrula embryos of the wild type strain of *Drosophila melanogaster* differentiated into various characteristic tissues or organs *in vitro*. When cells were cultured in the absence of ecdysterone, they differentiated into embryonic or larval-type cells such as muscle, epithelial and nerve cells(3,4,5). On the other hand, in the presence of ecdysterone, cells differentiated into structures very similar to adult organs such as the eye, the wing and the leg (6). In our previous studies, the tissue specificity and time specificity of the action of some sex-linked recessive lethal genes were elucidated (3,4). In the present study, the ultrastructures of cells differentiated *in vitro* from undifferentiated embryonic cells were examined under an electron microscope.

Materials and Methods

The wild-type Oregon-R strain of *D. melanogaster* was used. Newly laid eggs were dechorionated by treatment with 3% sodium hypochloride solution for 6 minutes, and then washed with distilled water. Embryos at the stage of post-gastrula were selected, and sterilized in 70% ethyl alcohol for 10 minutes, and washed three times with sterile physiological salt solution.

One hundred embryos were transferred to culture medium in a hollow slide, and torn into small fragments with a pair of fine needles under a binocular microscope. Then torn small fragments of tissues and cells were transferred into culture medium on a carbon-coated cover slip in a T-5 culture flask. The flasks were incubated at 28°C. The culture medium consisted of K-17 medium (7), supplemented with 0.1 mg/ml fetuin (Grand Island Biol. Co., Deutsch Method) and 15% fetal bovine serum (Microbiol. Assoc. Inc.). In some cultures, ecdysterone (Rhoto Pharmaceutical Co., Ltd.) was added to the medium at concentrations of 1-10 μg/ml.

After the flasks were incubated for several days, the cover slips to which tissue fragments and cells attached were taken out from the flasks, and fixed in 2.5% glutaraldehyde in 0.1 M phosphate buffer at pH 7.3, and then postfixed in 1% OsO_4 in the same buffer. After dehydration in an ascending ethanol series, the fixed materials were embedded in Epon 812, and split off the coverslips. Thin sections were cut with an LKB ultrotome and stained with uranyl acetate and lead citrate. They were examined with a Hitachi H-12, H-700 or JOEL 100C electron microscope operated at 75-175KV.

Results

Differentiation of embryonic or larval-type cells

When undifferentiated cells dissociated from post-gastrula embryos were cultured in medium without ecdysterone for several days, various characteristic embryonic or larval-type cells were differentiated. Muscle cells, epithelial cells and nerve cells were easily identified in their morphology and functions in culture.

Muscle cells were recognized by their characteristic spindle shape after cultivation for several days. They extended their cytoplasm over the glass surface and migrated actively. They were 50 to 100 µm in length and had an ovoid nucleus in the center. They grew, made contact with each other and fused to form syncytial complexes, some containing more than ten nuclei (3). Under the electron microscope, the unfused muscle cells elongated and contained many incompletely organized myofibrils in the cytoplasm after cultivation for 6 days. In some of these cells, thick and thin myofilaments aggregated parallel or nearly parallel, and these aggregates of filaments exhibited Z-band density. In other cells, well-developed myofibrils increased in an number and diameter. In multinucleated syncytial cells, nuclei were arranged longitudinally in the central axis of the cells and the A, I and Z bands in myofibril were clearly recognized (Fig. 1). Some nerve fibers were found to extend from fragments of nerve tissues or cells. They were fine and thread-like and began to extend after a few hours of cultivation. After several days, the nerve fibers formed branches in many directions from their thicker main axes (3). Under the electron microscope, many thin neurofilaments were running parallel in the nerve fibers (Fig. 2).

Flat polygonal epithelial-type cells were released from the peripheral regions of some tissue fragments after cultivation for several hours. They were about 5 um in diameter and had a relatively large nucleus with a prominent nucleolus. After 24 hours of cultivation, these cells gradually increased in size and number and formed a continuous monolayer sheet (3).

Under the electron microscope, these epithelial-type cells possessed many cylindrical processes or microvilli at their free, apical surface (Fig. 3). These processes were about 0.1 - 0.2 µm in diameter, and several micrometers in length. The cytoplasm within the microvilli was composed of fine filamentous material which showed a preferred orientation along the axis of the microvilli. The nucleus was located in the mid-basal part of the cell. Round or rod-shaped mitochondria were distributed in the cytoplasm. In some epithelial-type cells which connected with each other, interdigitations, gap junctions and desmosomes were found between adjacent cells (Fig. 4).

Differentiation of adult-type structures

When undifferentiated embryonic cells were cultured in the presence of ecdysterone, some characteristic structures which were very similar to those of imaginal discs or adult tissues developed. Among them the structures which were similar to the wing disc, antennal disc, and eye disc were remarkable (6,8).

Under the electron microscope, various types of cells developed in the presence of ecdysterone were observed. In the early period of cultivation, two distinct cell types were distinguished, i.e., dark and light cells. The former cells possessed a dark nucleus and a cytoplasm containing numerous reticula, mitochondria and ribosomes. The latter cells had a lightly stained cytoplasm with fewer endoplasmic reticula, mitochondria and ribosomes. After one week in culture, differentiation of muscle and nerve cells was evident. The most interesting feature was the observation of the development of tracheae in undifferentiated embryonic cells.

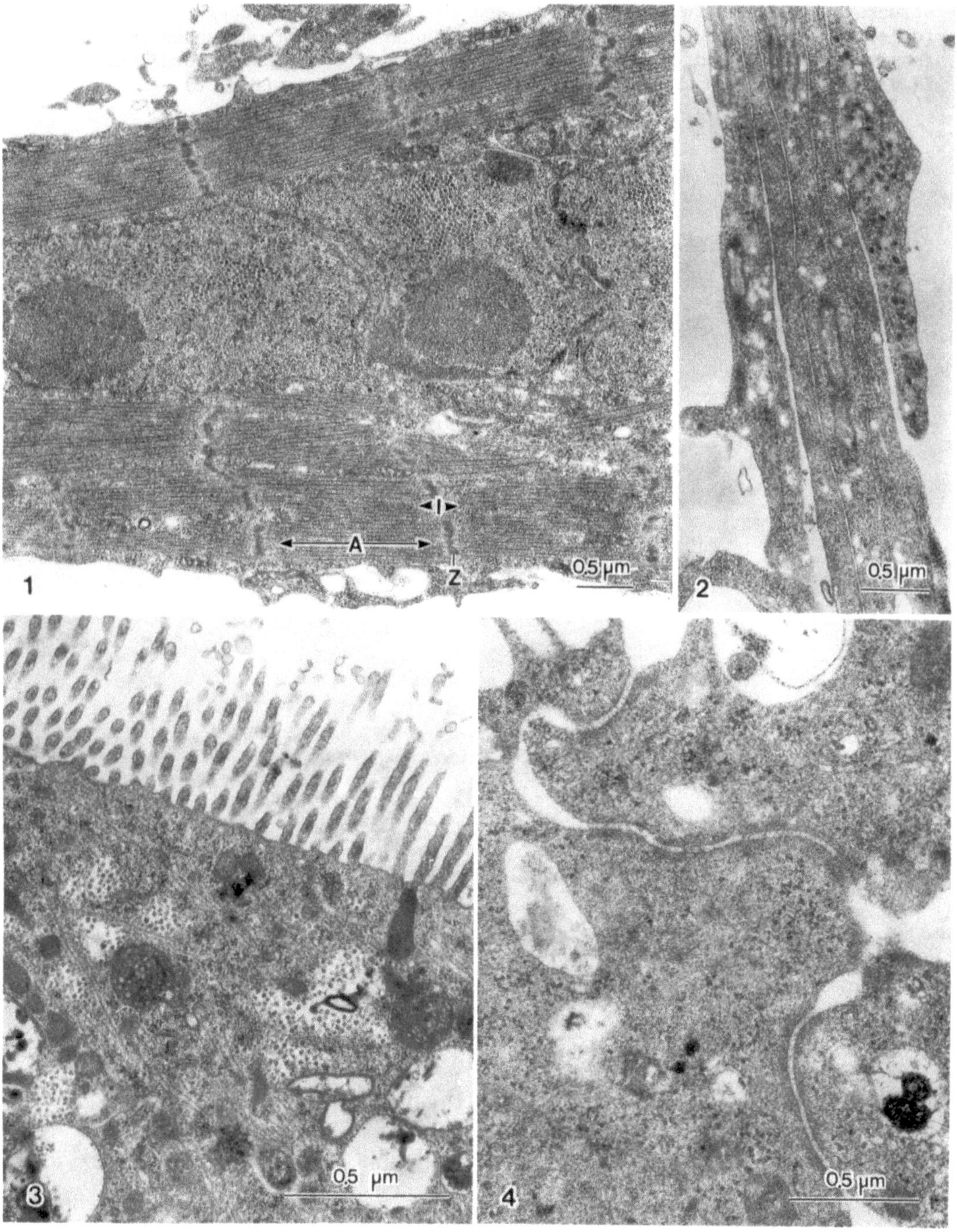

Figs. 1-4. Embryonic or larval-type cells differentiated in vitro from undifferentiated embryonic cells of D. melanogaster in the absence of ecdysterone. 1. Muscle cells, 2. Nerve fibers, 3. Epithelial-type cells, 4. Epithelial-type cells.

After cultivation for a few days, some trace and small structures of trachea developed in the cytoplasm of some cells (Fig. 5). These showed onion-like structures with many layers of unit membranes. They gradually grew up and elongated with more complex windings of unit membranes (Fig. 6). These structures extended in the cytoplasm and the inside space became more empty and formed tube-like structures (Fig. 7). The outer membranes extended in their longitudinal axes and their winding pitch and amplitude became more regular. Finally, they exhibited the mature structure of a trachea (Fig. 8).

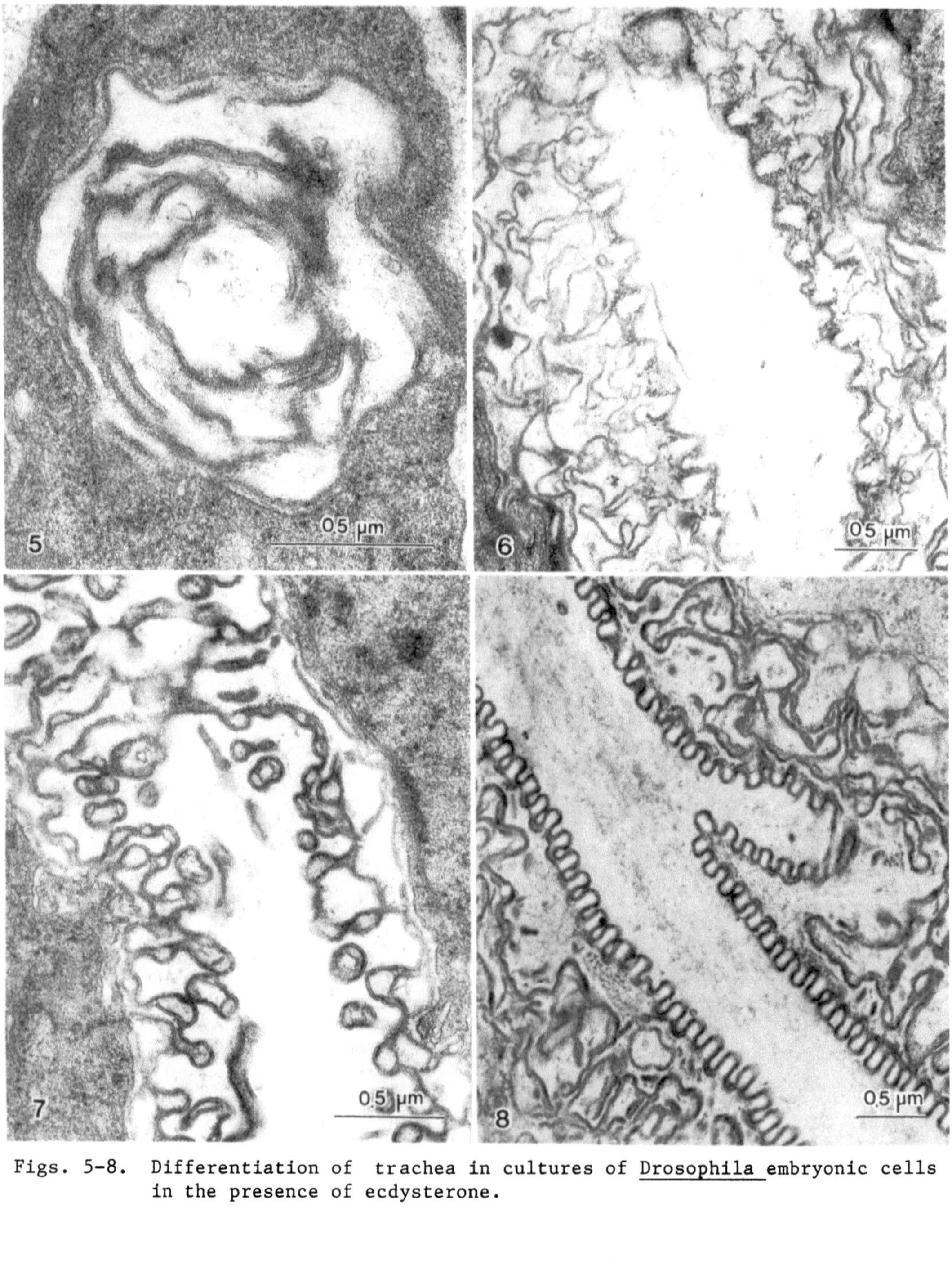

Figs. 5-8. Differentiation of trachea in cultures of Drosophila embryonic cells in the presence of ecdysterone.

Discussion

In the absence of ecdysterone, some other types of cells than those described in the present paper developed and differentiated. Small cells which were epithelial in morphology and about 3 μm in diameter were observed (7). Their origin and function are unknown. Some other types of epithelial cells were also observed. These were hemocytes and fat body cells. The hemocytes were polygonal and relatively large in size. They had a round nuclei in the center of a cytoplasm which had numerous granules. The fat body cells made aggregates consisting of more than 10 cells and had many granules in the cytoplasm. Both hemocytes and fat body cells are initially epithelial-type (9) and develop into these characterist types of cells (10, 11). In cultures of embryonic cells from a sex-linked recessive lethal mutant, rudimentary ^{39k}, the maturation of some epithelial cells was not observed and this defect in maturation was compensated for by addition of pyrimidine and purine nucleosides to the culture medium (5).

Cellular spheres (vesicles) were specific structures formed by insect cells in culture. They were balloon-like and consisted of a monolayer cell sheath of the outer membrane and many cells inside the spheres (7, 11). These cellular spheres are assumed to be the cell group of imaginal disc cells, since they differentiated into adult fly structures or organs when they implanted into larvae about to undergo metamorphosis (10, 12, 13). In the present experiment, they were found frequenctly in cultures without ecdysterone and remained in the state and morphology above described. In addition of ecdysterone into culture medium, they differentiated into various structures and tissues very similar to more advanced imaginal discs or adult organs (6, 8). Electron microscopic studies on these advanced stages of cellular spheres are now in progress.

Our culture system with Drosophila embryonic cells will be applicable to the examination of development and differentiation of various tissues and organs and the analysis of the action of specific genes on morphogenesis and organogenesis of these tissues and organs. Ecdysterone-independent and -dependent culture conditions may be useful for selecting the larval or adult tissues and organs to be analyzed.

Acknowledgements

The present work was supported by a Grant-in-Aid from the Ministry of Education, Science and Culture. This is contribution No. 1742 from the National Institute of Genetics, Mishima, Japan.

References

1. Poulson, D. F. Biology of Drosophila (Demerec, M., ed.), pp.168-274, Wiley, New York. 1950.
2. Hotta, Y. and Benzer, S. Nature, 240:527-535, 1972.
3. Kuroda, Y. Nature, 252:40-41, 1974.
4. Kuroda, Y. Develop. Growth Differ. 19:57-66, 1977.
5. Kuroda, Y. Invertebrate Cell Culture Applications (Maramorosch, K. and Mitsuhashi, J., eds.), pp.53-104, Academic Press, New York and London. 1982.
6. Kuroda, Y. Techniques in in vitro Invertebrate Hormones and Genes (Kurstak, E. and Oberlander, H., eds.), c215, pp. 1-5, Elsevier Biomedical, 1986.
7. Kuroda, Y. Develop. Growth Differ. 16:55-66, 1974.
8. Kuroda, Y. Ultrastructure and Functioning of Insect Cells (Akai, H., King, R. C. and Morohoshi, S., eds.), pp. 91-94, 1982.
9. Shields, G. and Sang, J. H. J. Embryol. Exp. Morphol., 23:53-69, 1970.
10. Shields, G., Dubendorfer, A. and Sang, J. H. J. Embryol. Exp. Morphol., 33:159-175, 1975.
11. Dubendorfer, A. and Eichenberger-Glinz, S. Invertebrate System in Vitro (Kurstak, E., Maramorosch, K. and Dubendorfer, A., eds.), pp. 169-185, Elsevier, Amsterdam. 1980.
12. Schneider, I. J. Embryol. Exp. Morphol., 27:353-365, 1972.
13. Dubendorfer, A., Shields, G. and Sang, J. H. J. Embryol. Exp. Morphol., 33:487-498, 1975.

DIFFERENTIATION OF THE *BOMBYX* WING DISC IN *IN VITRO* CULTURE

H. Kawasaki and Y. Iwashita

Faculty of Agriculture, Utsunomiya University, Utsunomiya, Tochigi 321, Japan

This study was undertaken to evaluate the methods of wing disc culture in the fifth larval instar of *Bombyx mori* and to observe the mode of differentiation of the wing disc in culture. The stainless mesh method proved the best for development of discs among the three organ setting methods tested. The high O_2 partial pressure and the addition of 10% *Bombyx* serum accelerated the wing development. Complete differentiation was observed on the discs exposed to 0.02-0.1 µg/ml ecdysterone. The mode of differentiation of the wing discs of larvae on the days 1 , 6 and 11 was identical. Mitoses, cuticle deposition and scale formation were observed in each case. From these results we concluded that ecdysteroid controlled the wing development.

In vitro differentiation of the Lepidopteran wing disc was reported(1), but the accurate doses of ecdysterone for wing differentiation including scale formation were not determined. And there is no observation on the differences after culture among the wing discs at the different larval stages. Therefore we studied to ascertain whether the wing disc of *Bombyx mori* differentiates *in vitro* and what kinds of hormonal conditions are required for complete differentiation, and to observe the mode of differentiation histologically.

Materials and Methods

The racial hybrid N124 x C124 of *Bombyx mori* was used. After being sterilized in ethanol, insects were dissected and the forewing discs were removed. After rinsing in the Grace medium, wing discs were transferred to conditioned Grace medium. Culture was done at 25°C. Medium was exchanged for fresh medium every two days.

Three kinds of tissue setting methods were tested; stainless mesh method, hanging drop and dipping method. The O_2 partial pressure and the addition of serum were examined. Moreover, the mode of differentiation was observed histologically. Paraffin sectioned samples were made according to ordinary method and stained with haematoxylin and aeosin. The mitoses of wing tissues were observed under the microscope in the whole mounted tissues stained with Feulgen's reaction after being fixed with Carnoy's solution.

Results

Complete differentiation including scale formation was observed on the wing tissue cultured with stainless mesh method, but it was not observed on the wing tissue with hanging drop or dipping method. On these tissues cultured with hanging drop or dipping method, mitoses perpendicular to the cell layer were observed. These are the mitoses of the scale stem cells. Tissues cultured with hanging drop or dipping method swelled and the wing disc cell layer became thinner. A cluster of the cells was observed in the wing cavity (Fig. 1,2).

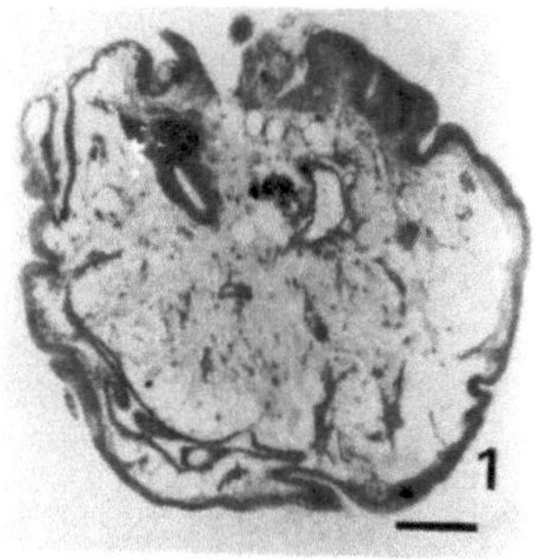

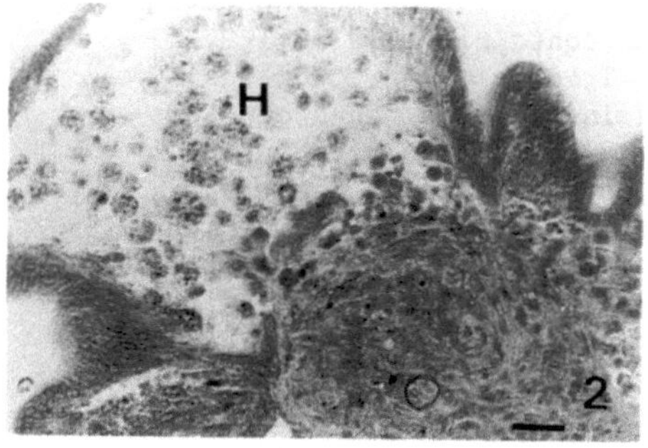

Figs.1,2 and Figs.4-7. Light micrographs of wing discs cultured *in vitro*. Bars equal 50 µm in Fig. 1 and 10 µm in Figs.2, 4, 5, 6 and 7.
Fig.1. Wing disc of day 3 larva was cultured with dipping method in a solution containing 0.05 µg/ml ecdysterone for 12 days.
Fig.2. Wing disc of day 3 larva was cultured by applying the hanging drop method in a solution containing 0.05 µg/ml ecdysterone for 16 days. H; haemocyte.

Table 1. Results of the examination of culture

	Cuticle deposition	Mitoses perpendicular to tissue	Scale formation
Air'	2	8	14
$95\%O_2+5\%CO_2$'	2	6	10
Hanging drop"	+	+	-
No serum added"	+	+	-
Bovine serum added"	+	+	±

Numbers are days before observation. + ; Observed, - ; Not observed, ± ; Observed in some cases, ' ; Discs of prepupa, " ; Discs of day 3 larva

Table 2. Disc reaction after treatments with different doses of ecdysterone

Doses of ecdysterone (µg/ml)	Cuticle deposition	Tracheole migration	Scale formation
0.5	Tanned	-	-
0.4	↓	-	-
0.3	↓	-	-
0.2	Transparent	+	-
0.1	↓	+	+
0.08	↓	+	+
0.02	↓	+	+
0.01	↓	+	-
0.005	↓	-	-

+ ; Observed, - ; Not observed

As indicated in Table 1. the high O_2 partial pressure accelerates the cell differentiation of wing tissue. The addition of *Bombyx* serum was necessary to induce the wing disc cell differentiation, while that of bovine serum did not necessarily induce. Scale formation was not observed when no serum was added. As indicated in Table 2, ecdysterone over 0.3 µg/ml induced the wing tissue tanned cuticle. The complete differentiation was observed in the medium containing 0.02-0.1 μg/ml ecdysterone. Tracheole migration was not observed in the wing discs exposed to ecdysterone lower than 0.008 μg/ml. Mitoses did not continue in the wing discs exposed to ecdysterone lower than 0.004 μg/ml.

Table 3 indicates the results of the culture of wing discs of the days 1 , 6 and 11 (prepupa) of the fifth larval instar. The mode of differentiation observed in the culture condition was the same as observed *in situ*. Mitoses (Fig. 4), cuticle deposition (Fig. 5), mitoses perpendicular to cell layer (Fig. 6) and scale formation were observed successively. The mode of differentiation of each cell in the same wing disc was less synchronious in the wing discs at early stages.

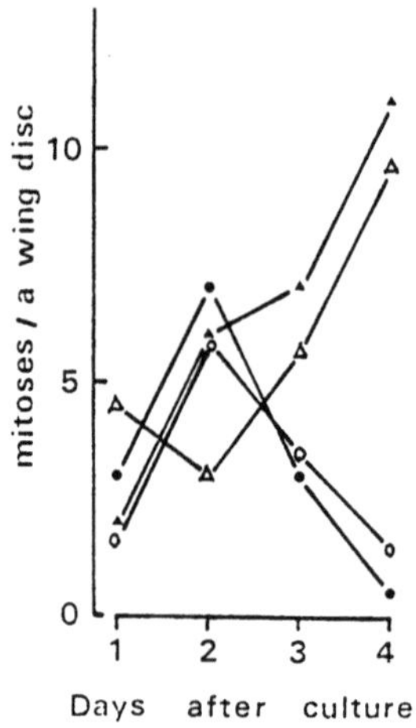

Fig. 3. Differences in the doses of ecdysterone resulted in different rates of mitoses in the wing discs of day 1 larvae. Each value repesents the mean of 3 samples. Discs were cultured for 24 hours in the hormone free medium and then they were transferred to a medium containing various doses of ecdysterone. ●-● ;0.001 μg/ml, ○-○ ; 0.004 μg/ml, △-△ ;0.008 μg/ml, ▲-▲ ; 0.015 μg/ml

Discussion

Complete differentiation was observed inthe wing discs cultured in the Grace medium with the stainless mesh method including 10 % *Bombyx* serum. The high O_2 partial pressure accelerated the disc development. These observations suggest that the disc development needs the substance supporting wing tissue and the O_2 supply. The concentration of ecdysterone between 0.02-0.1 μg/ml induced the complete differentiation on the wing disc. Mitoses did not continue in the medium containing ecdysterone lower than 0.004 μg/ml. As a result, wing disc development needs ecdysteroid and disc cells proliferate in the medium containing ecdysteroid of low concentration,followed by the differentiation of scale, socket forming cells and other epidermal cells. The high dose ecdysterone increases RNA and protein synthesis of the imaginal disc (2, 3). Therefore, a change of gene expression occurs in the wing disc cells with ecdysteroid of high concentration. As a result,the thick cuticle was deposited and the mitoses were inhibited.

Wing discs at three different stages could differentiate in the Grace medium containing 0.05 μg/ml ecdysterone. The discs at early stages, whichwere small in size, took fewer days to differentiate than those at late stages. This indicates that a relatively great deal of O_2 and nutrition were obtained in small discs. From

Table 3. Differences in the development of wing discs after culture depending on the larval stage

Stage	Day after culture		
	Mitosis	Cuticle deposition	Scale formation
V1	2-10	7	11
V6	1-3,6-8	3	10
V11	8		14

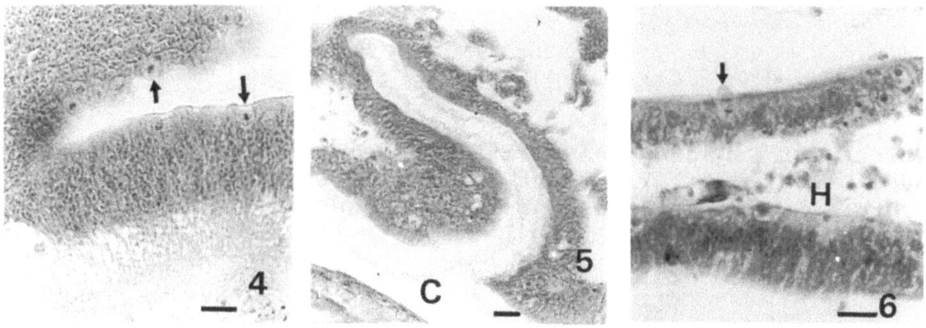

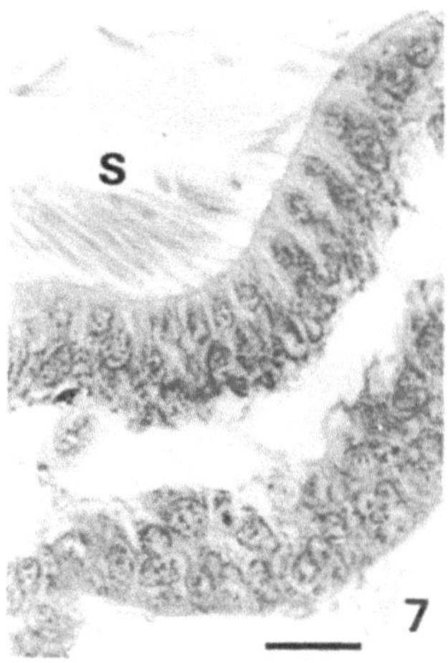

Figs.4-7. Light micrographs of wing disc cultured on the stainless mesh containing 0.05 μg/ml ecdysterone and 10 % *Bombyx* serum. Fig.4. Wing disc of day 6 larva ; Many mitoses were observed in the wing disc 1 day after culture (arrows). Fig. 5. Day 6 wing disc ; Cuticle deposition was observed in the disc 6 days after culture. C; cuticle. Fig. 6. Wing disc of prepupa ; Mitosis perpendicular to wing tissue (arrow) was observed in the wing disc cultured for 12 days. H; haemocyte. Fig. 7. Wing disc of prepupa ; Scale formation was observed in the wing tissue 10 days after culture under 95%O_2+5%CO_2 aeration condition.

these results we conclude that the wing discs of fifth larval instar can differentiate to imaginal wings in the medium containing ecdysteroid of a constant dose, but mitoses and cuticle deposition must be experienced before differentiation. We suppose that ecdysteroid sustains the disc development, but suppresses the cell proliferation and accelerates cuticle deposition at a high dose, and that the changes of ecdysteroid titer regulate the wing disc development by controlling cell cycle, and gene expression.

References

1. Nardi, J.B. and Willis, J.H. Dev. Biol. 68:381-395, 1979.
2. Raikow, R. and Fristrom, J.W. J.Insect Physiol. 17:1599-1614, 1971.
3. Fristrom, J.W., Gregg, T.L. and Siegel, J. Dev. Biol. 41:3ol-313,1974.

THE POSSIBLE ROLE OF THE Y CHROMOSOME DURING MALE GERM CELL DIFFERENTIATION IN *DROSOPHILA*

W. Liebrich* and N. Kociok

Institut für Genetik, Universität Düsseldorf, Universitätsstr.1 D-4000 Düsseldorf. *Present address: Universität Heidelberg, Chirurgisch Onkologisches Labor, Im Neuenheimer Feld 305, D-Heidelberg. F. R. Germany

The Y chromosomal factors (fertility genes) of *Drosophila hydei* have no function during early germ cell proliferation. They do not affect the formation of a cyst nor the phase of premeiotic mitoses. Their role is restricted to meiosis and spermiogenesis, where the single factors seem to be related to correct function of cytoskeletal elements.

The Y chromosome of *D. hydei* bears five factors (nooses, clubs, tubular ribbons, pseudonucleolus and threads), that control spermatogenesis (1). To learn more about their function, males are constructed, which are deficient in each one of the factors. Single cysts are isolated of the testes and cultivated *in vitro* or examined immediately after isolation. The aim of our studies is to exactly localize the phases in which for the first time defects occur and to describe the nature of the defects. From that we hope to draw some indirect conclusions about the possible function of a single loop and to localize the phases in which the special loop might have its function. Comparing cysts of testes of wild-type with Y deficient males, no morphologically visible differences become obvious. The same is even true for cysts isolated from males missing the whole Y chromosome (Fig. 1). It is, therefore , concluded that the Y chromosome has no effect on germ cell formation and early spermatogenesis: pole cell formation and movement; origin of cysts, number of premeiotic mitoses and the growth phases of cysts during prophase I of meiosis.
During spermatogenesis several processes occur in which cell surface properties are involved: cell recognition and adhesion play an essential role, e.g., during cyst formation. Cell surface glycoproteins are involved in many of such cellular processes. To study sugar containing components of the cell surface, lectins have been found to be a useful tool. We used wheat germ agglutinin (WGA) to study possible changes of the glycoprotein composition of cyst and germ cell membranes of wild-type and Y deficient males.

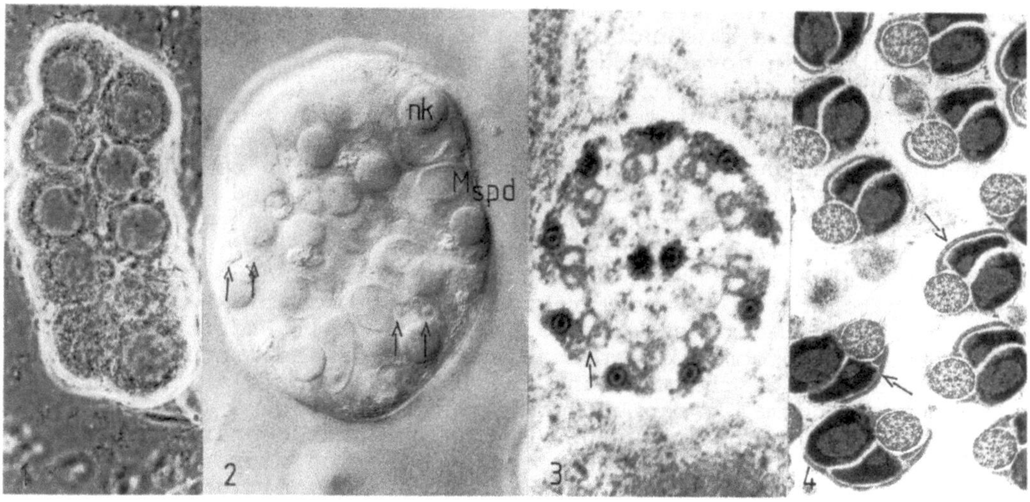

Figures 1 - 4. Drosophila hydei
Fig. 1. Phase contrast optics, x480, genotype: X/0, isolated spermatocyte cyst. The cyst is composed of 10 germ cells and 2 cyst cells.
Fig. 2. DIC optics, x480, genotype: Y chromosome without nooses, isolated and cultured cyst in onion phase. Mspd:single ring-like non-aggregated mitochondria of spermatid, Nk: nebenkern, arrow indicates spermatids with two or more nuclei.
Fig. 3. Transmission electron microscopy, x196.000, genotype: Y chromosome without the tubular ribbons. Axoneme in cross section, arrow indicates the "A" microtubulus without the outer dynein arm.
Fig. 4. Transmission electron microscopy, x33.250, genotype: Y chromosome without the pseudonucleolus. Cross section through the flagellum of spermatids; arrow indicates the mislocated thick part of the axonemal membrane.

In cysts isolated from wild-type flies WGA receptors are distributed in small patches over the cell surface up to the early elongation phase. The fluorescence diminishes in elongating spermatid bundles. The surface of isolated germ cells is more or less evenly stained. This pattern is found up to the phase of onion nebenkern.
The results obtained using cysts of Y deficient males do not differ visibly from that of the wild-type. We, therefore, conclude that the Y chromosome has no influence on events occurring during cell recognition and cell rearrangement.
Comparing squash preparations of testes of wild-type and Y deficient males, first visible morphological defects can be detected during meiosis as well as during spermiogenesis. In the following chapter the results of the single deficiences are summarized: spermatocytes of males deficient for the nooses start meiosis in vitro. The earliest disturbances are discovered during second meiotic division, where the single spermatocytes develop highly asynchronously: in one cyst secondary spermatocytes in meiosis and spermatids can be detected. After meiosis, at the beginning of spermiogenesis, some mitochondria are not included in the developing nebenkern (Fig. 2). Besidesthat,two or more dif-

ferent sized nuclei are found in differentiating spermatids. The results suggest a misfunction of the spindle apparatus and the microtubuli that govern aggregation of mitochondria into the spherical nebenkern (2).
Squash preparations of testes of males deficient in the tubular ribbons exhibit cysts at the terminal end, whose spermatids are individualized but show no motility. The same can be demonstrated if cysts of the respective region are isolated and cultured *in vitro*. Electron microscopy rereals that the "A" microtubuli of the axoneme lack the outer dynein arms (Fig. 3).
Analysing testes of males deficient for the pseudonucleolus we found cysts at the terminal region, whose spermatids were highly motile but failed to escape the head cyst cell. Using electron microscopy it becomes clear that spermatozoa are produced whose thickened part of the membrane is dislocated in relation to the axoneme: it normally can be seen around the axoneme (Fig. 4).
Germ cells of males deficient in the threads degenerate during the coiling phase. As demonstrated by electron microscopy the "A" microtubuli of the axoneme possess no outer dynein arms.
At this time it is not yet possible to construct males missing the clubs.
Our results demonstrate: The possible function of the Y chromosomal loops is restricted to meiosis and spermiogenesis. It could be related to the correct function of cytoskeletal elements during meiosis and spermiogenesis. - Supported by the Deutsche Forschungsgemeinschaft (Li 317/4-4).

References

1. Hess,O. Biology and Genetics of *Drosophila* (Ashburner, M. and Wright, T.R.F.,eds.), Vol. 2d, pp.1-32, Academic Press, London, 1980.
2. Liebrich, W. Cell Tissue Res. 224:161-168, 1982.

IN VITRO SPERMIOGENESIS OF THE DIAPAUSING CABBAGE ARMYWORM, *MAMESTRA BRASSICAE*

T. Shimizu and S. Yagi

The Institute of Physical and Chemical Research, Wako, Saitama 351-01, Japan and Faculty of Agriculture, Tokyo University of Agriculture and Technology, Fuchu, Tokyo 183, Japan

The cultured pupal testis of *M. brassicae* released 2-deoxyecdysone, ecdysone and 20-hydroxyecdysone into the medium during 24hr of cultivation. Disc electrophoretic analysis showed that two major protein bands were obtained from the medium after cultivation. Ecdysteroid in the hemolymph of the nondiapausing pupae was gradually increased after pupation, and cAMP level in the testis was higher than that of the diapausing pupal testis. It was concluded that unknown heat-stable factors from the testicular peritoneal sheath are required to promote spermiogenesis of the diapausing pupal "naked" spermatocysts.

In our first paper on *in vitro* spermatogenesis we reported that 20-hydroxyecdysone stimulated the elongation of spermatocysts of the rice stem borer, *Chilo suppressalis*(1). We also observed that in the absence of the hormone some spermatocysts developed into elongated spermatids within several days after they had been liberated from ruptured testes. Later, Kambysellis and Williams(2, 3) published papers on the hormonal control of spermatogenesis in silkworms. They found that naked spermatocysts showed no response to ecdysone, but when a macromolecular factor (MF) contained in hemolymph was present or was added to an ecdysone-free medium, a rapid differentiation to spermatozoa took place. Since spermatocysts within intact testis only responded when both MF and ecdysone were present, they concluded that the function of ecdysone was to facilitate the entry of MF into spermatocysts. In our experiments in 1969, we used CSM-2F(4) as a culture medium. This did not contain any insect hemolymph, but it had a mammalian serum which may have acted as a MF. Using Grace's medium without any insect hemolymph or serum, we succeeded in cultivating testes of the cabbage armyworm, *M. brassicae*, provided ecdysone or active prothoracic glands were present(5, 6). We observed that spherical spermatocysts in intact testis elongated during one week's cultivation. Furthermore, without any ecdysone spermatocysts elongated when the testis walls were ruptured and the cysts were allowed to migrate into the medium. Spherical spermatocysts developed into elongated cysts in Grace's medium after only 7 days if they were cultivated with ruptured or intact testis. Spherical cysts in intact testis did not develop although they survived well. Naked spermatocysts degenerated into darkened cysts or small fragments within 3 days. In the presence of fat body or brain, spermatocysts showed no further development *in vitro* and degenerated within 7 days. Only co-culturing with an intact testis was effective in stimulating the elongation of spermatocysts. This suggests the testis may produce some unknown substance which facilitates development or maintenance of the spermatocysts *in vitro*. Next, the role of the testes themselves in the development of spermatocysts was examined. Naked spermatocysts were cultured for 5 days with testicular fluid. This fluid was called "pre-conditioning medium". Newly added spermatocysts were cultivated in this preconditioning medium without Grace's medium. When naked spermatocysts were cultivated in the preconditioning medium, the cysts developed into elongated ones after 4 days; however, the cysts degenerated in control experiments. Thus, the testis seems to produce a MF-like substance(s) that stimulate(s) spermatocysts development(7). Since the pre-condioning medium was effective in stimulating the elongation of naked spermatocysts, we decided to analyze the medium by electrophoresis. Although Grace's medium did not show any protein band, the culture medium after cultivation of testes for 3 days showed two major protein bands, a and b. Therefore, it seems likely that these proteins were derived from the cultivated testes themselves. Further experiments showed that similar patterns of proteins were also present in the supernatants of the homogenates of testes from diapausing pupae, or from the culture medium in which ruptured testes had been cultivated for 6 days. These proteins must therefore originate from the testes and be liberated into the medium. From these results, we suggested that the testis in M. brassicae synthesizesproteins and that cells of the testis wall may be one of the sources of proteins with MF-activity(8). Until now there has been no chemically defined medium for the observation of spermatogenesis in lepidopteran diapausing pupal spermatocysts. In the present study, using Grace's medium which is chemically defined and contains neither insect hemolymph nor mammalian serum, we investigated the factors and preparation methods of media for inducing spermiogenesis in vitro. In medium II, the added spermatocysts became gradually brownish. This medium is not suitable for spermatocyst culture. In the heat treated medium II(mediumIII), spermatocysts survived for 5 days in good condition. Media I, III and IV have the ability to promote the elongation of spermatids. The MF is heat-labile substances(proteins), but medium III did not contain heat-labile substances or MF-like proteins. Thus, unknown factors released from the testis may be a heat-stable substance(9).

Next, we measured ecdysteroid level released from the cultured testis. Within 1 day after cultivation, ecdysteroids released into the culture medium reached about 700pg/testis. The conditioned medium containing a similar level of ecdysteroids showed effectiveness in inducing spermiogenesis. We also measured ecdysteroid levels of diapausing and nondiapausing day 3 pupal testes. Ecdysteroid level of the nondiapausing pupal testis was 4.4-fold higher than in the diapausing pupal one(10). As a possible hormonal effect, we found changes in the testicular surface which had been cultivated for 7 days in the presence of 20-hydroxyecdysone. A part of the testis wall enlarged and granules appeared in the surrounding nuclei. As shown in Fig.1, a cell sheet formed on the surface on the glass. Large and small granules can be observed in the cells. The degenerated testis was removed from the cell sheet, the cell sheet was maintained in Grace's medium containing 20-hydroxyecdysone and was cultivated for one month by replacing the medium with fresh one every 3 days. We then investigated the formation of cell-sheet buds from the testis of various stages. The best formation of cell-sheet buds was from pharate pupae, and the buds were formed in about 50% of the cultures. Left and right testes fused and formed a single body at this stage. Thus, cell-sheet buds may relate to the fusion of testes. Using this cell sheet, we cultured reproductive spermatocysts on the somatic cell sheet on the glass, or in the conditioning medium with a

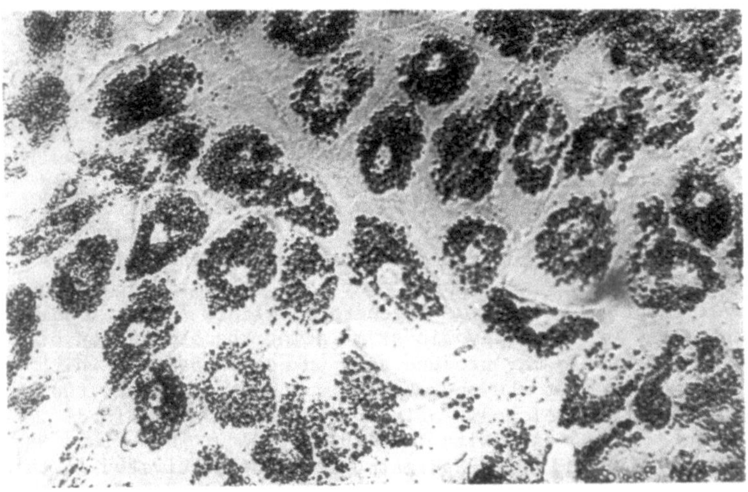

Fig. 1. A cell sheet formed on the surface of a glass. (x 315)

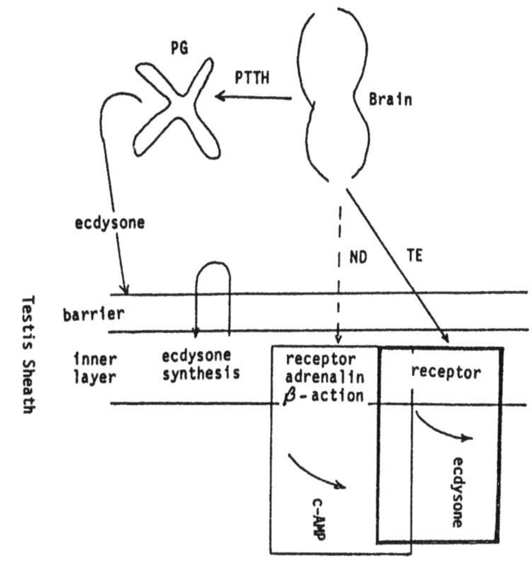

Fig. 2. Schematic illustration of the relationships between testicular ecdysone, ecdysone synthesized by the prothoracic gland and c-AMP.

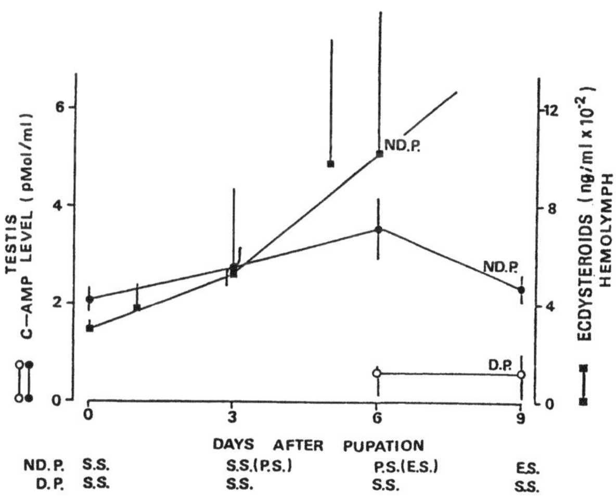

Fig. 3. C-AMP level and ecdysteroid titer during spermatogenesis of M. brassicae pupae.
■—■ ecdysteroid titer of nondiapausing pupal hemolymph (from Agui & Hiruma),
●—● c-AMP level of the testis of nondiapausing pupae,
○—○ c-AMP level of the testis of diapausing pupae (≦ 0.62 pMol/ml). Vertical bars at each point indicate standard deviation in three separate samples. ND.P.;nondiapausing pupae,D.P.; diapausing pupae. S.S.;spherical cysts,P.S.; pyriform cysts, E.S.;elongated cysts.

newly formed cell sheet. When the diapausing naked pupal spermatocysts were cultured for 3 days in conditioned medium (medium I), they developed into pyriform spermatocysts and consequently became elongated spermatocysts, while the cysts cultured in Grace's medium containing only 20-hydroxyecdysone degenerated within 3 days. After removing the testis the cell sheet was cultivated for 5 days. Medium II in which the cell sheet alone had been cultured was collected, and naked spermatocysts were cultivated in this medium with the cell sheet. Subsequently, spermatids were formed and spermiogenesis was observed. This medium seemed to be the best one for spermatocyst culture(11).

Since Loeb et al.(12, 13) suggested in Heliothis that testicular ecdysteroid level was increased by TE(testis ecdysiotropin, a peptide) from the brain, we investigated testicular ecdysteroid, ecdysteroid synthesized by the prothoracic gland and cAMP. Testicular cAMP level was increased in the nondiapausing pupae. Therefore we speculated that increased cAMP level is dependent upon peptide like a TE from the brain. These receptors of the inner layer of the testis seem similar to the relationship of the prothoracic gland and PTTH. A low level of ecdysteroids was demonstrated by Loeb et al.(14, 15) and synthesized by the inner layer of the testis sheath(16) (Fig. 2). We investigated whether this peptide-like factor(TE) increases the cAMP level of target testis, and found that neither cAMP level nor ecdysteroid of the nondiapausing pupae was increased. Elongated spermatocysts appeared when cAMP reached the maximal level and the ecdysteroid level increased approximately 1.0μg/ml. Both the values of the testis and ecdysteroid titer of hemolymph act as indicators for the time of occurrence of spermiogenesis in M. brassicae testis. On the contrary, in the diapausing pupae, cAMP level was low(Fig. 3).

As shown in Fig. 4, in vivo, ecdysteroid

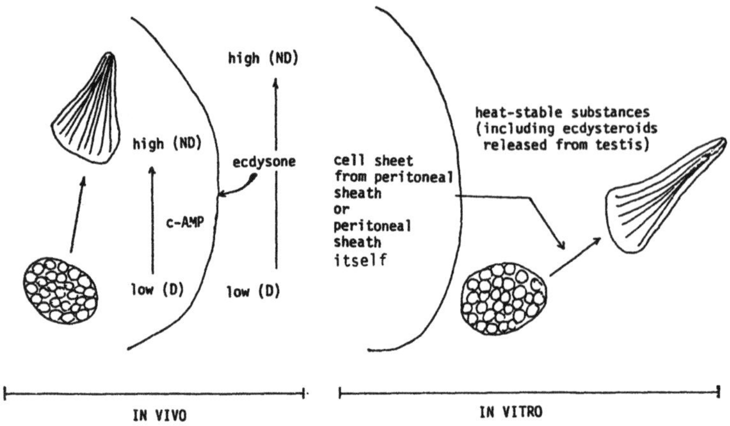

Fig. 4. Schematic summary of the relationships between spermatogenesis and environmental information obtained in vivo and in vitro.

titer in the hemolymph is gradually increased in the nondiapausing pupae. Cyclic AMP level of the testis is also increased in the nondiapausing pupae. In vitro, heat-stable substances including ecdysteroids released from testis intervene in spermiogenesis. Effective substances were also obtained from the medium conditioned with cell sheet from the peritoneal sheath or with the peritoneal sheath itself. That is to say, unknown heat-stable factors are required to promote spermiogenesis of the diapausing pupal naked spermatocysts.

In conclusion, spermiogenesis appears from these results to be promoted by the material from the testis itself. The testicular peritoneal sheath plays an important role in the differentiation of spermatocysts.

References

1. Yagi,S.,Kondo,E. and Fukaya,M. Appl. Ent. Zool.,4:70-78,1969.
2. Kambysellis,M. and Williams,C.M. Biol. Bull.,141:527-540,1971.
3. Kambysellis,M. and Williams,C.M. Biol. Bull.,141:541-552,1971.
4. Mitsuhashi,J. Appl. Ent. Zool.,3:1-4,1968.
5. Fukushima,T. and Yagi,S. Appl. Ent. Zool.,10:220-225,1975.
6. Shimizu,T. and Yagi,S. Appl. Ent. Zool.,13:278-282,1978.
7. Shimizu,T. and Yagi,S. Appl. Ent. Zool.,17:385-392,1982.
8. Yagi,S. and Shimizu,T. In "The Ultrastructure and Functioning of Insect Cells" (Akai,H.,King,R.C. and Maramorosch,K. eds.),pp.57-60, The Society for Insect Cells Japan, Tokyo,1982.
9. Shimizu,T. Zool. Sci.,3:203-206,1986.
10. Shimizu,T.,Moribayashi,A. and Agui,N. Appl. Ent. Zool.,20:56-61,1985.
11. Shimizu,T. Zool. Sci.,3:309-313,1986.
12. Loeb,M.J.,Woods,C.W.,Brandt,E.P. and Borkovec,A.B. In "Insect Neurochemistry and Neurophysiology,1986" (Borkovec,A.B. and Gelman,D.B. eds.),pp.367-370, Humana Press Inc.(Clifton),1986.
13. Loeb,M.J.,Brandt,E.P. and Woods,C.W. In "Techniques in the Life Science", Cell Biology, C2; In vitro Invertebrate Hormones & Genes(Kurstak,E. ed.),pp.1-11, Elsevier Scientific Publishers, Ireland Ltd.,Contry Clare,1986.
14. Loeb,M.J. and Woods,C.W. J. exp. Zool.,240:75-82,1986.
15. Giebultowicz,J.M.,Loeb,M.J. and Borkovec,A.B. Inter. J. Invert.Rep.& Develop.,11 :211-224,1987.
16. Loeb,M.J. Arch. Insect Biochem. Physiol.,3:173-180,1986.

POSSIBLE CORRELATION BETWEEN TRANSPOSITION OF COPIA-LIKE NOMADIC ELEMENTS AND ESTABLISHMENT IN VITRO OF CONTINUOUS DROSOPHILA CELL LINES

G. Echalier[1] and N. Junakovic[2]

1) UA CNRS 1135, Université Paris VI, 7 quai Saint-Bernard
75005 Paris, France

2) Centro Acidi Nucleici CNR, Università "La Sapienza"
Depte di Genetica e Biologia Molecular - Piazzale A. Moro 5
00185 Roma, Italia

Amplification and redistribution of copia-like elements, as previously described in the genome of cultured Drosophila cells, seem to be limited to the first period of the cultures.
According to Hayward's "promoter insertion model", a moving retrotransposon might activate some proto-oncogene and lead to in vitro "immortalization" of cultured cells.

Everyone with some experience in the field of cell culture knows that there still remains something mysterious about the process of in vitro establishment of continuous cell line.

In a suitable medium supplemented with fetal calf serum, many insect tissues can be grown in vitro for a few weeks. With rare exceptions (the most noteworthy and, no doubt, meaningful examples are the Drosophila leukemic cell lines obtained by E. Gateff (1)), those cells from first explantation, even if they look healthy and multiply, will degenerate more or less rapidly. After a phase of stagnation, if the medium is regularly renewed, another wave of cell proliferation may develop, then regress, and the cycle may be repeated several times. Finally, in some rare culture flasks, a new cell type, usually with a fibroblast-like morphology, will suddenly emerge and, henceforth, can be subcultured for years. A new cell line (2) is born. This capacity of indefinite multiplication in vitro is called cell "immortalization".

This phenomenon is now actively investigated in vertebrate cells. Many current studies point out the crucial role, in such cell "perennization", of deregulations of some cellular proto-oncogenes, especially c-myc or related genes : their products were localized in the cell nucleus by immunofluorescence and can interact with double-stranded DNA. So, it is widely thought that they might participate in the control of the normal multiplication cycle of cells, more precisely for the passage from stage G_0 into G_1 and S (that is the entry into the "competence" phase, according to Pledger's terminology (3).

Among the multiple possible causes of such deregulations, probably the best analyzed mechanism is the cis-activation of proto-oncogenes by retroviral insertion in their close vicinity. That is the "promoter insertion model" propounded by Hayward et al. (4 - see also 5-6), when studying the slow transforming ALV, responsible for chicken B-cell lymphomas. They showed that "as a rare event, the ALV provirus integrates (into the host-cell genome) adjacent to the c-myc gene and that transcription, initiating from a (strong) viral promoter (either of its LTRs), causes enhanced expression of c-myc, leading to neoplastic transformation".

DNA sequences homologous to the main vertebrate proto-oncogenes were identified in the genome of Drosophila melanogaster (7, 8, 9). In spite of the huge evolutionary distance between dipterous insects and higher vertebrates, the structure of those Drosophila proto-oncogenes display a high degree of sequence homology compared to their vertebrate counterparts. Nine proto-oncogenes and two growth factor receptors have so far been characterized and, with the method of in situ hybridization, could be assigned to their approximate positions on Drosophila melanogaster polytene chromosomes. They correspond to all three major families of vertebrate oncogenes (i.e. src, ras and myc/myb). Concerning this last group, no authentic c-myc could be found, but c-myb (which is c-myc related) does exist and is expressed throughout the course of Drosophila development (10).

Besides, the genome of Drosophila melanogaster shelters an uncommon abundance and variety of mobile genetic elements (up to 5-10 % of total DNA)(11). The most frequent transposons (a dozen families, each represented by 20 to 50 scattered repetitions) are collectively referred to as "copia-like elements" (from the first detected example named copia, a Latin word meaning abundance). They are characterized by their long terminal direct repeats (LTRs) and display striking similarities, in their structure as well as in their biology, with the proviruses of vertebrate retroviruses : They exist under circular DNA replicative forms and can generate RNA virus-like particles (VLPs) associated with reverse transcriptase activity, which are non-infectious and may be interpreted as defective virions (12, 13, 14). So, for a similar transposon in yeast, Boeke et al. (15) put forward the quite justified denomination of "retro-transposons".

It is well known from the work of Potter et al. (16), and it was confirmed by several authors (17, 18), that **the number of retrotransposons copies is generally increased in Drosophila cultured cell lines and that they can be dispersed** to new genomic sites. For instance, from data of DNA reassociation kinetics, they estimated the number of copia elements per haploid genome to be 60 in embryos from a Drosophila melanogaster Oregon R strain and to **increase** to 170 in Schneider's cell line S2 (19) derived from a comparable strain. Moreover, they demonstrated, after restriction enzyme analysis of the genomic sequences flanking the mobile elements, that the distribution of retrotransposons may differ, not only from embryos to cultured cells but also among individual cells of the cell line. Potter et al. assumed a uniform rate of transposition in cell cultures and (putting together the number of additional copies of copia, i.e. 110, in Schneider's line 2 and the 10 years during which it had been grown in vitro at that time), they even approximated this rate of transposition to be as high as $10^{-3} - 10^{-4}$ transpositions per element per generation ...

We decided to re-investigate the stability, meaning the presumed unstability of retro-transposons,under more rigorous conditions of cell culture : no Drosophila cell line is indeed, and probably from the beginning, a homogeneous population. So, in order to study the possible movement of mobile genetic elements, one has imperatively to work with cloned cell sublines. Such a cloned subline from our Kc cells (20, 21) was followed for years (8 years on the whole) and, at intervals, sub-cloned again to get an idea of the genomic repartition in individual cells. All clones or sub-clones were partly frozen in liquid nitrogen just after their isolation and partly propagated under standard culture conditions until the time of our analysis.

We used the Southern approach : all genomic DNAs were digested with the restriction enzyme Hind III, then the blots were successively hybridized with internal fragments from 6 different copia-like elements (copia, 1731, 412, 297, B104, mdgl). For each cell sample DNA, many restriction fragments ware detected by the probes (which is the visualization of the repetitive and scattered characteristics of transposons), but the autoradiographic patterns of hybridization should differ if the genomic distribution of the transposons has varied (16, 22, 23).

Detailed results of our analysis will be published elsewhere, but we could not escape the amazing conclusion : only minor pattern heterogeneity was observed among the successive cell samples, reflecting the basic stability of six different retro-transposons in Drosophila Kc cells throughout eight years of culture.

A totally different picture emerged from the study of short term cultures: the genomic distribution of the same six retro-transposons was analysed in four recently established cell lines (less than one year old), by comparison with not only the same Drosophila strain but the very fly laboratory stock from which they derived. By contrast with the preceding situation, numerous quantitative and qualitative differences were observed, in general agreement with Potter's original findings.

We may conclude that the amplification and important redistribution of retrotransposons, as described in Drosophila cell lines, seem to occur essentially during the first times of the culture. and are perhaps induced by the stress of in vitro environment (23, 24).

So, it is extremely tempting to correlate such initial burst of transposition in cell cultures with the "spontaneous" establishment of Drosophila continuous cell lines and to suggest that, according to the "promoter insertion model" of Hayward, some moving retro-transposon might haphazardly integrate in the proximity of some Drosophila proto-oncogenes(such as c-myb or any other) and, enhancing its expression, lead to cell "immortalization".

Let us admit that it will be rather difficult to get direct proof of this assumption, because Drosophila primary cell cultures are too delicate and heterogeneous to be good controls, if one wants to measure and compare the transcripts of any proto-oncogene in established line cells. As indirect arguments, we will point out that most, if not all, available Drosophila cell lines display several features which, in vertebrate cells, are considered as typical of a "transformed" state : a roundish morphology, a poorly developed cyto-skeleton, many surface microvillosities and peculiar sensitivity to lectins (as concanavalin), weak adherence to the culture flasks and no "contact inhibition", and, above all, the ability (at least for our KcO sub-line) to grow without any serum supplementation (which is highly suggestive of "autocrinic" or paracrinic" capacities (25).

Acknowledgements

The participation of one of us (G.E.) in the 7th International Conference on Invertebrate Tissue Culture was supported by grants from the J.E.C. Fund and the Japan Shipbuilding Industry Foundation.

References

1. Gateff, E., Gissmann, L., Shresta, R., Plus, N., Pfister, H., Schröder, J. and Zur Hausen, H. Invertebrate Systems In Vitro (Kurstak, E., Maramorosch, K. and Dübendorfer, A., eds), pp. 517-533, Elsevier/North Holland. 1980.
2. Hayflick, L. and Moorhead, P. S. Exp. Cell Res., 25:585-621, 1961.
3. Pledger, W. J., Stiles, C. D., Antoniades, H. N. and Scher, C. D. Proc. Natl. Acad. Sci. U. S. A., 74:4481-4485, 1977.
4. Hayward, W. S., Neel, B. G. and Astrin, S. M. Nature, 90:475-480, 1981.
5. Nusse, R. Trends in Genet., 2:244-247, 1986.
6. Bishop, J. M. Science, 235:305-311, 1987.
7. Shilo, B. Z. and Weinberg, R. A. Proc. Natl. Acad. Sci. U. S. A., 78:6789-6792, 1981.
8. Bishop, J. M., Drees, B., Katzen, A. L., Korberg, T. B. and Simon, M. A. Cold Spring Harbor Symp. Quant. Biol.,50:727-731, 1985.
9. Shilo, B. Z. Trends in Genet., 3:69-72, 2987.
10. Katzen, A., Kornberg, T. B. and Bishop, J. M. Cell.,41:449-456, 1985.
11. Finnegan, D. J. and Fawcett, D. H. Oxford Surveys on Eukaryotic Genes, 3, pp. 1-62 + Appendix. 1986.
12. Siba, T. and Saigo, K. Nature, 302:119-124, 1983.

13. Finnegan, D. J. Nature, 302:105-106, 1983.
14. Temin, H. Cell, 21:599-600, 1980.
15. Boeke, D., Garfinkel, D., Styles, C. A. and Fink, G. Cell, 40:491-499, 1985.
16. Potter, S. S., Brorein, W. S., Dunsmuir, P. and Rubin, G. M. Cell, 17:415-427, 1979.
17. Ilyin, Y. V., Chmeliavskaite, V. G., Ananiev, E. V. and Georgiev, G. P. Chromosoma, 81:27-53, 1980.
18. Tchurikov, N. A., Ilyin, Y. V., Skryabin, K. G., Ananiev, E. V., Bayev, A. A. Jr, Krayev, A. S., Zelentsoya, E. S., Kulguskin, V. V., Lyubomirskaya, N. V. and Georgiev, G. P. Cold Spring Harbor Symp. Quant. Biol., 45:655-665, 1980.
19. Schneider, I. J. Embryol. Exp. Morphol., 27:353-365, 1972.
20. Echalier, G. and Ohanessian, A. C. R. Acad. Sci., Paris, 268:1171-1173, 1969.
21. Echalier, G. and Ohanessian, A. In Vitro, 6:162-172, 1970.
22. Strobel, E., Dunsmuir, P. and Rubin, G. M. Cell, 17:429-439, 1979.
23. Junaković, N. and Angelucci, V. J. Mol. Evol., 24:83-88, 1986.
24. Junaković, N., Di Franco, C., Barsanti, P. and Palumbo, G. J. Mol. Evol., 24:89-92, 1986.
25. Sporn, M. B. and Todaro, G. J. N. Engl. J. Med., 303:878-880, 1980.

PRODUCTION OF RETROVIRUS-LIKE PARTICLES BY TRANSPOSONS IN GENOME OF DROSOPHILA SIMULANS

T. Miyake,[1] N. Mae,[1] T. Shiba,[2] S. Kondo,[1] K. Saigo,[3] and K. Hattori[2]

1) Mitsubishi-Kasei Institute of Life Sciences, Machida, Tokyo 194, Japan
2) Kitasato University, Sagamihara, Kanagawa 228, Japan
3) Kyushu University, Fukuoka 812, Japan

The genomic *copias* of *Drosophila simulans* were isolated and analyzed for the structure and ability to produce retrovirus-like particles (RVLP). Based on the structure, the isolates were classified in 3 types. Among these 3 types, the longest one (5 kb) could produce RVLP while the middle one (3 kb) could not. The RVLP produced by 5 kb *simulans copia* in the transfectants were very similar to those produced by a *melanogaster copia* (cDm 2055) in various aspects such as size, morphology and chemical constitution. They were similar also to RVLP spontaneously appearing in *melanogaster* cell lines and *simulans* cell lines, except for the activity of a reverse transcriptase.

Previously, we reported (1) that the transposon *copia* in cDm 2055, which was isolated from genomic DNA of *Drosophila melanogaster* and ligated to pBR 322 by G.M. Rubin could produce the retrovirus-like particles (RVLP) when transfected into the *copia*-free cell line KUN-DH-33 (2) of *Drosophila hydei*. This finding strongly indicated the causal relation between the transposon (*copia*) and the retrovirus or RVLP.

Utilizing the system, we have next examined whether or not (i) the *copia* transposon in another *Drosophila* species, *D. simulans* also could produce RVLP, (ii) all *copias* in the genome could produce RVLP and (iii) the RVLP were similar to those produced by cDm 2055. Here, we describe the results concerning these questions briefly.

Isolation and Characterization of *Copias* from the Genome of *Drosophila simulans*

DNA was prepared from embryos of *Drosophila simulans*, and partially digested with *Eco*RI. The fragments were then ligated to Charon 4A λ phage and the *copia* hybridizable fragments were selected by plaque hybridization using, as the probe, cDm 2055 containing a *melanogaster copia*. The fragment were analyzed by examining heteroduplexes formed between each fragments and *melanogaster copia* clone. As shown in Table 1, they were classified in 3 types; type I, about 5 kb in length with LTR at both ends; type II, about 3 kb with LTR at the right end; type III, about 0.3 kb with only LTR. The restriction mapping for one of the type I, cDs-1 and of type II, cDs-10 showed that the latter corresponded to the 3' region of the former. The map of cDs-1 was quite similar to that of *melanogaster copia* in cDm 2055 (Fig. 1).

Table 1. Genomic copias of D. simulans

Plasmids or phage DNA	Fly source	Length of copia region	LTR
cDs-1	D.simulans	5.2 Kb	R,L
" -3	"	3.0	R
" -7	"	5.0	R,L
" -10	"	2.9	R
" -13	"	5.1	R,L
" -16	"	0.3	R
" -45	"	2.7	R
" -48	"	2.7	R
cDm 2055	D.melanogaster	5.0	R,L

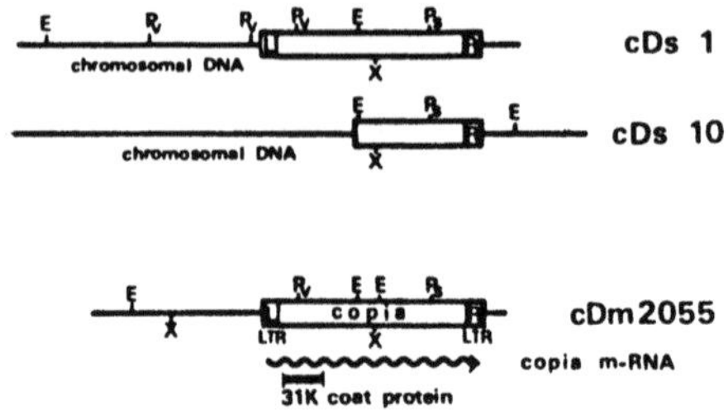

Fig. 1. Restriction map of simulans copias and melanogaster copia.

Transfection of the Simulans Copias into the Cell Line KUN-DH-33

The procedure for transfection was similar to that described previously (1). Briefly, the simulans copia in Charon 4A was co-transfected with the plasmid pSVC^{neo}-1 by the calcium-phosphate procedure into the cell line KUN-DH-33. The plasmid pSVC^{neo}-1 constructed by T. Todo of Osaka University carried the prokaryotic Neomycin-r gene under the control of copia promoter, and provided G418 resistance to the cells. The cell line KUN-DH-33 was established by Sondermeijer et al. from embryos of D. hydei (2). This fly, and hence the cell line contained no sequence hybridizable to copia. After the expression time of 48 hrs, the transfected cells were selected in the medium M3(BF) containing heat inactivated fetal bovine serum (10%) and Geneticin (G418-sulfate, 1.5 mg/ml, GIBCO). Small colonies appeared at the frequency of 10^{-4}~10^{-6} after 4-6 weeks at 25°C.

Analysis of the Transfectant DNA

By Southern blot analysis of the EcoRI digested genomic DNA, all these transfectants were found to contain copia sequences as well as Neo-r sequences. The patterns of copia hybridizable bands also suggested that there were copia sequences integrated in intact form.

Production of RVLP in the Transfectants

The presence or absence of RVLP was examined by Northern blot analysis and electronmicroscopy. For the former, the supernatant of nuclear sonicate was run by electrophoresis on 0.8% agarose. RVLP on the agarose were disrupted by treating with 1% SDS for 60 minutes at 50°C, and the RVLP-RNA was transferred onto the filter and hybridized with the ^{32}P labelled copia DNA, and autoradiographed. Hybridization signals were observed for the nuclear lysate of the transfectants with cDs-1 (5 kb simulans copia). No such signals were observed for cDs-10 (3 kb simulans copia) transfectants or for the recipient KUN-DH-33. The results were perfectly supported by electronmicroscopic observation. Virus-like structures with dia 50 nm were observed only in the nuclei of the transfectants with cDs-1 (Fig. 2). Results are summarized in Table 2.

Table 2. Retrovirus-like particles (RVLP) in various transfectants

Cells	No. of cells examined	*copia*-RVLP type		Other type
		Nucleus	Cytoplasm	Cytoplasm
NC^{sim}-3*	560	159	-**	-
NC^{sim}-7*	550	3	-	-
NC-1*	500	440	2	58
KUN-DH-33 (Recipient)	1000	0	0	190

* NC^{sim}-3 = one of cDs-1 transfectants ** not scored
NC^{sim}-7 = one of cDs-10 transfectants
NC -1 = one of cDm 2055 transfectants

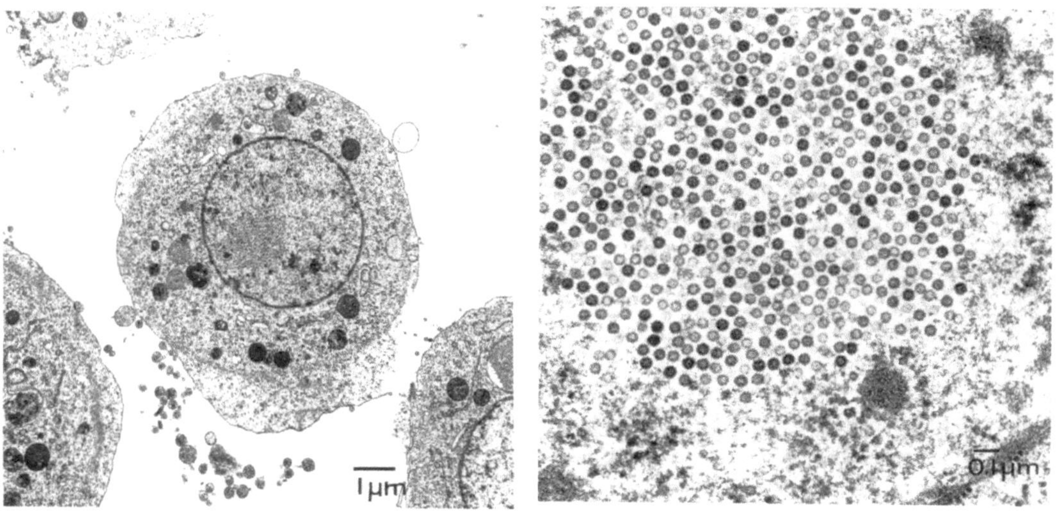

Fig. 2. Retrovirus-like structures in cDs-1 transfectant

RNA in RVLP

As shown previously (3), RVLP spontaneously appearing in nuclei of *melanogaster* cultured cells contained low molecular weight RNAs (4S, 4.5S, 5S. and 6S) and high molecular weight RNA (H-RNA), and only the latter hybridized to *copia* DNA. RNAs in RVLP produced in the transfectant with cDs-1 were similarly analyzed. The same 4 species of low molecular weight RNA and H-RNA were detected and the latter hybridized to the *copia* probe.

RVLP Protein

RVLP protein was electrophoresed on 12.5% polyacrylamide gel containing SDS-urea, and stained with Coomassie brilliant blue. The result showed that RVLP in the cDs-1 transfectant contained 31 kd protein as the major component, similarly to RVLP in the cDm 2055 transfectant and those spontaneously appearing in the simulans cell line. These 31 kd proteins immunoreacted to the anti-RVLP rabbit serum prepared with melanogaster RVLP. There seemed to be, however, a little difference in the species of minor protein components of RVLPs from these three sources.

Reverse Transcriptase Activity Associated with RVLP

It has been reported that RVLP spontaneously appearing in melanogaster cell lines showed reversetranscriptase activity (4). RVLP produced in the cDm 2055 transfectants also showed similar activity although at significantly low level (1). RVLP in cDs-1 transfectants were examined for this activity, and the result was positive. The level of the activity, however, was much lower compared with that of RVLP spontaneously appearing in simulans cell lines or in melanogaster cell lines. The activity was in the same order as that of RVLP in cDm 2055 transfectants.

RVLP RNA and RVLP Protein in the Transfectants with cDs-10 (3 kb Simulans Copia)

As mentioned in the previous section, RVLP were not detected in the cDs-10 transfectants. Therefore, we next examined the synthesis of RVLP protein and RVLP RNA in the transfected cells. The major protein component of 31 kd was detected in cDs-1 transfected cells but not detected in cDs-10 transfected cells. When probed with copia DNA, hybridization was observed with 5 kb cellular RNA from cDs-1 transfected cells. With cellular RNA from cDs-10 transfected cells, very low level of hybridization was observed after a long exposure. However, the size of the hybridizable RNA was about 3 kb, much shorter than that in cDs-1 transfected cells.

Conclusion

Thus, it was revealed that (i) at least some copias in the genome of Drosophila simulans could produce RVLP, (ii) there seemed to be incomplete copias in the simulans genome and (iii) the RVLP produced in the transfectants by simulans copia were similar to those by melanogaster copia. They were also similar to those spontaneously appearing in melanogaster cell lines and simulans cell lines except for the activity of reverse transcriptase and the species of minor protein components.

References

1. Miyake, T., Mae, N., Shiba, T. and Kondo, S. Mol. Gen. Genet., 207: 29-37, 1987.
2. Sondermeijer, P.J.A., Derksen, J.W. and Lubsen, N.H. In Vitro, 16; 913-914, 1980.
3. Saigo, K., Shiba, T. and Miyake, T. Invertebrate Systems In Vitro (Kurstak, E., Maramorosch, K. and Dubendorfer A., eds.) pp.411-424, Elsevier,Amsterdam, 1980.
4. Shiba, T. and Saigo, K. Nature, 302: 119-124, 1983.

STUDY ON A *SARCOPHAGA* LECTIN GENE ACTIVATING FACTOR IN THE HEMOLYMPH

A. Shiraishi[1)], Y. Nakanishi[2)], and S. Natori[1)]

1) Faculty of Pharmaceutical Sciences, University of Tokyo, Tokyo 113, Japan
2) Faculty of Pharmaceutical Sciences, Kanazawa University, Kanazawa 920, Japan

Sarcophaga lectin gene was activated in the fat body of the third instar larvae of *Sarcophaga peregrina* when the body wall was injured. Hemolymph from injured larvae activated the lectin gene when it was injected into the posterior part of ligated larvae and added to *in vitro* cultured fat body. Therefore, hemolymph from injured larvae was found to contain a factor that activates *Sarcophaga* lectin gene in the fat body.

Selective gene expression, in response to endogenous or exogenous signals, is one of the most important features of life. Research is being carried out in different laboratories to elucidate the signal pathway and the regulation of gene expression. In most cases, molecular mechanism of the signal mediation is not yet known.

Sarcophaga lectin was induced in the third instar larvae of *Sarcophaga peregrina* when the body wall was injured with a hypodermic needle. *Sarcophaga* lectin was purified (1) and an antibody was raised against it. Using this antibody, it was found that the lectin participates in the elimination of sheep red blood cells introduced into the body cavity of larvae (2). Therefore, at least one role of *Sarcophaga* lectin is to eliminate foreign substances possibly invading through the injured body wall.

On the other hand, an interesting fact was observed that the same lectin was induced in embryos and pupae in their normal development without any treatment (1,3). This result suggests that *Sarcophaga* lectin does not only participate in a defence mechanism but also plays a role in these two stages of development.

Recently, we isolated cDNA clone of *Sarcophaga* lectin (4). Using this clone as probe, it was found that induction of the lectin occurred at the mRNA level (3). Thus, *Sarcophaga* lectin gene has a unique feature of being expressed in different physiological stages. Therefore, it may be a good model to study the regulation of gene expression in response to different signals. As a first step, the regulation of lectin gene expression was studied on injury. This paper describes results suggesting that hemolymph from injured larvae contains a factor which activates the lectin gene expression in the fat body.

Expression of *Sarcophaga* Lectin Gene on Injury

Third instar larvae were injured with a hypodermic needle. Fat body was isolated at 1.5, 3, 6, 12, and 24 h after injury. Then the amount of lectin mRNA was determined by Northern blot hybridization. As shown in Fig. 1, the lectin mRNA was induced on injury and reached the maximum level at 3 to 6 h after injury. Next, the tissue specificity of the lectin gene expression on injury was examined. Although the fat body contains a large amount of the lectin mRNA, almost no lectin mRNA was detected in the other tissues. This result indicates that the lectin gene is expressed specifically in the fat body. From these results, we believe that injured larval hemolymph contains a factor which activates the expression of lectin gene in the fat body.

Effect of Ligation on the Expression of Lectin Gene on Injury

In consideration of the involvement of the endocrine system in lectin gene expression, the larvae were ligated at the boundary of the abdomen and the head where most of the endocrine glands are located. Therefore, the fat body was divided into anterior and posterior parts. Then, the posterior part of the larvae was injured and the expression of the lectin gene in the anterior and posterior parts was examined. As shown in Fig. 2, when larvae were not ligated, the lectin mRNA was detected on injury in both parts of the fat body. However, when larvae were ligated, no significant lectin mRNA was detected in the posterior part of fat body, whereas the anterior part of fat body contained a significant amount of lectin mRNA. The same result was obtained when the anterior part of the ligated larvae was injured (data not shown). These results suggested that the anterior part of larvae is important for lectin gene expression. On the other hand, the lectin mRNA was found to be induced in the anterior part of fat body when the animal was ligated without injuring the body wall. In the posterior part of the fat body, almost no lectin mRNA was detected. This lectin gene expression seems to be induced by an internal injury caused by the ligation.

Expression of *Sarcophaga* Lectin Gene by Transfusion of Injured Larval Hemolymph

If an activating factor of the lectin gene is present in the hemolymph of the injured larvae, transfusion of this hemolymph to the posterior part of ligated larvae may induce the lectin mRNA. In order to prove this, larvae were ligated and the anterior part cut off. Then, the injured larval hemolymph was injected to posterior part of the larvae. The hemolymph was collected by cutting off the anterior tip of the larvae and centrifugation. As shown in Fig. 3, when the injured larval hemolymph was injected, the lectin mRNA was induced, whereas insertion of a hypodermic needle had almost no effect on the lectin gene expression. These results suggested that injured larval hemolymph contains the activating factor of *Sarcophaga* lectin gene. Injection of normal larval hemolymph also induced the lectin mRNA. Induction of the lectin mRNA by normal larval hemolymph may be due to the activating factor appearing in the process of isolation of the hemolymph.

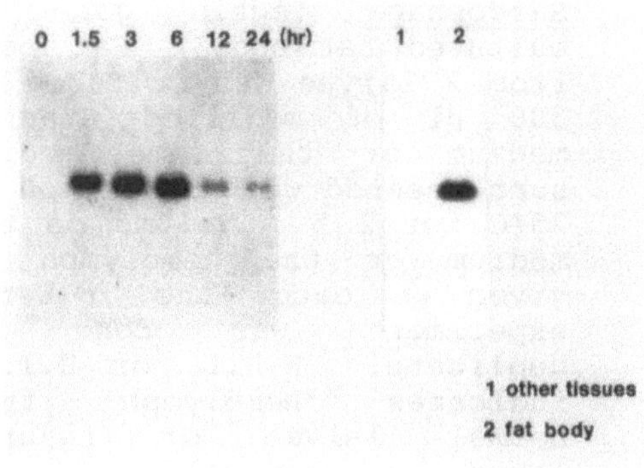

Fig. 1 Expression of Sarcophaga lectin gene on injury. In all results presented in Fig. 1-4, the amount of lectin mRNA was determined by Northern blot hybridization using cDNA clone of Sarcophaga lectin, pLE10, as probe. The time course of expression of the lectin after injury is shown in the left part of this figure. Fat body was isolated at the time mentioned on each lane. Tissue specificity of expression of the lectin gene after injury is shown in the right part of this figure. Fat body and other tissues were isolated from larvae 3 h after injury.

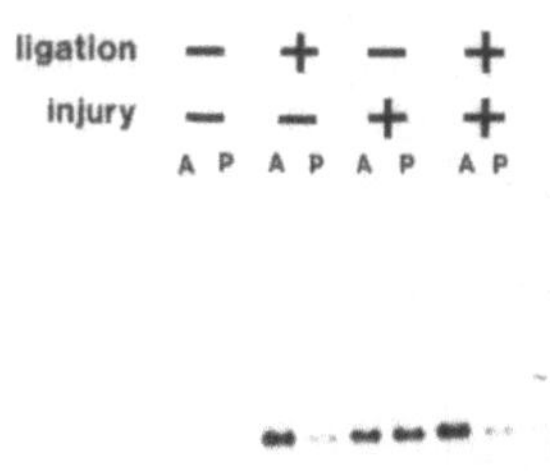

Fig. 2 Effect of ligation on the expression of Sarcophaga lectin gene. A or P indicates the anterior or the posterior part of larvae, respectively.

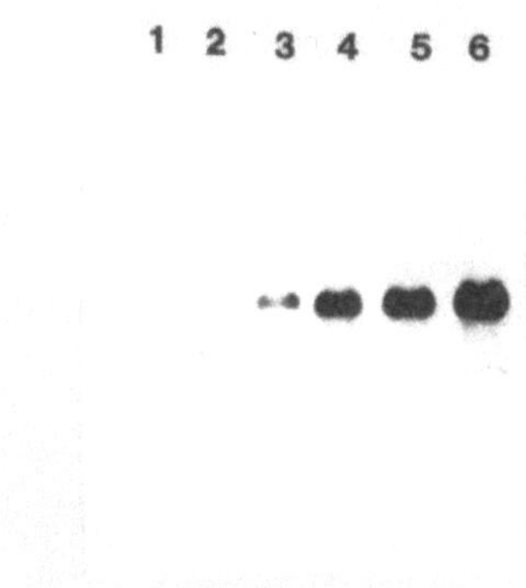

Fig. 3 Expression of Sarcophaga lectin gene by transfusion of injured larval hemolymph. Control larvae (1), the posterior part of ligated larvae was treated in the following way: insertion of hypodermic needle (2); injection of 2 μl of water (3); injection of 2 μl of the hemolymph from normal larvae (4); injection of 2 μl of the hemolymph from larvae 1 h after injury (5) and injection of 2 μl of the hemolymph from larvae 3 h after injury (6).

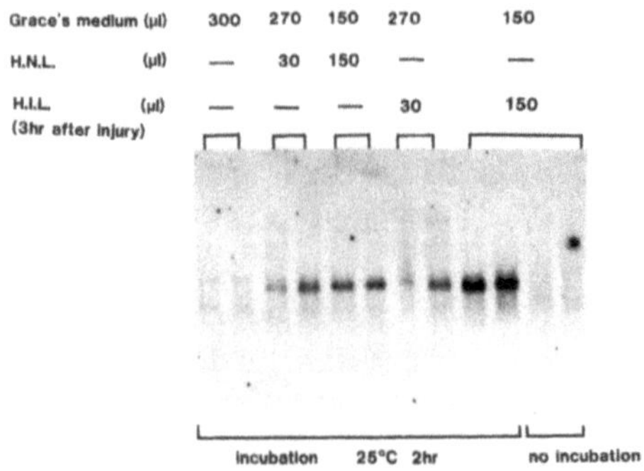

Fig. 4 Expression of Sarcophaga lectin gene in cultured fat body. Fat body from 2 larvae was cultured in 300 µl of modified Grace´s medium or the same medium supplemented with hemolymph at 25°C for 2 h. Volume of the medium or the hemolymph is given on each lane. Every experiment was done in duplicate. H.N.L. or H.I.L. indicates hemolymph from normal larvae or injured larvae, respectively.

Expression of Sarcophaga Lectin Gene in Cultured Fat Body

We next tried to induce the lectin mRNA in cultured fat body by adding injured larval hemolymph to the medium. Fat body was isolated from larvae and was cultured in modified Grace´s insect medium or the same medium supplemented with hemolymph at 25°C for 2 h. The induction of lectin mRNA was examined by Northern blot hybridization. As shown in Fig. 4, no lectin mRNA was detected in the fat body cultured in Grace´s medium alone. However, the lectin mRNA was induced in the fat body cultured in the medium supplemented with injured larval hemolymph according to dose. Addition of normal larval hemolymph also induced the lectin mRNA. However, the amount of lectin mRNA induced by normal larval hemolymph was less than that induced by injured larval hemolymph. These results suggested that a factor which activates Sarcophaga lectin gene expression in the fat body was present in the injured larval hemolymph.

We next tested heat stability of the activity of inducing lectin mRNA in the hemolymph. The injured larval hemolymph was heated at 60 °C for 15 min and centrifuged. The resulting supernatant was further heated at 80°C or in boiling water for 15 min and then centrifuged to obtain supernatant. The heat treatment at 80°C did not decrease the activity of inducing lectin mRNA. Even in boiling water, the activity was not lost (data not shown). Thus the activity of inducing lectin mRNA is stable against heat treatment.

Finally, from these results presented here, the following speculation is possible. Namely, signal of injury to a body wall is not transmitted to the fat body directly, but first to some organ in the anterior part of larvae, and from there a factor which activates Sarcophaga lectin gene expression in the fat body may be released. We are now trying to purify and to characterize the activating factor.

References

1. Komano, H., Mizuno, D. and Natori, S. J. Biol. Chem. 255:2919-2924, 1980.
2. Komano, H. and Natori, S. Devel. Comp. Immunol. 9:31-40, 1985.
3. Takahashi, H., Komano, H. and Natori, S. J. Insect Physiol. 32: 771-779, 1986.
4. Takahashi, H., Komano, H., Kawaguchi, N., Kitamura, N., Nakanishi, S. and Natori, S. J. Biol. Chem. 260:12228-12233, 1985.

GENETIC PROGRAMMING OF *DROSOPHILA* CELLS BY HEAT SHOCK AND BY ECDYSTERONE

E. M. Berger

National Science Foundation, Washington D.C., U.S.A.

Several established *Drosophila* tissue culture lines respond to the addition of the moulting hormone, ecdysterone, by initiating the transcription and translation of the four small heat shock protein (hsp) genes. These genes, and several others, are also induced in cells by chemical, or physical (heat) stress. In order to analyze the molecular mechanism of small hsp gene regulation, we have developed and employed DNA mediated gene transfer techniques for these cells. In this brief report we summarize the results of our lab and others in identifying the nature and location of cis-acting regulatory elements, in the 5' flanking region of each gene, that are involved in regulation.

In the VIth International Conference on Invertebrate Tissue Culture, held in St. Augustine, Florida, we discussed our progress in identifying specific genes whose expression (transcription) was induced in cultured cells, by the addition of physiological concentrations of the insect moulting hormone, ecdysterone, into the medium. The most intriguing case involved the transcriptional activation of the four small heat shock protein genes. These genes are tightly linked, and are found clustered within a 10 kilobase pair region located in the cytological band 67B1, on chromosome 3. The genes encode four proteins, designated hsp22, hsp23, hsp26, and hsp28. Small hsp synthesis is also induced in these cells by physical (heat), or chemical stress. We found that ecdysterone induction occurred in several independently derived cell lines, including Schneider 3, Kc, and a haploid line call-D, but did not occur in hormone insensitive cell lines, such as MDER and F6 (10). Induction of small hsp synthesis by heat shock was found in every cell line tested, suggesting that heat induction and hormone induction involve independent mechanisms. We and others subsequently found that small hsp gene induction was developmentally regulated in the animal. Expression is abundant in the imaginal discs of prepupae, and, for two of the gene, expression also occurs in ovary tissue during oogenesis. Induction can be produced in imaginal discs of immature, third larval instar animals by incubation in medium containing hormone. Our conclusion, then, was that the expression of the small hsp genes in *Drosophila* is under a complex pattern of control. Gene expression could be independently activated by stress of various sorts, or by the hormone ecdysterone, in a stage and tissue specific manner. Finally, we provided evidence that each of the four genes appeared to be regulated independently, rather than as a cluster. Several other genes embedded within the small hsp cluster, while showing developmentally regulated expression, are not inducible by either heat shock or ecdysterone, in any cell line (10).

Functional Assays

The hypothesis we wished to test next was that each of the four small hsp genes is flanked, at their5' end, by one or several cis-acting regulatory elements that function as target sites for trans-acting transcription factors. These factors would include the so-called heat shock transcription factor, HSTF, discovered and purified by Parker's group (11), the putitive ecdysterone-receptor complex (12), and other protein factors yet to be identified. Our initial suspicion was that there would turn out to be several, functionally specific, and non-overlapping control elements. Each element (heat shock or hormone) might itself be made of sub-elements that interact additively, or cooperatively. In order to map the various elements,several approaches were considered. The first was simply to look at the DNA sequence of each of the four cloned genes and identify regions of homology among all four, in the 5' upstream region. This approach was flawed, however, because we could not distinguish regions of sequence homology based on common function from regions of sequence homology based on common ancestry. These four small hsp genes appear to have arisen from a single ancestral gene that underwent a series of duplications and mutations.

A second approach was to isolate mutant cell lines that were defective in small hsp gene regulation, and then examine the nature and location of alterations in the DNA sequence. This approach was abandoned,as well, because we had no means of screening or selecting for such mutants. In fact, we have no idea what the phenotype of small hsp defective cells,or flies would be. We proceded to the third approach, which involves the development of a functional assay for regulatory elements, based on the technology of DNA mediated gene transfer. The reasoning behind the approach goes as follows. The first step is to develop a technique for introducing cloned small hsp genes into responsive cells, under conditions that lead to the acquisition of normal regulation in the cell. The next step would be to modify the cloned gene, by in vitro mutagenesis, reintroduce the mutated construct into a responsive cell, and then assay for regulated expression. If the altered DNA region is not necessary for regulation, then the mutation will have no effect. If the altered or deleted region is required for specific regulation, then the mutated gene construct will show abnormal regulation. Expression could be turned off permanently, or remain constitutive if, for example, a repressor binding site is lost. One problem that had to be overcome involved distinguishing between the regulated expression of genes introduced into the cell,from the regulated expression of endogenous genes. The solution was to replace the protein coding region of each small hsp gene with a "reporter" gene sequence, such as herpes virus thymidine kinase (tk), or bacterial chloramphenicol acetyltransferase (CAT). The expression of these foreign genes could be assayed unambiguously by analyzing RNA composition, using radiolabelled "reporter" gene DNA as a hybridization probe, or by carrying out a CAT enzyme assay. The assumption which needed to be proven is that the regulatory elements of interest lie outside of the protein coding sequence. The evidence to date is consistent with this assumption.

Developing the System

We (13) and others (14) have adapted techniques for introducing cloned genes into Drosophila cells, under conditions that lead to the acquisition of normal regulation and expression. For resaons of efficiency we have chosen to use a transient expression assay system, in which several to many genes are introduced into each cell, but replication does not occur. In several of the reports discussed below gene transformation assays were utilized. Here, 1 or a few gene copies are present in each cell, integrated into the chromosomes, and undergoing replication.

Once the gene tranfer procedures were developed we went on to show that recombinant DNA constructs containing the 5' upstream region of each gene, and part of the transcribed but untranslated leader, ligated to a tk or CAT gene did, in fact, acquire the normal pattern of heat and hormone regulated expression when introduced into responsive Drosophila cells(15). Similar results were obtained using cultured mosquito or moth cell lines (16). The final step involved the construction of mutated plasmids, using restriction endonucleases or Bal 31 exonuclease to progressively resect the hsp 5' region, from the 5' end. Each of these constructs was then assayed.

Location of Regulatory Elements

Figure 1 summarizes the results obtained from several labs, with respect to the position of regulatory elements required for heat shock or ecdysterone control. In several cases there are unresolved discrepencies between two labs concerning the location or limits of an element. These differences might be based on differences in the assay protocol (transformation versus transient expression), or they could reflect the specific types of constructions used. Space restrictions prevent me from describing these results in detail, or from discussing the biases and potential artifacts intrinsic to these kinds of studies. There are several general conclusions, that are summarized below.

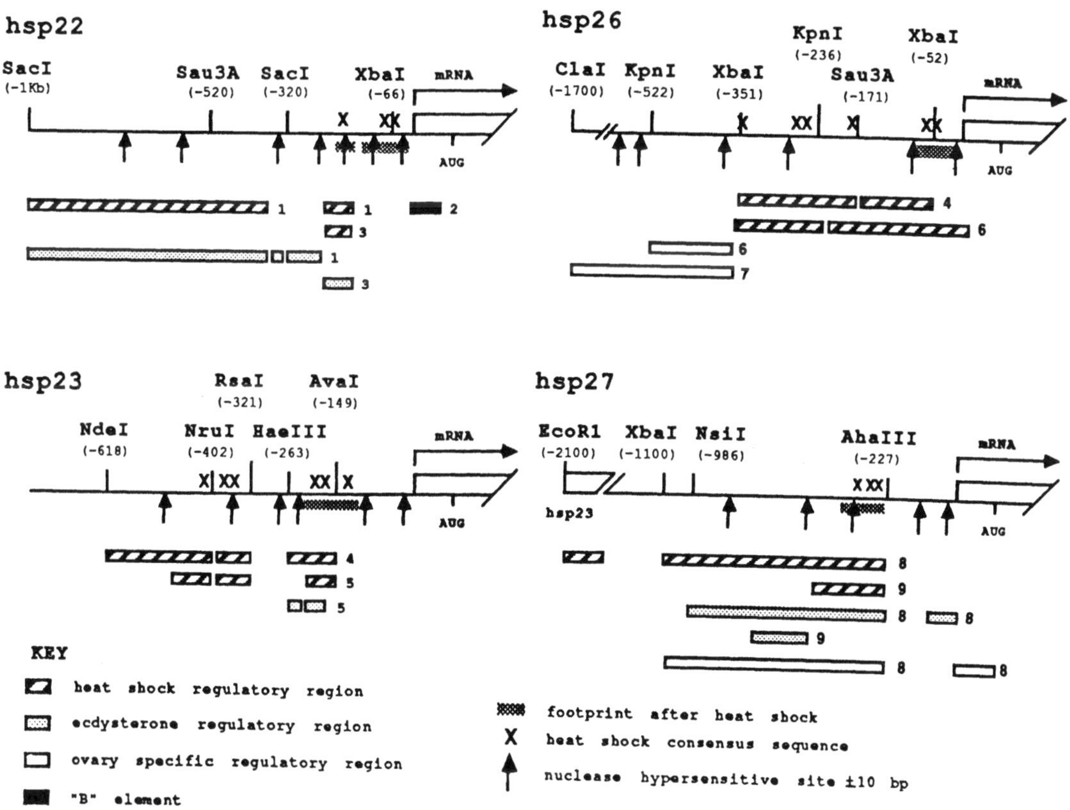

Figure 1. The location of regulatory elements near small heat shock protein genes. The sites indicated are described in the Key. The "B" element is a transcription site in the leader, and the arrows point to DNAse hypersensitive sites. The "footprint" regions become resistant to nuclease digestion following heat shock. References are indicated to the right of each mapped element.

1. The heat shock regulatory element contains a consensus sequence called the Pelham box (17), that is CT-GAA--TTC-AG. The element itself is composed of 2 or more such sequences, arrayed in tandem, which function cooperatively. They are HSTF binding sites (11). Using synthetic DNA constructs, we have shown that a functional element contains at least two Pelham boxes (18).

2. The ecdysterone and heat shock elements are distinct. In the case of hsp22, studied in our lab (1), there is a region between -1,000 and-320 that, when deleted, reduces the level of both heat and hormone induction, while not affecting the on-off switch. It behaves like an enhancer element. The specific ecdysterone element appears to lie between positions -320 and -239, while the specific heat shock element maps to a position between -208 and -146.

References

1. Morganelli, C.M. Ph.D. Thesis, Datrmouth College, 1986.
2. Hultmark, D., Klemenz, R., Gehring, W.J. Cell 44:429, 1986.
3. Klemenz, R., Gehring, W.J. Molec. Cell Biol. 6:2011, 1986.
4. Pauli, D., Spierer, A, Tissieres, A. EMBO J.5:755, 1986.
5. Mestril, R., Schiller, P., Amin, J., Klapper, H., Anathan, J., Voellmy, R. EMBO J. 5:1667, 1986.
6. Cohen, R., Meselson, M. Cell 43:737, 1985.
7. Glaser, R., Wolfrer, M., Lis, J. EMBO J. 5:747, 1986.
8. Hoffman, E., Gehring, S., Corces, V. in press, 1986
9. Riddihough, G., Pelham, H.R.B. EMBO J. 5:1653, 1986.
10. Berger, E.M., Vitek, M.P., Morganelli, C.M. in: Techniques in In Vitro Invertebrate Hormones and Genes (eds. E. Kurstak, H. Oberlander) C220, pp.1-14,Elsevier Biomedical, 1986.
11. Topol, J., Ruden, D., Parker, C. Cell 43:527, 1985.
12. Yund, M.A., King, D.S., Fristrom, J. Proc. Nat. Acad. Sci. US 75:6039, 1978.
13. Morganelli, C.M., Berger, E.M. Science 224:1004, 1984.
14. DiNocera, P.P., Dawid, I.B. Proc. Nat. Acad. Sci. US 80:7095, 1983.
15. Morganelli, C.M., Berger, E.M., Pelham, H.R.B. Proc. Nat. Acad. Sci. US 82:5865, 1985.
16. Berger, E.M., Marino, G., Torrey, D. Somat. Cell and Molec. Genet. 11:371, 1985.
17. Pelham, H.R.B. Cell 30:517, 1982.
18. Berger, E.M., Torrey, D., Morganelli, C.M. Somat. Cell and Molec. Genet. 12:433, 1986.

ULTRASTRUCTURAL EFFECTS OF ECDYSTEROID ON GROWTH AND DIFFERENTIATION OF SILK GLAND CELLS OF BOMBYX MORI

H. Akai and M. Kiuchi

Sericultural Experiment Station, Yatabe, Ibaraki 305, Japan

The authors found various specific ultrastructural changes induced by ecdysteroid treatment with relation to the ecdysteroid titer. First, ecdysteroid titers of silkworm larvae in 4th to 5th instar were measured. Next, two types of histological change in the silk gland were revealed by ecdysteroid treatments. Ecdysis of cuticular intima of the posterior silk gland was also confirmed electron-microscopically by these treatments. In the processes of histological changes of the silk gland cells induced by ecdysteroid treatment, the massive nucleoli were completely changed into strand-like structures. The treatments also accelerated lysosome formation.

The silk gland of the larva of Bombyx mori is a typical exocrine gland designed to secrete silk proteins. At the completion of larval maturation during the fifth instar, the silk glands are so large that they reach a length of about 25 cm and make up about 40% of the body weight. During larval development the cells in the middle and posterior regions increase about 86,000 and 57,000 times, respectively. During the larval period, the DNA content of the average cell in the posterior silk gland increases 250,000-fold. The nuclei in the posterior silk gland undergo 18-19 endomitotic cycles, the middle silk gland undergoes 20 cycles, and the anterior silk gland only 13.

Concerning the ultrastructural characteristics of the posterior silk gland cells, the fibroin which is mainly silk is synthesized in endoplasmic reticulum (ER) and secreted into the silk layer via Golgi complexes and fibroin globules(1), and finally into the lumen passing through the cuticular intima(2). During the larval molting stage also, the gland cells undergo characteristic histological changes, digestion of the silk materials, disappearance of ER, etc.(3). In this paper we report various specific ultrastructural changes induced by ecdysteroid treatments with relation to the ecdysteroid titers.

The ecdysteroid titers of the silkworm larvae from just after the 3rd ecdysis to the pupation were measured at 12 hr intervals by radioimmunoassay(4). As shown in Fig. 1, the ecdysteroid titer during the 4th instar increased gradually, and reached a peak at three days after the 3rd ecdysis, then decreased to a much lower level for five or six days after the 4th ecdysis, and then increased and reached a peak on the tenth day (Fig.2). These results basically coincide with the ecdysteroid levels which were published by Kiguchi and Agui(5).

In the second experiment, the effect of ecdysteroid treatment on the histological changes of the silk gland cells were observed light- and electron-microscopically. Ecdysterone(20 ug/1 g. of body weight) was injected into four groups of animals of different developmental stages : larvae just after the 4th ecdysis, and 48 hrs, 96 hrs, and 120 hrs after the 4th ecdysis. Histological changes of the posterior silk gland after the treatments were observed at 6 hr intervals, as shown in

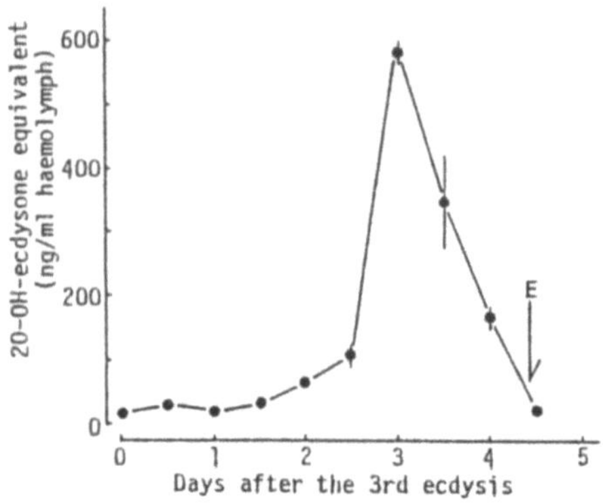

Fig 1 Ecdysteroid titer of haemolymph during 4th instar.

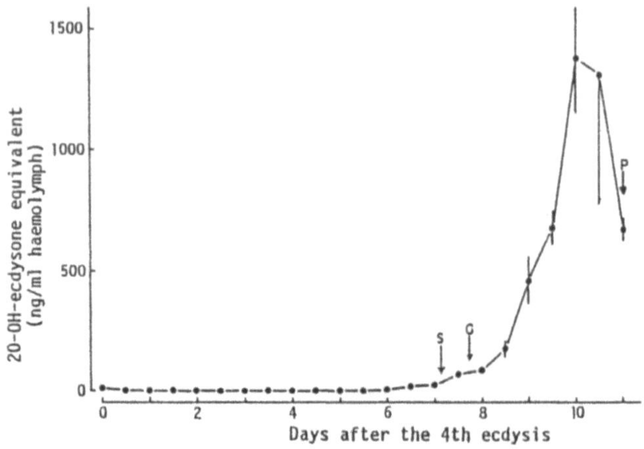

Fig. 2 Ecdysteroid titer of haemolymph during 5th instar.

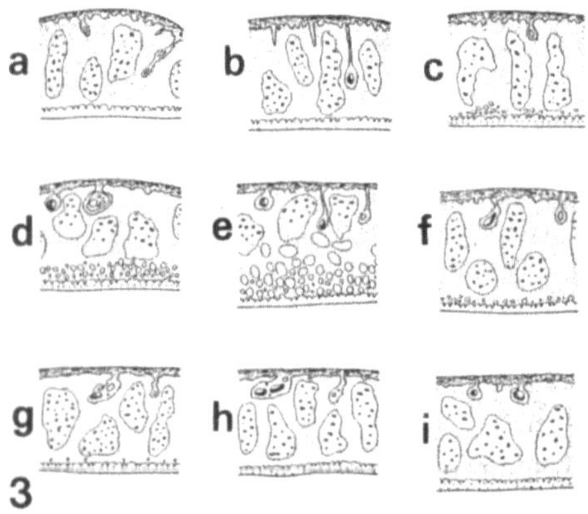

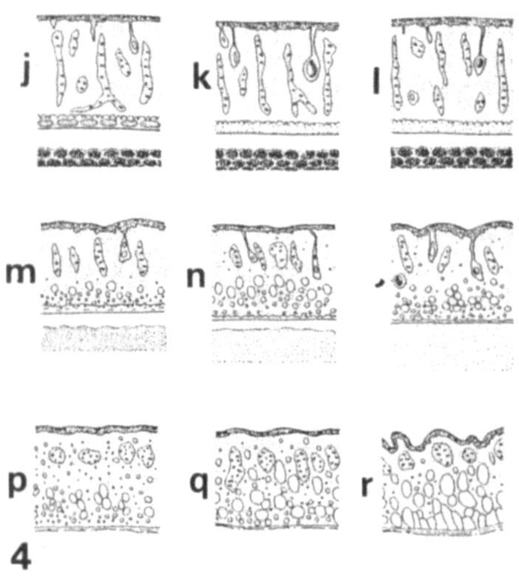

Fig. 3 (a-i) Structural changes of posterior silk gland cells(6 hr intervals) were induced by ecdysteroid treatments at beginning of 5th instar. Cytoplasmic vacuoles induced disappear and are recovered the same as in the normal cell during the molting stage.

Fig. 4 (j-r) Similar changes of the cells (6 hr intervals) were induced by ecdysteroid treatments at 48 hrs of the 5th instar. Cytoplasmic vacuoles are greatly increased and finally are histolyzed.

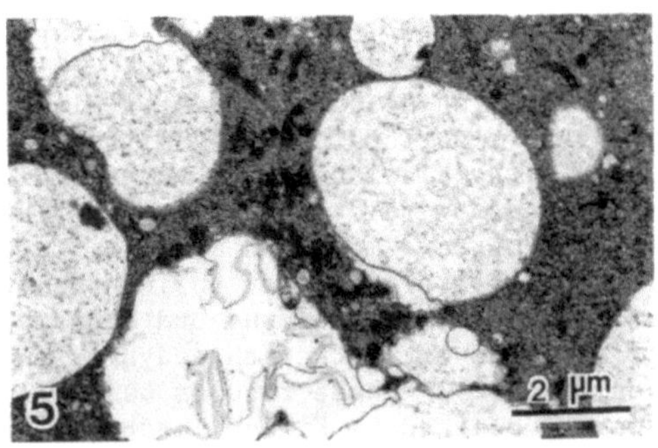

Fig. 5 Various sizes of vacuoles in the posterior silk gland cells, induced by ecdysteroid treatment during the 5th larval instar. The vacuoles contain fine fibrous materials.

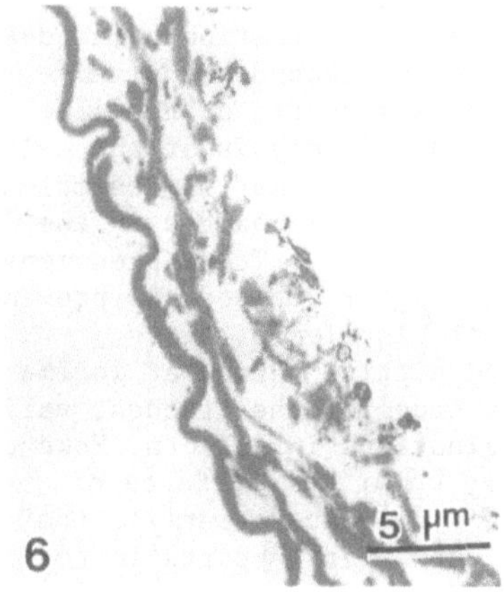

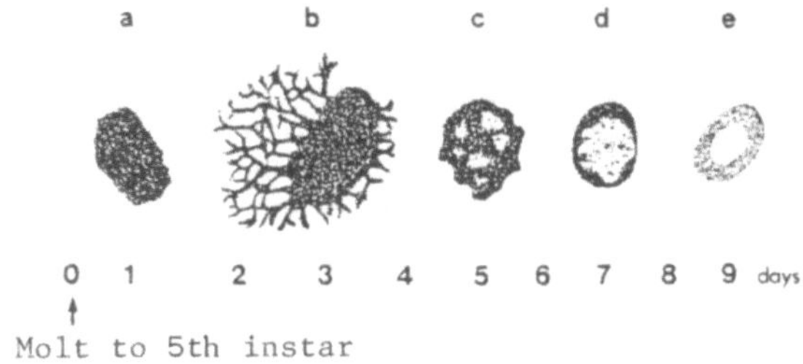

Fig. 7 Changes in the nucleoli of the cells of the posterior silk gland during the 5th instar. (a) Early 5th instar (1 day after 4th molt); (b) stage of active RNA synthesis (2-3 days); (c) middle 5th instar (5 days); (d) mature to spinning stage (7 days); (e) prepupal stage (9 days).

Fig. 6 Ecdysis of the intima in the posterior silk gland induced by ecdysterone treatment.

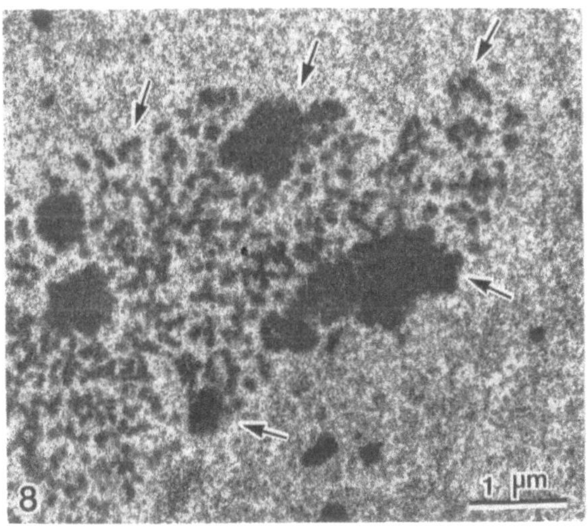

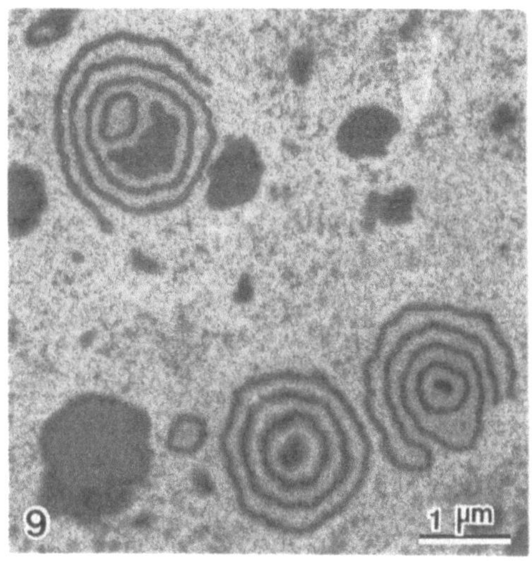

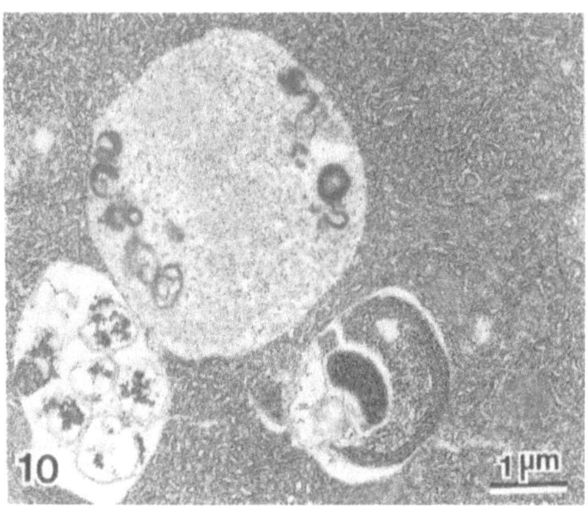

Fig. 8 Well-developed nucleoli(arrows) in the stage of active RNA synthesis.

Fig. 9 Strand-like nucleoli which appeared in the posterior silk gland cells during the molting stage, induced by ecdysterone treatment after 48 hrs of the 5th instar.

Fig. 10 Lysosomes induced by the same ecdysterone treatment as mentioned above.

Figs. 3 and 4. The most interesting point of cellular change induced is the vacuole formation in the apical cytoplasm (Fig. 5). In the first group(ecdysterone treatment just after 4th ecdysis), the vacuole formation occurred at 18 hrs after the treatments and disappeared at 30 hrs. This type of vacuole formation coincides with the normal cellular changes in larval molting(3). In the second group(ecdysterone treatment at 48 hrs after 4th ecdysis), the vacuoles appeared at the same time but increased in both size and number, and preceded the histolysis of the cells without any disappearance of the vacuole. This type of vacuole formation resembles normal histolysis during larval-pupal metamorphosis. Histological changes in the third and fourth groups were similar to these in the second group. These phenomena indicate that the ecdysterone treatments later than 48 hrs after 4th ecdysis prevent normal development of the silk gland due to the process of histolysis.

The Bombyx silk gland is an ectodermal organ, and a thick cuticular intima lines the inner surface of the anterior silk gland as is true for the trachea, salivary glands and other organs present as ectodermal invaginations in insects. However, it has been claimed that there are no cuticular membranes on apical surfaces of thecells of the middle and posterior silk glands(6). In a previous paper(7), the first author found several thin cuticular membranes and cuticular supports in the silk layer of the posterior silk gland. Even by electron microscopic observation, however, it is difficult to determine whether such thin membranes are the true cuticular membrane or not. The ecdysterone treatment in this experiment gave clear evidence that these membranes are renewed by the injections(Fig. 6).

It is well known that the nucleolus is a site of RNA synthesis in the nucleus(1). In posterior silk gland cells, the nucleolus changes to various sizes and shapes in conjunction with the synthetic activity of RNA. Ultrastructural changes of nucleoli in the cells of the posterior silk gland during the 5th instar were confirmed as shown in Fig. 7. The most active synthesis of RNA occurs 2-3 days(b) after the 4th ecdysis. The nucleolus develops enormously in size and shape, forming numerous strand-like structures(Fig. 8). Ecdysteroid treatment induced very unique morphological changes of nucleoli in the posterior silk gland cells as shown in Fig. 9. Such strand-like structures from the nucleolus are undocumented, not only in insect cells but even in general animal cells. It generally has been claimed that a nucleolus consists of a pars amorpha and a nucleolonema which is strand-like shape. Such nucleolar changes may be an uncommon phase of a basic ultrastructure of the nucleolus which appeared because of a special hormonal condition.

Lysosomes are popularly formed in silk gland cells during the process of larval-pupal metamorphosis. Ecdysterone treatment accelerates the lysosome formation which is one of the markers of metamorphosis, when applied after than 48 hrs after the 5th instar(Fig. 10).

References

1. Akai, H. Insect Ultrastructure 2 (King, R.C. and Akai, H., eds.),pp. 323-364, Plenum Press, 1984.
2. Akai, H. J. Seric. Sci. Jpn., 55 : 163-167, 1986.
3. Akai, H. Bull. Sericul. Exp. Stat., 19 : 375-483, 1965. (in Japanese with English summary).
4. Takeda, S., Kiuchi, M. and Ueda, S. Bull. Sericul. Exp. Stat., 30 : 361-374, 1986. (in Japanese with English summary).
5. Kiguchi, K. and Agui, N. J. Insect Physiol., 27 : 805-812, 1981.
6. Tanaka, Y. Lecture of Silkworm Anatomy, pp. 80-102, Meibundo, Tokyo, 1928. (in Japanese).
7. Akai, H. J. Seric. Sci. Jpn., 53 : 72-76, 1984.

EFFICIENT EXPRESSION OF FOREIGN GENES IN CULTURED DROSOPHILA MELANOGASTER CELLS USING HYGROMYCIN B SELECTION

A. van der Straten, H. Johansen, R. Sweet, and M. Rosenberg

Smith Kline & French Laboratories, Dept. of Molecular Genetics, 709 Swedeland Road, King of Prussia, PA 19406, USA

A dominant hygromycin B selection system was developed to stably co-introduce and express foreign genes in the Schneider S2 cell line. Cells were transfected with a vector carrying the *E. coli* hygromycin resistance gene under the control of the COPIA 5'LTR and selected for 2-3 weeks at 300μgr/ml of hygromycin B. This system was used to co-introduce and overexpress the *E. coli galK* gene and several other genes of mammalian origin. The transfected gene is stably integrated in the host DNA and cell lines with up to 10^3 gene copies/cell were obtained. This system was also utilized to compare the efficiency of several promoter signals.

Cultured insect cells have become useful hosts for overexpression of heterologous gene products (1). These cells grow in suspension at room temperature and to a high cell density (1-2x10^7 cells/ml) (2). Also, higher eucaryotic processing signals such as those used for secretion and glycosylation are recognized (3,4). Expression of heterologous genes in insect cells has been restricted primarily to the lytic baculovirus system. However, the copious uptake of transfected DNA by *Drosophila* Schneider S2 cells (2,5) suggested the use of these cells for the stable overexpression of foreign genes. To this end we sought a rapid and efficient system for transfection and overexpression using the S2 cell line.

Results

Selection systems for introduction of foreign genes in insect cells.

We have developed a hygromycin B (hygro) selection system and compared it with those previously described for methotrexate (mtx)(8), and G418 (10,11). Genes encoding resistance to mtx, G418 and hygro were placed under the transcriptional control of the COPIA 5'LTR (6) and flanked downstream by the SV40 early polyadenylation signal (7),(Fig. 1). The pHGCO vector (8) carries a bacterial dihydrofolate reductase (DHFR) gene, pCO*neo carries a neomycin phosphotransferase gene (9) and pCOdhygro carries the *E. coli* hygromycin B phosphotransferase gene (12). S2 cells were transfected with these vectors by Ca-PO_4 co-precipitation

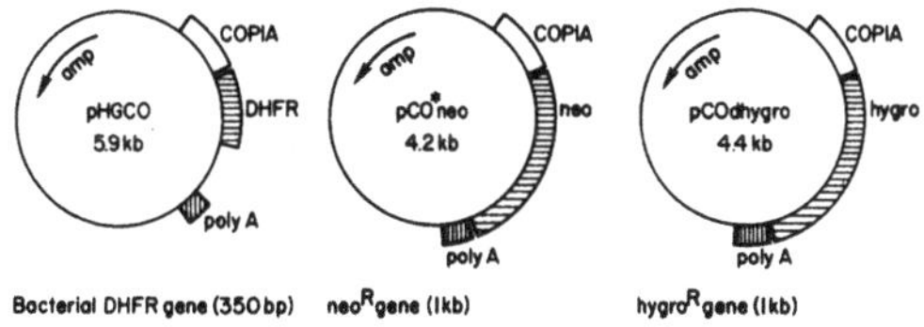

Figure 1.

(13,14) and then selected with the appropriate drug. A resistant cell population arose within a few weeks and all subsequent analyses were carried out on these polyclonal populations. As summarized in Table 1, selection with hygro is completed within 2-3 weeks, which is twice as rapid as with mtx. Moreover, it does not suffer from the high background of spontaneous resistance observed with G418 selection.

Table 1.

DRUG	CONCENTRATION	VECTOR	SELECTION TIME	BACKGROUND RESISTANCE
Methotrexate	0.1 µg/ml	pHGCO	6 weeks	0
G418	1 mg/ml	pCO'neo	3 weeks	+
Hygromycin B	300 µg/ml	pCOdhygro	2-3 weeks	0

Co-introduction and Expression of E. coli Galactokinase in S2 Cells.

The *E. coli* galactokinase gene (*galK*)(7) was placed under the control of three different promoters: the constitutive COPIA 5'LTR (6), pCOdGALK; the inducible *Drosophila* metallothionein (MT) promoter (15), pDMTGALK; or the mammalian SV40 early promoter (7), SV40GALK, (Fig. 2). Each of these *galK* expression vectors was co-transfected into S2 cells with pCOdhygro vector (20 µgr DNA total), and populations of stably transfected cells were obtained upon selection in hygromycin B.

a. Gene copy number correlates with the amount of DNA in the transfection.

The copy number of the integrated *galK* gene in the resulting resistant cell population was modulated by varying the ratio between the *galK* vector and pCOdhygro in the transfection (Fig. 3, upper panel, lanes A->C and Table 2). Transfectants carrying more than 1000 copies of *galK* gene were obtained (data not shown).

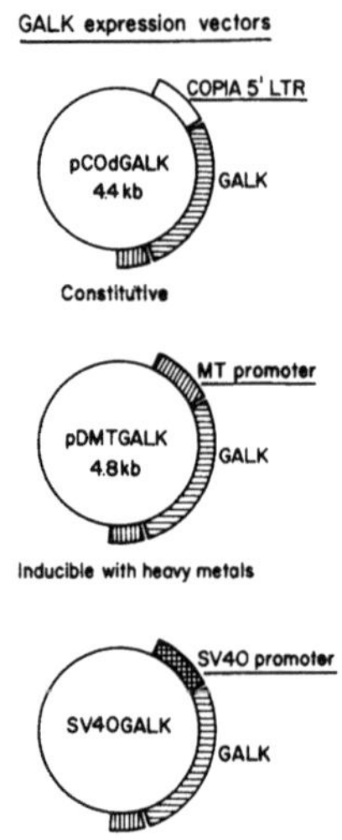

Figure 2.

b. Transcription and expression of *galK* in S2 cells.

Expression of *galK* RNA and protein was measured in each of the resistant cell lines transfected with the three *galK* vectors. By RNA blot analysis (16), the COPIA and MT vectors gave *galK* transcripts of the expected size, 1.5Kb, and RNA levels increased with increasing gene copy number (Fig. 3, middle panel, lane A->C). For the MT promoter, *galK* expression was regulated and could be selectively induced by the addition of heavy metals (e.g. cadmium, Fig. 3, lanes F(-) and (+)). Very low levels of *galK* RNA were observed with the SV40 promoter (Fig. 3, lane E).

The expression of *galK* protein was monitored by immunoblot analysis (17). The results indicated that the *galK* product has the appropriate size, 41 Kdal, and the protein levels paralleled the RNA expression results (Fig. 3, lower panel). The results on the expression of *galK* in S2 cells are summarized in Table 2. Using this same hygromycin B selection system, several other genes have been expressed. In each case, the recombinant products are efficiently expressed, properly localized and biochemically active. Examples include the intracellular human H-*ras*, the mouse glucocorticoid receptor and a secreted serine-protease of human origin.

Figure 3: DNA, RNA and protein analysis of stably transfected cell lines with the various *galK* expression vectors.
Upper panel: Southern blot analysis (16).
High MW DNA was extracted from the cell cultures, digested with restriction endonucleases which excise the *galK* coding sequence from the vectors, run on an 0.8% agarose gel and transferred onto a nitrocellulose filter. The presence of stably integrated genes was detected with a ^{32}P-nick-translated *galK* probe. Copy number

was deduced by comparison to known amounts of digested plasmid DNA. Low (A), medium (B), and high (C) copy number of galK gene were obtained upon co-transfection with the pCOdGALK and pCOdhygro vectors at ratios of 2 to 18µgr, 10 to 10 µgr and 18 to 2 µgr, respectively. D:Control cell line transfected with pCOdhygro alone. F: cell line co-transfected with pDMTGALK and pCOdhygro vectors. E: cell line co-transfected with SV40GALK and pCOhygro vectors.
Middle panel: Northern blot analysis (16). Total RNA was extracted from the cell cultures, run on a 1.2% formaldehyde-agarose gel and transferred onto a nitrocellulose filter. Specific mRNAs were detected with a ^{32}P-nick-translated GALK DNA probe. Lane F: RNA extracted from the MT GALK cell line before (-) or after (+) induction for 24-48 hrs. with 10µM $CdCl_2$.
Lower panel: Western blot analysis (17). 30 µgr of protein from total cell lysate were electrophoresed on a 12% SDS-polyacrylamide gel and transferred onto a nitrocellulose filter. The filter was then treated with rabbit anti-galK antiserum followed by ^{125}I-protein A. Lane F (-) and (+): proteins from the uninduced (-) and induced (+) MT GALK cell line.

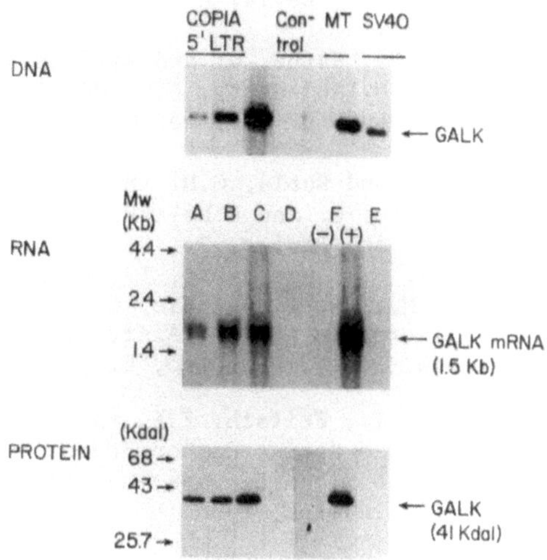

Figure 3.

Conclusions

A rapid hygromycin B selection system was developed for the introduction of foreign genes into Drosophila S2 cell line. These cells can acquire over a thousand copies of the transfected gene(s), and gene copy number can be modulated by varying the amount of vector DNA in the transfection. Cell lines expressing high levels of the transfected gene product(s) can be obtained within 3 weeks. Expression increases with gene copy number but not proportionally. Vectors for constitutive (COPIA 5'LTR) and inducible (MT promoter) expression were developed. The MT promoter is 2-5 fold more efficient than the COPIA 5'LTR and the mammalian SV40 early promoter works very poorly in Drosophila cells. Finally, the MT promoter is tightly regulated by cadmium; RNA and protein levels are very low in the absence of metal but both are induced 30-50 fold within 24 hrs. after the addition of cadmium.

Table 2.

EXPRESSION of E. coli GALK in DROSOPHILA CELLS.

		DNA		RNA		PROTEIN
		Gene copy # (copies/cell)		level[a] (pg/µg)	RNA/DNA[b] (x10)	level (ng/µg)
COPIA	A	43		4	≈1	0.9
	B	150		6	0.4	1.3
	C	480		8	0.16	2.3
MT	F	210	(-)Cd	≈1	0.047	0.03
			(+)Cd	44	2.0	3.0
SV40 EARLY	E	87		<0.6	—	—

(a) GalK mRNA levels were measured by dot blot analysis of total cellular RNAs and compared to known amounts of digested plasmid DNA, using a ^{32}P-nick-translated GALK DNA probe.

(b) RNA/DNA ratio normalizes the amount of GALK RNA per gene copy in each of the cell cultures.

References

1. Fallon, A.M. and Willis, J.H. Trends in Biotech., 3:217-218, 1985.
2. Schneider, I. J. Embryol. Exp. Morph., 27:353-365, 1972.
3. Smith, G.E., Summers, M.D. and Fraser, M.J. Mol. Cell. Biol., 3:2156-2165, 1983.
4. Smith, G.E., Ju, G., Ericson, B.L., Moshera, J., Lahm, H-W., Chizzonite, R. and Summers, M.D. Proc. Nat. Acad. Sci. USA, 82:8404-8408, 1985.

5. Moss, R., Cherbas, L., Koehler, M., Bunch, T. and Cherbas, P. In Vitro,21, 3 pII:42A, 1985.
6. Flavell, A.J., Levis, R., Simon, M.A. and Rubin, G.M. Nucl. Ac. Res., 9:6279-6291, 1981.
7. Pfarr, D.S., Sathe, G. and Reff, M. DNA, 4:461-467, 1985.
8. Bourouis, M. and Jarry, B. EMBO J., 2:1099-1104, 1983.
9. Beck, E., Ludwig, G., Auerswald, E.A., Reiss, B. and Schaller, H. Gene, 19:327-336, 1982.
10. Rio, D.C. and Rubin, G.M. Mol. Cell. Biol., 5:1833-1838, 1985.
11. Maisonhaute, C. and Echalier, G. FEBS Lett., 197:45-49, 1986.
12. Gritz, L. and Davies, J. Gene, 25:179-188, 1983.
13. DiNocera, P.P. and Dawid, I. Proc. Nat. Acad. Sci. USA, 80:7095-7098, 1983.
14. Wigler, M., Sweet, R., Sim, G.K., Wold, B., Pellicer, A., Lacy, E., Maniatis, T., Silverstein, S. and Axel, R. Cell, 16:777-785, 1979.
15. Lastowski-Perry, D., Otto, E. and Maroni, G. J. Biol. Chem., 260:1527-1530, 1985.
16. Maniatis, T., Fritsch, E.F. and Sombrook, J. Molecular Cloning, A Laboratory Manual, Cold Spring Harbor Press, Cold Spring Harbor, New York. 1982.
17. Griswold, D.E., Hillegass, L., Antell, L., Shatzman, A. and Hanna, N. J. Immun. Meth., 91:163-168, 1986.

Part 5

APPLICATION OF CELL CULTURES TO PATHOLOGY AND MICROBIOLOGY

SUSPENSION CULTURE OF AEDES ALBOPICTUS, CLONE,C6/36, CELLS AND ITS APPLICATION TO THE PRODUCTION OF FLAVIVIRUSES

A. Igarashi and A. K. Srivastava

Department of Virology, Institute of Tropical Medicine, Nagasaki University, Sakamoto, Nagasaki 852, Japan

Aedes albopictus, clone C6/36, cells were adapted to grow in Eagle's medium in Earle's saline supplemented with 0.2 mM each of nonessential amino acids and 10 % heat-inactivated fetal calf serum. The cells were then adapted to grow in suspension and attached to microcarrier. The cells were infected with Japanese encephalitis virus and infected fluid was (A) continuously kept, or (B) daily harvested and replaced with fresh maintenance medium. The virus yield in the fluid was assayed by infectivity titration, ELISA, as well as the peak height of the purified virion in the sucrose gradient, finding that the infective virus yield was almost the same in both methods.Virus yield varied according to the virus used in the experments.

Japanese encephalitis (JE) is existing in East, Southeast, and South Asia(1). Although the epidemics of JE in Japan and the Republic of Korea appear to have been controlled, frequent epidemics in other parts of Asia are still causing large numbers of deaths and disabilities (2). Formalin-inactivated and highly purified JE vaccine from infected mouse brains has been developed in Japan and widely used to control the disease (3). Similar mouse brain-derived vaccine has been used also in the Republic of Korea. On the other hand, China has developed their own formalin-inactivated JE vaccine from hamster kidney cell cultures(4). However, these vaccines are rather costly and limited in amount for wide-scale vaccination in present epidemic areas, most of which are in developing countries. Live attenuated JE vaccine has been developed in China and applied for humans and horses (5). However, it will take some time more before the live JE vaccine will be internationally accepted for human use. In order to make a breakthrough on the problem of JE in the developing countries, the World Health Organization (WHO) proposed development of a second generation JE vaccine using recombinant DNA technology. Toward this objective, we tried to analyze the entire nucleotide sequence of JE virus RNA and to localize the gene encoding envelope glycoprotein (E), which has been considered to play an important role in the initial step of virus infection to host cells(6),hoping that the study will eventually lead to the expression of the E protein in appropriate systems. The analysis of the genome and structural protein requires quantities of the purified virus. For this purpose, we set up a laboratory-scale mass production system of high virus-yielding clone C6/36 cells of Aedes albopictus cell line (7).

Materials and Methods

The cell suspension of C6/36 clone was made from the stationary culture by pipetting and placing it in a spinner flask with rotating blade at 60 rpm at 28°C. The same cell growth medium was used for both stationary and spinner cultures, that is, Eagle's medium in Earle's saline supplemented with 0.2 mM each of nonessential amino acids (8) and 10 % heat-inactivated fetal calf serum. After a few days of lag phase, the cells entered into the logarithmic phase of growth, and the cultures could simply be divided by diluting with fresh medium for further growth. For production of the virus, we used 1 liter volume spinner flasks. When the cell density reached around 10^6 cells/ml, microcarrier (Cytodex 1, Pharmacia) was added to a final concentration of 2 mg/ml in order to attach the cells onto it. After 6-12 hours of incubation, the flasks were kept on the table for a while to settle down the cells on the microcarrier. The supernatant was removed by decantation, and the cells were infected with 40 ml volume of seed virus, JaOAr S982 strain of JE virus isolated from field-caught mosquitoes. Virus adsorption was carried out at room temperature for 1-2 hours, and the cells were supplied by the maintenance medium (cell growth medium from which serum concentration was reduced to 1 %), to a final volume of 750 ml/flask. Then, the flasks were kept at 28°C as before with rotating speed of 60 rpm.

In experimental series (A), infected culture fluid was kept as a whole and only a part of it was sampled every day for virus assay. In experimental series (B), all infected fluid was harvested and replaced with fresh maintenance medium every day.

Virus infectivity was assayed by plaque formation on BHK21 cells, and virus antigen titer was determined by the sandwich ELISA (9).

The virus was purified from the infected culture fluid. The fluid was first clarified by filtration and high-speed centrifugation at 10,000 x g for 15 min. To the supernatant, polyethylene glycol (PEG) 6000 and NaCl were added to final concentrations of 6 % and 0.5 M, respectively. When PEG was completely dissolved, the solution was centrifuged at 10,000 x g for 30 min. The supernatant was removed and the precipitate was resuspended in STE buffer (0.1 M NaCl, 0.01 M Tris-HCl, 0.001 M EDTA, pH 7.4) using 1/50 - 1/60 volume of the initial infected culture fluid. The specimen was again centrifuged at 20,000 x g for 15 min. The supernatant was layered on 30-45 % sucrose gradient plus 15 % sucrose layer in STE buffer, and centrifuged in a SW 28 rotor of a Beckman model L8M ultracentrifuge at 21,000 rpm for 17 hours at 4°C. Fractions were collected by an ISCO gradient fractionator model 640 and peak fractions of OD_{254} were used as purified virions.

Results

No significant difference was observed between the virus yield in experimental series (A) and (B), as examined by the infectivity and ELISA assay. Therefore, the method in series (B) was applied for later studies of virus production.

When the purified JE virion peak fractions were examined by SDS-PAGE, a negligible amount of contaminating proteins was found besides structural proteins of JE virus (10). Besides the purified virion peak, slow-sedimenting ELISA antigen was found in the upper fractions. These antigens might be breakdown products or envelope fragments of the JE virion.

Similar strategy was applied to prepare other flaviviruses. Murray Valley encephalitis virus was shown to give quite high yield. In contrast, West Nile virus and Saint Louis encephalitis virus did not give sufficient amount of purified virions, although they could give slow-sedimenting ELISA antigens with comparable titers with JE virus. Recent studies using other strains of JE virus showed that virus yield or recovery was dependent on the combination of the virus strain and host cells. We have also used the same method to obtain dengue virus type 3 in order to produce sufficient quantities to analyze its nucleotide sequence and structural proteins.

Discussion

Flaviviruses consist of almost 60 members, of which many are counted as causative agents of severe manifestations of human diseases, such as encephalitis and haemorrhagic fevers (12). However, a significant proportion of flavivirus-infected humans remain undetected or subclinical infections. It is considered important to analyze the pathogenesis of the flavivirus infections at molecular levels. Analysis of the nucleotide sequence and structural proteins of the virus are the most direct way to show the strain difference, besides the development of the second generation vaccine. The establishment of the suspension culture of high virus-yielding clone C6/36 cells was the first step in studies of JE virus at its molecular level. However, this system did not always give high virus yield for all the flaviviruses. Although the system worked well for JE, dengue and Murray Valley encephalitis viruses, critical preliminary experiments appear to be necessary before the same system is applied to other flaviviruses.

References

1. Miles, J. A. R. Bull. WHO., 22: 339-371, 1960.
2. Umenai, T., Krzysko, R., Bektimirov, A. and Assaad, F. A. Bull. WHO 63: 625-631, 1985.
3. Hammon, W. McD., Kitaoka, M. and Downs, M. D. (eds.). Immunization for Japanese Encephalitis, Igaku Shoin, Tokyo, 1971.
4. Yin Min Wang. Virus Vaccines in Asian Countries (Fukai, K., ed.), pp105-115, University of Tokyo Press, Tokyo, 1984.
5. Li Ho Min. Virus Vaccines in Asian Countries (Fukai, K., ed.), pp159-165, University of Tokyo Press, Tokyo, 1984.
6. Kimura-Kuroda, J. and Yasui, K. J.Virol., 45: 124-132, 1983.
7. Igarashi, A. J. Gen Virol., 45: 124-132, 1983.
8. Eagle, H. Science, 130: 432-437, 1959.
9. Voller, A., Bidwell, O. and Bartlett, A. Manual of Clinical Immunology (Rose, N. R. and Friedman, N., eds.), pp506-512, ASM, Washington,D. C., 1976.
10. Srivastava, A. K., Aira, Y., Mori, C., Kobayashi, Y. and Igarashi, A. Arch. Virol., 96: 97-107, 1987.
11. Hori, H. Trop. Med., 28: 179-190, 1986.
12. Monath, T. P. Virology (Fields, B. N., Knipe, D. M., Chanock, R. M., Melnick, J. L., Roizman, B. and Shope, R. E., eds.), pp955-1004, Raven Press, New York, 1985.

INVESTIGATION OF CELL-CELL FUSION IN SEMLIKI FOREST VIRUS (SFV) INFECTED C6/36 (MOSQUITO) CELLS

H. Koblet, A. Omar, U. Kohler, and Ch. Kempf

Institute of Medical Microbiology of the University of Bern,
Faculty of Medicine, Bern, Switzerland
Friedbühlstrasse 51, CH - 3008 Bern, Switzerland

Semliki Forest virus (SFV) infected Aedes cells undergo cell-cell fusion at a pH below 6.2. This syncytium formation is a fusion from within (FFWI) and can be inhibited by a wide variety of reagents. Using protein modifying agents which inhibit cell-cell fusion we showed that the viral spike protein E_1 is predominantly labeled. Thus, this protein might be involved in the fusion process. Furthermore, we have demonstrated that FFWI requires a specific expenditure of energy and can be blocked by inhibitors of oxidative phosphorylation. In addition, an exposure of SFV-infected cells to a pH below 6.2 results in the opening of a "proton channel" in the plasma membrane. Implications to viral infection by the endosomal pathway are discussed.

Infection of certain Aedes albopictus cell lines with Semliki Forest virus (SFV) induces a strong cytopathic effect (CPE) (1). The CPE is discerned as cell-cell fusion, leading to giant syncytia. This finding differed from the consensus that Togavirus-infected mosquito cells rarely exhibit morphological changes. We and others have found that there is a strong correlation between the time of seeding of the cells and the time of infection for the CPE to occur (2). This is due to the fact that certain cell lines acidify the culture medium relative to the time of seeding and that at a critical pH of 6.1 fusion is triggered. Thus, Togavirus-infected mosquito cells show a cytopathic effect in the form of syncytium formation if the cell line itself lowers the pH of the culture fluid by metabolic activity or if the pH is intentionally lowered by acidification to 6.1 or below. We have shown that this fusion corresponds to a fusion from within (FFWI) (3, 4). FFWI is defined as a late cell-cell fusion occurring after the viral eclipse. In contrast to fusion from without (FFWO) which takes place upon addition of virions to cells, FFWI can be blocked by inhibitors of viral nucleic acid- or protein synthesis (5, 6). In other words, FFWO is a fusion of the virion envelope with the plasma membrane whereby the plasma membrane becomes competent to fuse with the plasma membranes of neighbouring cells, while FFWI is based on the insertion of newly synthesized viral proteins into the plasma membrane. These proteins are transported by vesicles from the endoplasmic reticulum through the Golgi into the plasma membrane, and this integration from the inside of the cell, "from within", renders the plasma membrane fusion competent.

SFV induced FFWI takes place in two steps (7). The first step is very fast and pH-dependent, but not temperature dependent. An exposure of the infected cells to pH 6 for a few seconds at any temperature between 0 and 28°C is sufficient to trigger the FFWI. The second step is pH-independent, but temperature-dependent. Cells exposed to pH 6 can be left at this pH or be brought into medium of pH 7 and fusion will proceed under both conditions provided the temperature is higher than 17°C. This has allowed us to apply protein modifying reagents at pH 7 and examine whether they acted as inhibitors of fusion before or after exposure of the cells to pH 6. We have shown that a considerable number of reagents such as tetrathionate, sulfite, Ellman's reagent or butanedione inhibit the FFWI (7-9) of the infected C6/36 _Aedes_ cells (10) only if added at pH 7 after the exposure of the cells to pH 6. From these results we anticipated that a conformational change of a cell surface protein takes place at pH 6, exposing -SH, -S-S-, amino- and guanidino-groups which are essential for the fusion to occur. We tried, with appropriate controls, to identify the protein which changes its conformation by labeling the cells and at the same time inhibiting the fusion with $[^{35}S]SO_3^{2-}$ at pH 7 after a short exposure to pH 6. The membrane proteins were isolated and analyzed by chromatography and gel electrophoresis. Based on the finding that the SFV-spike protein E1 is preferentially labeled when the reagent was applied after pH 6 exposure, we conclude that this protein is the one which changes conformation and might be involved in triggering the fusion (manuscript in preparation).

The two fusion types mentioned, FFWO and FFWI, are related to a third fusion phenomenon, the fusion of enveloped virions with the endosomal membrane during endocytosis (11). The prevailing consensus, based mainly on the work of Helenius and coworkers, is that viruses are bound to the cell surface and taken up in coated pits (12-14). These pits then form endosomes, the contents of which are acidified by the action of proton pumps in the endosomal membrane (15). When the critical pH is reached, the viral spike proteins change their conformation and trigger the fusion of the viral envelope with the endosomal membrane. The nucleocapsid carrying the viral information is ejected into the cytoplasm and the infection has formally come to an end; replication then starts.

Thus, there are three viral fusion phenomena, which share a common biochemical bias: (i) FFWO (ii) fusion at the endosomal level and (iii) FFWI. It is reasonable to speculate that the endosomal fusion and the cytopathic effects of FFWI are the expression of similar mechanisms at different locations of the cell. In all of these cases low pH leads to a conformational change of a viral protein which then initiates fusion.

However, there is certain experimental evidence which seems to question the whole concept. Whereas the FFWI of the Togavirus infected mosquito cells proceeds to completion at both pH 6 and pH 7, it had been shown that it is mandatory for infected BHK-cells to be brought back to pH 7 after a short exposure to pH 6; otherwise fusion is not possible (16 - 18). This finding is in conflict with the conditions prevailing in endosomes. No simple mechanism can explain why in endosomes the pH should drop and then suddenly rise again. Thus, either the concept is wrong or it is a peculiar behavior of BHK cells giving rise to a special situation which is not amenable to a general concept. From previous results, we have been aware of the fact that sodium azide (8) and potassium cyanide - potent inhibitors of oxidative phosphorylation - prevented the fusion of SFV-infected _Aedes albopictus_ cells. Also Okada had shown that dinitrophenol inhibited giant cell formation by Sendai virus (19, 20). Therefore we considered the following possibilities: Mosquito cells are able to provide enough ATP to drive the fusion and concomitantly maintain the cytoplasmic pH (pH_i) at its normal level even if the external pH (pH_e) is at 6. In contrast, BHK cells cannot perform both duties simultaneously and their energy is exhausted to keep the pH_i constant if the pH_e is at 6; thus, FFWI would be impeded. An answer to this hypothesis would confine the conflict to a question of energy stores without doubting the mechanism of endosomal infection. We therefore investigated the ATP levels in C6/36 cells, which were either mock infected or infected 16 hrs before (21). ATP was found to be constant in mock infected cells at pH 7.2 or pH 6 and also in SFV-infected cells at pH 7.2 and amounted in all cases to about 1-3 mM. However, initiating the fusion in SFV-infected cells by lowering the pH to 6 results in a slow decrease of the intracellular ATP over 60 min to 10-15% of its initial value. An identical result was obtained at pH 7 after a brief exposure to pH 6. This shows that the consumption of ATP during fusion is not entirely due to the imposed pH gradient across the plasma membrane and it also supplies good evidence for an energy requirement for polykaryocyte formation itself. Moreover, if medium of pH 6 containing either 1 mM KCN, 1 mM NaN_3 or 10 µM dinitrophenol is added, the infected cells are rapidly and totally depleted of ATP and polykaryocyte formation is prevented. However, if infected cells - in presence of the inhibitors - were exposed for only a short time to pH 6 (e.g. 1 min) the intracellular ATP recovered to approximately 50% of the initial value which was sufficient for polykaryon formation to occur. In general it can be stated that SFV-induced syncytium formation of _Aedes_ cells occurs when the intracellular ATP level is above 10 to 20% of its initial value. This can be explained as follows: (i) blockage of de novo ATP synthesis at pH 6 prevents polykaryocyte formation. (ii) Rapid depletion of ATP in infected cells at pH 6 in the presence of inhibitors is probably due to the requirement of ATP to deal with the large artificial pH gradient across the membrane (22-25). If no inhibitors are present at pH 6 sufficient ATP is synthesized to cope with this gradient and simultaneously support fusion. (iii) Similarly, if in the presence of inhibitors, the pH stress is released, the ATP concentration can be partially restored and fusion takes place. Apparently, _Aedes_ cells possess massive stores of energy-rich compounds which allow generation of ATP by transphosphorylations. (iv) Therefore, two main systems compete for ATP during fusion at pH 6: all biochemical reactions which maintain the internal pH on one hand, and the fusion per se on the other hand. It is not yet known what ATP is used for during the fusion process. However, we have been able to show that ATP is consumed for an early process during membrane-membrane fusion (26). By microinjection of the highly fluorescent, easily diffusible, membrane-impermeant dye lucifer yellow (27) into SFV infected C6/36 cells and arresting the fusion at any given stage with glutaraldehyde, cell-cell fusion can be observed already 1 min after exposure to pH 6 by the spreading of the dye through microscopically small connections between neighbouring cells. If KCN is added concomitant to lowering the pH, this early process is not observed. Performing similar analysis with SFV-infected BHK-cells showed that these cells behave differently (manuscript submitted). As soon as the pH_e is lowered the ATP concentration collapses, even in the absence of inhibitors. Only if the pH is raised again to 7.4, ATP is restored; then fusion proceeds. Thus, these cells behave as mosquito cells to which inhibitors of oxidative phosphorylation have been added. All these data fit very well into the working hypothesis, namely that C6/36 cells can provide enough energy to drive the fusion at pH 6 and to sustain the pH gradient across the membrane, whereas BHK cells cannot.

However, a fundamental question has not yet been answered. $[H^+]$- or $[H_3O^+]$-ions do not easily diffuse through lipid bilayers (28). Why should therefore the ATP-level of BHK-cells break down as soon as the external pH is lowered? It does not collapse in uninfected control cells. There is only one possible and completely new explanation for this phenomenon: in infected cells, at pH 6, H^+-ions stream into the cytoplasmic compartment. To prove this hypothesis, the cytoplasmic pH was examined as a function of the extracellular pH (29). To this end an uncharged non-fluorescent derivative of a fluorescent dye was added to cells which diffuses into the cytoplasm where it is enzymatically cleaved. The cleavage product is now charged and membrane-impermeant and becomes trapped in the cytoplasm. pH-dependent fluorescence intensities were recorded at the appropriate excitation and emission wavelengths. The intracellular pH was then calculated from standard curves. Using fluoresceine diacetate as a dye in the case of SFV-infected BHK-cells, pH_i was found to drop immediately if the external pH reaches the critical value of 6.1 (manuscript submitted), at which the conformational change of the E_1-protein takes place (30). In C6/36-cells, where Quene 1 was used as pH indicator (31), this drop of pH_i at pH_e 6.1 could also be detected; however, an inhibitor of oxidative phosphorylation had to be present (manuscript submitted). Thus, the different findings fit together. At pH 6.1, in SFV-infected cells, a proton channel in the plasma membrane is irreversibly opened, and H^+-ions rapidly stream into the cytoplasm. ATP is consumed by the pumps in the plasma membrane to exchange H^+-ions for Na^+-ions and Na^+-ions for K^+-ions. Whereas an influx of Na^+

cannot be shown under the experimental conditions tested, the efflux from preloaded infected C6/36-cells at a pH_e 6.1 can well be demonstrated. Correspondingly, an influx of K^+-ions can be shown (manuscript submitted). The opened proton channels seem to be quite selective. These phenomena explain why mosquito cells can fuse at pH 6, whereas BHK-cells have to be brought back to pH 7 in order to fuse. However, the following questions remain unanswered: (i) is one of the viral transmembrane spike proteins the proton channel which opens at an identical pH where the conformational change is induced? Or does the conformational change of a viral spike influence a potential channel-forming host protein? (ii) Is the channel opening biochemically directly related to FFWI or indirectly via the ATP levels?

References

1. Späth, P.J. and Koblet, H. Invertebrate Systems In Vitro (Kurstak, E., Maramorosch, K. and Dübendorfer, A., eds.), pp. 375-388, Elsevier Biomedical. 1980.
2. Pudney, M., Leake, C.J. and Buckley, S.M. Invertebrate Cell Culture Applications (Maramorosch, K. and Mitsuhashi, J., eds.), pp. 159-194, Academic Press. 1982.
3. Omar, A., Flaviano, A., Reigel, F., Kohler, U. and Koblet, H. Invertebrate Cell Systems in Applications, Vol. 1 (Mitsuhashi, J., ed.), CRC Press, in press.
4. Omar, A., Flaviano, A., Kohler, U. and Koblet, H. Arch. Virol.,89:145-159, 1986.
5. Poste, G. and Nicholson, G.L. Membrane Fusion, Cell Surface Reviews, Vol. 5, Elsevier Biomedical. 1978.
6. Bratt, M.A. and Gallaher, W.R. Proc. Natl. Acad. Sci. USA, 64:536-543, 1969.
7. Koblet, H., Kempf, C., Kohler, U. and Omar, A. Virology, 143:334-336, 1985.
8. Koblet, H., Omar, A. and Kempf, C. Arbovirus Cultivation in Arthropod Cells In Vitro (Yunker, C.E., ed.), CRC Press, in press.
9. Omar, A. and Koblet,H. Invertebrate Cell Systems in Applications, Vol. 2 (Mitsuhashi, J., ed.), CRC Press, in press.
10. Igarashi, A. J. gen. Virol., 40:531-544, 1978.
11. Besterman, J.M. and Low, R.B. Biochem. J., 210:1-13, 1983.
12. Helenius, A., Marsh, M. and White, J. J. gen. Virol., 58:47-61, 1982.
13. Marsh, M., Bolzau, E. and Helenius, A. Cell, 32:931-940, 1983.
14. Goldstein, J.L., Anderson, R.G. and Brown, M.S. Nature, 279:679-685, 1979.
15. Tycko, B. and Maxfield, F.R. Cell, 28:643-651, 1982.
16. Edwards, J., Mann, E. and Brown, D.T. J. Virol., 45:1090-1097, 1983.
17. Mann, E., Edwards, J. and Brown, D.T. J. Virol., 45: 1083-1089, 1983.
18. Cassell, S., Edwards, J. and Brown, D.T. J. Virol., 52:857-864, 1984.
19. Okada, Y. Exp. Cell Res., 26:98-107, 1962.
20. Okada, Y. Exp. Cell Res., 26:119-128, 1962.
21. Kempf, Ch., Kohler, U., Michel, M.R. and Koblet, H. Arch. Virol., in press.
22. Grinstein, S. and Rothstein, A. J. Membrane Biol., 90:1-12, 1986.
23. Schulten, Z. and Schulten, K. Meth. Enzymol., 127:419-438, 1986.
24. Scheiner, S. Meth. Enzymol., 127:456-465, 1986.
25. Moolenaar, W.H. Trends Biochem. Sci., 11:141-143, 1986.
26. Kempf, C., Michel, M.R., Kohler, U. and Koblet, H. Arch. Virol., in press.
27. Stewart, W.W. Nature, 292:17-21, 1981.
28. Deamer, D.W. and Branhall, J. Chem. Phys. Lipids, 40:167-188, 1986.
29. Roos, A. and Boron, W.F. Physiol. Reviews, 61:296-434, 1981.
30. Kielian, M. and Helenius, A. J. Cell Biol., 101:2284-2291, 1985.
31. Rogers, J., Hesketh,T.R., Smith, G.A. and Metcalfe, J.C. J. Biol. Chem., 258:5994-5997, 1983.

THE ROLE OF C6/36 CELL CYTOSKELETON IN THE REPLICATION OF KUNJIN VIRUS

M.L. Ng and S.S. Hong

Department of Microbiology, Faculty of Medicine, National University of Singapore, Lower Kent Ridge Road, Singapore 0511, Republic of Singapore

In Kunjin virus-infected Vero cells, there was interesting rearrangement of the microtubulin into paracrystals. There is evidence that these virus-induced microtubulin paracrystals are important for the virus replication . However, when Kunjin virus infection was performed in C6/36 cells these paracrystals were observed only rarely. Coincidental with the low concentration of the tubulin paracrystals, the virus titre from C6/36 cells was about 100-fold lower than that from infected Vero cells.

The cell cytoskeleton is a complex network of protein fibres found in all eukaryotic cells. Work done on a variety of tissues and cells grown in culture have revealed 3 major fibre systems which make up the cytoskeleton; the microfilaments (5-6 nm in diameter), the intermediate filaments (7-10 nm) and the microtubules (20-25 nm). Distinct functions are associated with the different cytoskeletal components. The microfilaments determine cell morphology (15) and the microtubules are involved in intracellular organelle movement (14). The functions of the intermediate filaments however, are still unclear.

Much work has been published reporting the involvement of the cytoskeleton or components of the cytoskeleton in different virus-host systems (5, 7, 9). Investigations into the possible functions of the cytoskeleton in virus replication had not been done in invertebrate hosts and thus it was interesting to compare cytoskeletal involvement during Kunjin virus replication in an invertebrate and a vertebrate host cell.

Materials and Methods

Cell Culture : The vertebrate and invertebrate cell lines used in our work were the Vero (African Green Monkey Kidney) (8) cells and the C6/36 (a clone of the mosquito line *Aedes albopictus*) (6) cells.

Virus Infections : The Kunjin virus strain used was MRM 61C (16). Confluent monolayers of cells were infected with Kunjin virus at a multiplicity of infection of 10. In vinblastine-treated infected samples, the drug was supplemented to the infection media at a concentration of 10 µg/ml.

Electron Microscopy : The cell monolayer was washed with phosphate buffered saline (PBS) and fixed with 2.5% glutaraldehyde - 1% paraformaldehyde in 0.1 M sodium cacodylate at 4°C overnight. The cells were further post-fixed for 2 hours in 1% OsO_4 in 0.1 M cacodylate buffer at 4°C. After the post-fixation, the pellets were dehydrated through a graded series of acetone and finally embedded in epoxy resin. Ultrathin sections were cut, stained with uranyl acetate and lead citrate before viewing in the Philips T400 transmission electron microscope.

Cytoskeleton Preparation : To compare the cytoskeletal networks of vertebrate and invertebrate cells, we adopted a cell extraction procedure (3) which extracts the cell cytoplasm leaving the cytoskeleton intact. Cells grown on formvar-covered, carbon-coated gold grids were extracted with Triton X-100. The extracted cells were fixed for 30 min with 0.5% glutaraldehyde in Triton-extraction buffer and post-fixed for 1 min with 0.1% osmium tetraoxide in Triton-extraction buffer. Finally, the cells were dehydrated through an ethanol series of 50, 75, 90, 95 and 100% ethanol and then were allowed to air-dry before viewing under the transmission electron microscope.

Results

The initial study on the dependence of cytoskeleton in the replication of Kunjin virus was conducted in Vero cells. Thus a brief description of the observation is given as an introduction to this work.

In Kunjin virus-infected Vero cells, several virus-induced structures were observed (Fig. 1). The rough endoplasmic reticulum (RER) was very much extended. In certain sections of the RER the distended lumen was observed to contain vesicles (Ve) of uniform sizes. At earlier stages of the infection, these vesicles were seen to contain thread-like structures but by 30 hr p.i. (Fig. 1), most appeared empty. Mature virus particles (Vi) were also seen within the lumen of the RER. An interesting observation was the presence of microtubulin paracrystals. A drug vinblastine, which causes dissolution of the tubulin also decreased the virus yields quite drastically (50%). Thus it showed that Kunjin virus replication in Vero cells is dependent on at least one component of the cell cytoskeleton, most likely the microtubulin.

Since this is an interesting observation, we were curious to see if Kunjin virus growing in C6/36 cells also requires some components of the cell cytoskeleton. At low magnification in the Kunjin virus-infected C6/36 cells, similar pronounced ultrastructural changes were observed in these cells (Fig. 2A). There was extensive proliferation of the RER and also clusters of vesicles were in the distended lumen of the RER. However, there was a distinct absence of paracrystals like those seen in the Vero cells. Even at higher magnification (Fig. 2B), the microtubulin paracrystals were not seen.

A clearer picture of the thread-like structures within the vesicles is shown in Fig. 3A. The vesicles and virus particles were consistently observed within the distended lumen of the RER. Occasionally, bundles of stress fibres (SF) were seen in these infected cells (Fig. 3B). The stress fibres are actually microfilament bundles which is another component of the cytoskeleton. Although paracrystals were not seen in the infected cytoplasm, microtubulin filaments were visualised. These filaments were not in the crystalline fashion but appeared in a normal linear configuration (Fig. 3C).

Thus unlike infection in Vero cells, microtubulin filaments did not seem to play a role in the infected C6/36 cells. To add further evidence to this deduction, Kunjin virus-infected C6/36 cells were treated with vinblastine. There were insignificant changes in the virus titres. The observation again differs from the Kunjin virus-infected Vero cells. The presence of almost normal quantity of virus particles was confirmed visually by electron microscopy (Fig. 4A). All the expected virus-induced structures were present in the vinblastine-treated cells. The very distended portions of the RER are given the term amorphous structures (AS) and are often observed to contain numerous virus particles. These amorphous structures are only observed in infected mosquito cells and not in Vero cells.

In the presence of vinblastine, the infected C6/36 cells appeared more vacuolated and the presence of stress fibres was more obvious (Fig. 4B). At higher magnification, all the virus-induced structures similar to the untreated cells were clearly seen. Again the linear arrangement of the microtubulins was observed in these drug-treated cells (Fig. 5A). In Fig. 5B, a high

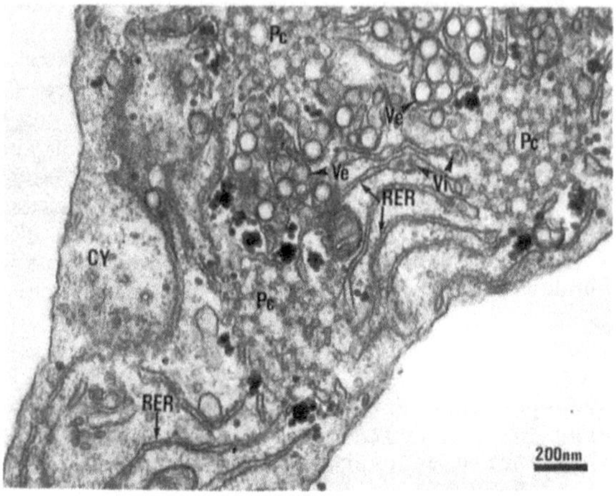

FIGURE 1. Kunjin virus-infected Vero cells at 30 hr post infection (p.i.). The obvious lengthening of the rough endoplasmic reticulum (RER) is seen in the infected cytoplasm. Vesicles (Ve) are seen enclosed within the distended portions of the endoplasmic reticulum. Virus (Vi) particles can also be observed in the lumen of the RER. The RER appears to lead into clusters of microtubulin paracrystals (Pc). On close observation, these crystals are made up of hexagonal subunits with a tubulin molecule at each apex. In close association with the paracrystals are the vesicles and virus particles.

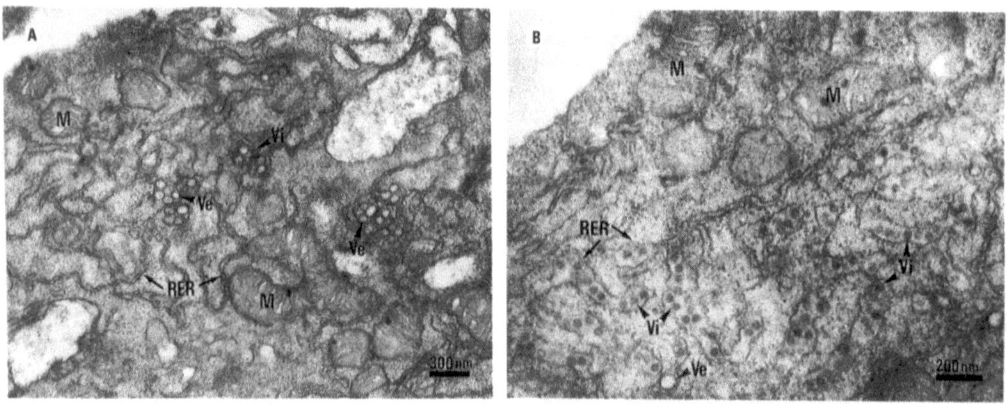

FIGURE 2. Kunjin virus-infected C6/36 cells at 30 hr p.i. (A).Similar to the Kunjin virus-infected Vero cells, there is proliferation of the rough endoplasmic reticulum (RER). Clusters of vesicles (Ve) were seen in the distended portion of the RER. Occasionally virus (Vi) particles are seen among the clusters of vesicles. (B).At higher magnification, numerous virus (Vi) particles are seen in the lumen of the rough endoplasmic reticulum (RER). Vesicles (Ve) are again seen in close association with the virus. There is a distinct absence of the microtubulin paracrystals as seen in Figure 1. The lettering M denotes mitochondria and V - vacuoles.

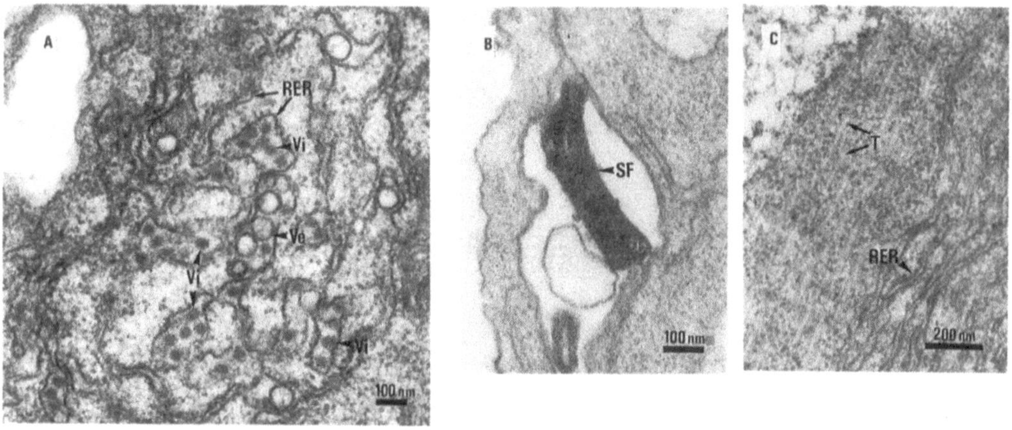

FIGURE 3. Kunjin virus-infected C6/36 cells at 30 hr p.i. (A).Fine thread-like structures can be observed within some of the vesicles (Ve). It can clearly be seen that the vesicles and virus (Vi) are enclosed within the lumen of the rough endoplasmic reticulum (RER). Again the vesicles and virus are in close association. No microtubulin paracrystals are observed in the infected C6/36 cell cytoplasm. (B).In the infected cells, occasionally, bundles of stress fibres (SF) are observed. (C).Although microtubulin paracrystals are not seen in infected C6/36 cells, microtubulin filaments (T) in their normal linear configuration can be seen.

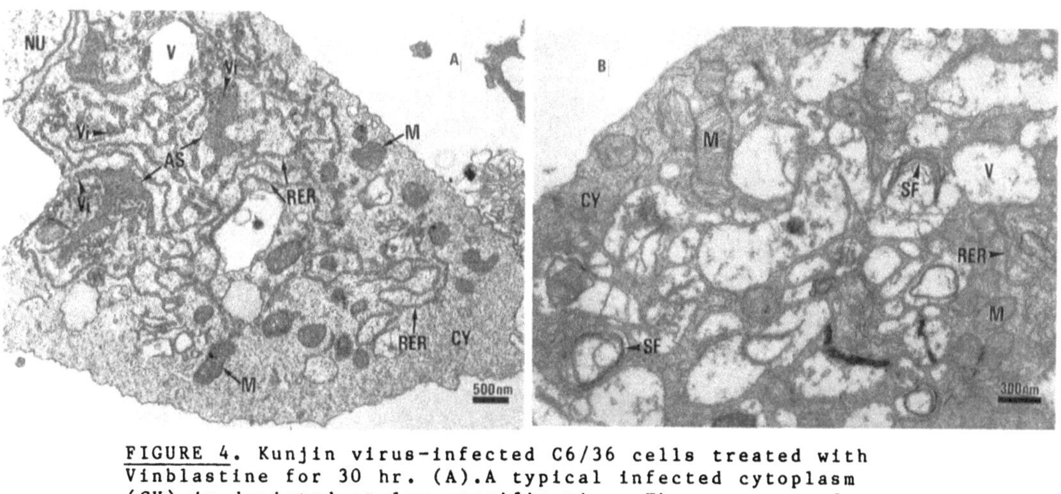

FIGURE 4. Kunjin virus-infected C6/36 cells treated with Vinblastine for 30 hr. (A).A typical infected cytoplasm (CY) is depicted at low magnification. The presence of Vinblastine did not seem to affect the progression of the infection. The expected proliferation of the rough endoplasmic reticulum (RER) is seen over the whole cytoplasm. Within the lumen of the RER, virus (Vi) particles are present. Certain portions of the RER have become very distended and are termed as amorphous structures (AS). Numerous virus particles are found in the amorphous structures. (B).With Vinblastine treatment, the cells become more vacuolated and the presence of the stress fibres (SF) is more obvious. Letterings denote: V - vacuoles, Nu - nucleus, CY - cytoplasm and M - mitochondria.

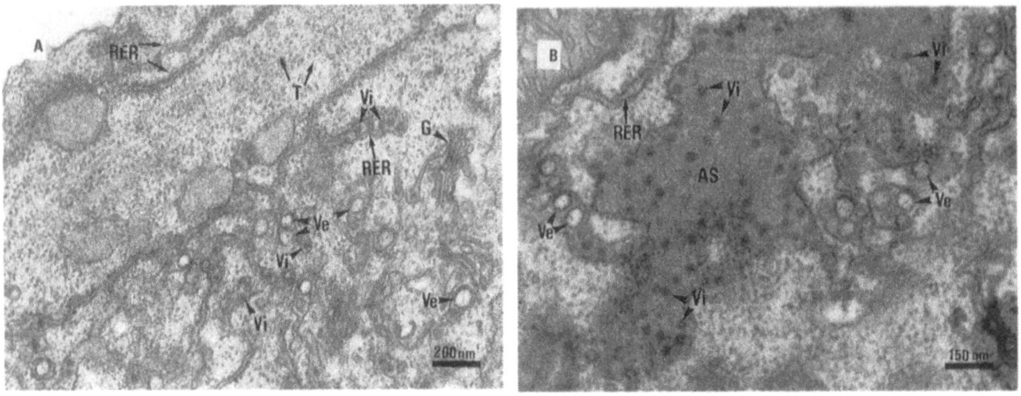

FIGURE 5. Kunjin virus-infected C6/36 cells treated with Vinblastine for 30 hr. (A).Vesicles (Ve) with thread-like enclosures are seen scattered among virus (Vi) particles. Both are enclosed within the lumen of the rough endoplasmic reticulum (RER). Microtubulin (T) filaments are again seen in the linear arrangement. The lettering G denotes Golgi apparatus. (B).An amorphous structure (AS) is shown at high magnification. Within this structure numerous virus (Vi) particles are seen. Vesicles (Ve - with thread-like enclosures) are observed in close contact with the virus particles (marked by arrow). The letterings denote : RER - rough endoplasmic reticulum and M - mitochondria.

FIGURE 6. Kunjin virus-infected mosquito cells. (A).Infected C6/36 cells. On very rare occasions paracrystal structures are seen in infected C6/36 cells. The architecture of these structures is different from those seen in the infected Vero cells (see Figure 1). The subunit structures are tightly packed. (B).Infected C6/36 cells. Higher magnification of the paracrystal structures seen in (A). (C).Infected *Aedes pseudoscutellaris* (AP61) cells. The crystal structures seen in infected C6/36 cells are more frequently seen in the AP61 cells. The subunit structures seem to be packed in a similar honey-comb fashion as the crystals seen in the infected C6/36 cells.

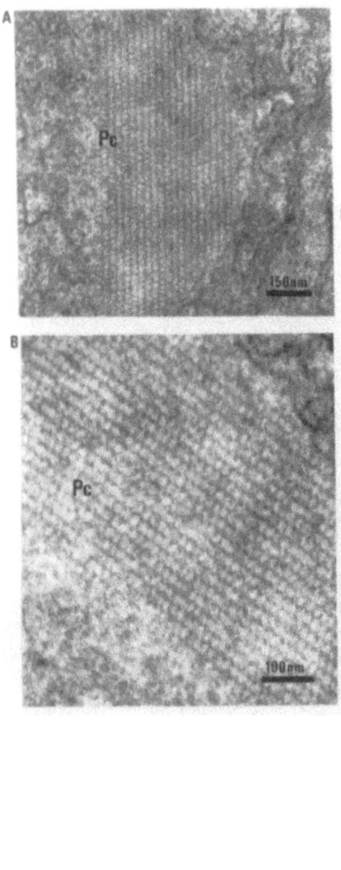

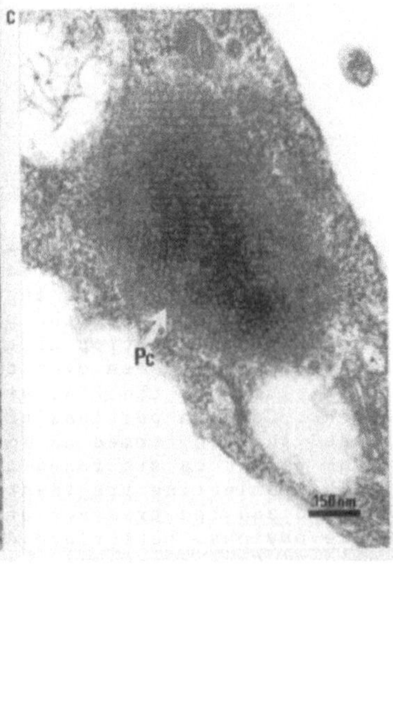

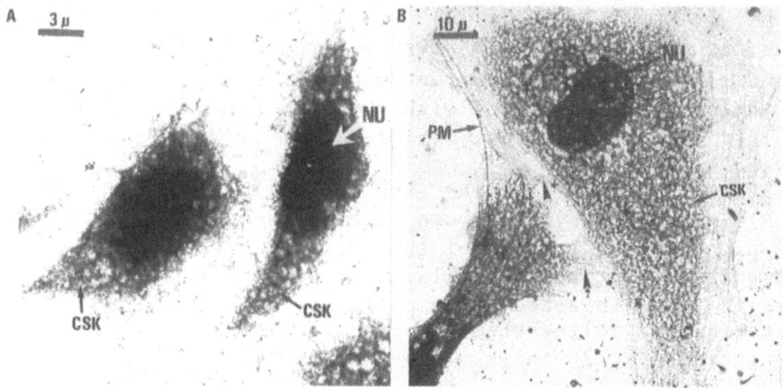

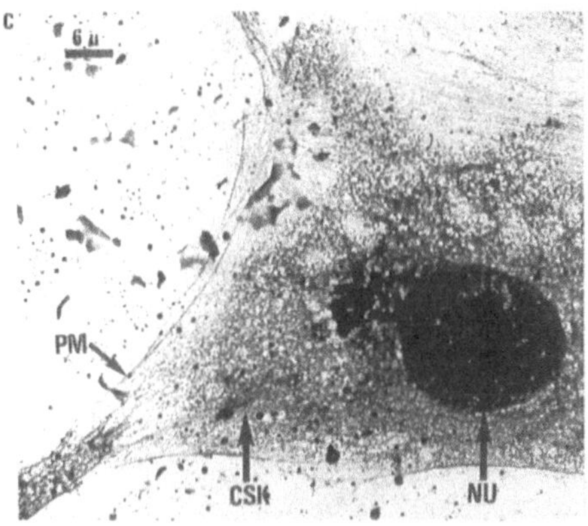

FIGURE 7. Triton X-100 extracted C6/36 and Vero cells. (A).Extracted C6/36 cells. The cytoskeleton (CSK) seemed rather compact and is closely associated with the cell nucleus (NU). The extracted cells also appeared to be smaller than the extracted Vero cells (see B & C). (B).Extracted Vero cells. The cytoskeleton (CSK) of the extracted cells seems better laid out. There is a coarser area of cytoskeleton nearer the nucleus (NU). The cytoskeleton becomes finer as it proceeds to the cell plasma membrane (PM), appearing as very fine straight filaments. The cell junctions seem to be connected by very fine filaments. (C).Extracted Vero cells. A higher magnification of a portion of the extracted Vero cells. The magnification is similar to (A). The letterings denote similar connotation as in (A) and (B).

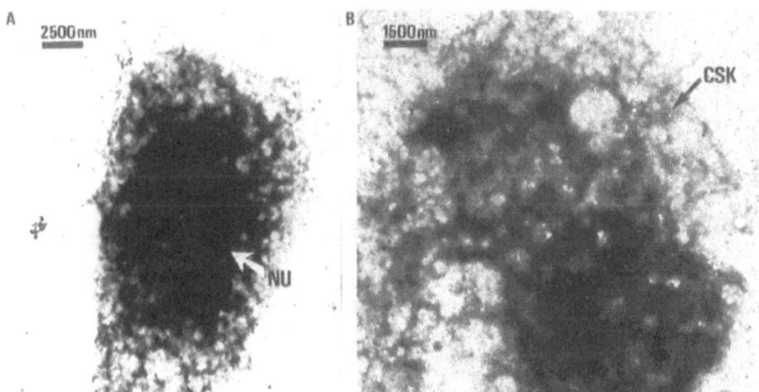

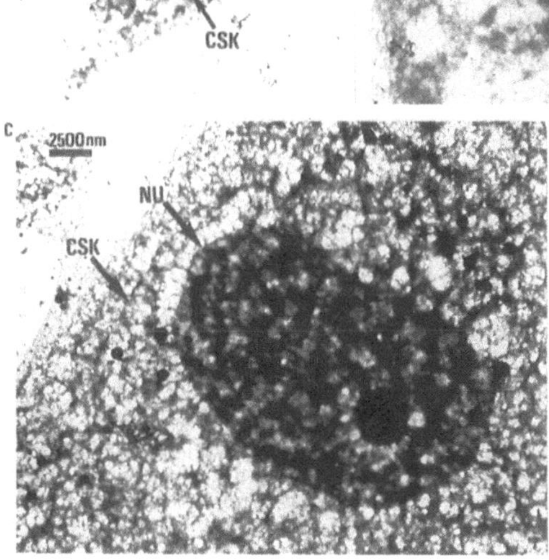

Figure 8.
Triton X-100 extracted C6/36 and Vero cells.
(A) and (B).High magnifications of extracted C6/36 cells. The cytoskeletal (CSK) lattices are not clearly defined as in the extracted Vero cells (C). Even the plasma membrane boundaries of the cells are not seen clearly. (C).Extracted Vero cells. A high magnification of extracted Vero cells. The cytoskeletal (CSK) lattice can be seen radiating from the nulceus (NU) in a reticulate fashion. As it approaches the edge of the cell, filaments are clearly seen.

magnification of an amorphous structure filled with mature virus particles is illustrated. The vesicles enclosing thread-like structures can be clearly seen in close association with the virus particles within the amorphous structures. In fact on the left of the figure the vesicles can be seen lining next to virus particles (marked by arrow).

On very rare occasions, a form of paracrystal was seen in Kunjin virus infected C6/36 cells (Fig. 6A). The paracrystals were tightly packed in a honey-comb configuration. A higher magnification of this architecture is shown in Fig. 6B. Just as a point of interest, similarly structured paracrystals were seen in Kunjin virus-infected Aedes pseudoscutellaris (AP61) cells (Fig. 6C). In the infected AP61 cells, these crystals were seen at a higher frequency than in the infected C6/36 cells. However, at present the nature of these crystals is uncertain. They could be microtubulin paracrystals as in the infected Vero cells or some other skeletal components which are unique to invertebrate cells.

Preliminary study was carried out to investigate if the infrequent observation of paracrystal formation in infected C6/36 cells was due to some intrinsic characteristic of the cell cytoskeleton. An extraction procedure using Triton X-100 was performed. It was surprising to find that the cytoskeleton pattern of the C6/36 and Vero cells differed drastically (Fig. 7 & 8). The cytoskeleton network of the C6/36 cells was seen to be about 3 to 4-fold smaller than Vero cells. The cytoskeleton of C6/36 cells was thick (Fig. 7A and Figs. 8A & B) and not in fine filamentous structure as in the Vero cells (Figs. 7B & C and 8C). In the extracted Vero cells, the cytoskeletons radiating from the perinuclear region were denser and became very fine at the edge of the cells. The junction between cells was connected by very fine filaments (arrowheads - Fig. 7B).

Discussion

The flaviviruses in general do induce some similar ultrastructural changes in vertebrate and invertebrate cells, e.g. proliferation of the RER and vesicles with thread-like enclosures (4, 11 & 12). However, data have shown that Murray Valley encephalitis and Japanese encephalitis viruses have preferential membrane fractions involved in their translation processes in the two host systems (Ho & Ng, manuscript submitted for publication). Also, when some series of human sera were reacted against the Dengue-2 virus grown in Vero and C6/36 cells, the Western blot profiles were different (unpublished data, L. Corner personal communication.). Previous study by Ng and Westaway (13) also observed that the progeny Kunjin virus grown in Vero and A. albopictus cells differs phenotypically.

Here in this study, we again encountered some differences on the dependence of cell cytoskeleton by Kunjin virus when replicating in Vero and C6/36 cells. We found that during Kunjin virus replication in Vero cells, in addition to the expected virus-induced structures, another virus-induced change was the formation of microtubulin paracrystals (Fig. 1). The formation of these paracrystals by the virus seemed important in the morphogenesis process since disruption with vinblastine caused a significant reduction in the infectious virus titres. Microtubulin has been observed to play some roles in several virus replications (1, 9 & 10).

From this brief study, we were able to deduce that Kunjin virus when grown in C6/36 cells did not project a need for the cell microtubulin during its replication, unlike in infected Vero cells. Since infectious virus was still produced from C6/36 cells one might postulate that there is probably a substitute cellular component which could take over the role of the microtubulin paracrystals in these cells. At present, we have no real evidence to support this postulation.

Acknowledgements

This work was supported by a grant from the National University of Singapore

(Grant No. RP305/86) and Shaw Foundation (Grant No. GR05303N). We thank Mrs. Josephine Howe for printing the photographs.

References

1. Dales, S. and Chardonnet, Y. Virology 56 : 465-483, 1973.
2. De Brabander, M. Endeavour, New Series 6(3) : 124-134, 1982.
3. Fulton, A.B., Wan, K.M. and Penman, S. Cell 20 : 849-857, 1980.
4. Ho, E.S.P., Chezian, S. and Ng, M.L. Tropical Medicine (in press).
5. Howard, J.M., Eckert, B.S. and Bourguignon, L.Y.W. J. Gen. Virol. 64 : 2379-2385, 1983.
6. Igarashi, A. J. Gen. Virol. 40 : 531-544, 1983.
7. Lenk, R. and Penman, S. Cell 16 : 289-301, 1979.
8. Liebhaber, H., Riordan, J.T. and Horstman, D.M. Proc. Soc. Exp. Biol. Med. N.Y. 125 : 636-643, 1967.
9. Luftig, R.B. J. Theor. Biol. 99 : 173-191, 1982.
10. Miles, B.D., Luftig, R.B., Weatherbee, J.A., Weihing, R.R. and Weber, J. Virology 105 : 265-269, 1980.
11. Ng, M.L. J. Gen. Virol. 68 : 577-582, 1987.
12. Ng, M.L. and Ho, E.S.P. In "Invertebrate Cell Systems in Applications" (Jun Mitsuhashi,ed.) (in press).
13. Ng, M.L. and Westaway, E.G. J. Gen. Virol. 64 : 1715-1723, 1983.
14. Schliwa, M. In "Cell and Muscle Motility" (R.W. Dowben and J.W. Shay,eds.) Vol. 5, pp. 1-81, Plenum Press, 1984.
15. Tilney, L.G. Modern Cell Biology 2 : 163-199, 1983.
16. Westaway, E.G. Am. J. Epidermiol. 84 : 439-459, 1966.

CLONING AND PASSAGES OF A CYTOPLASMIC POLYHEDROSIS VIRUS (CPV) IN INSECT CELL CULTURE

S. Belloncik and M. Arella

Centre de recherche en virologie, Institut Armand-Frappier, Université du Québec, Laval, Qc, Canada H7N 4Z3

Several insect cell lines support the replication of Euxoa scandens cytoplasmic polyhedrosis virus (CPV). Different methods were compared for their efficiency in the extraction of virus cell-associated particles. Although sonication of the cells gave higher virus recovery than freeze-thawing, the titer of the virus extracted was relatively low and inadequate for serial passages of CPV in cell cultures. Treatment of infected cells with a solution of Trypsin-EDTA yielded high titers of virus from cells and was efficiently used for the demonstration of genomic stability after serial passages of CPV in cell culture. Moreover, this methodology could be used for cloning of infected cells as well as for selection, type purification and amplification of CPV.

Cytoplasmic polyhedrosis viruses (CPV s) of insects are members of the Reoviridae family. One of the characteristics of the virus replication is the production in the cytoplasm of infected cells of protein inclusion bodies named polyhedra. Up till 1976, CPV of Bombyx mori was the most studied amongst the CPVs for its replication and molecular and biochemical characteristics. However, of all the CPV types studied, none was demonstrated to infect efficiently insect cell lines. Quiot and Belloncik (1) isolated in 1976 from the tobacco cutworm, Euxoa scandens, a CPV (EsCPV) classified as type 5 (2) and which has the important characteristic to replicate at high level in different established insect cell lines as Euxoa scandens, Lymantria dispar, Bombyx mori, Heliothis armigera, Choristoneura fumiferana (3).

Using EsCPV, several experiments were conducted in our laboratory in relation with the virus cell interactions and quantification of viral replication in permissive or non permissive cell lines. Particularly, the role of the nature of the cell lines and the condition of the culture of the cells were demonstrated to be important for an efficient virus and polyhedra production (4).

Both in permissive and non permissive cell lines, the virus particles penetrate the cell by viropexis. In permissive cells, the replication of the virus leads to a complete cycle of infection and production of certain amount of non occluded virus and polyhedrin, the major compound of polyhedra. Some characteristics of the viral replication in permissive cells are important to point out for the cloning of the infected cells and CPV particles. For example, E. scandens, L. dispar and B. mori cell lines infected by Es CPV gave high polyhedra and non occluded virus production. However, the same CPV could replicate in other systems as C. fumiferana with a low production of polyhedra and virus particles and in H. armigera cell line with a higher non occluded virus synthesis. Recently, different authors reported the viral replication of B. mori (5, 6), (Furuichi, personal communication) and H. armigera (Su, D.M., personal communication) CPV in insect cell lines. However, in both cases few polyhedra were found and crystallization of the polyhedrin showed some abnormal features during replication of H. armigera CPV in a homologous cell line. A close relationship between CPV replication and cell physiology could explain the variability in the viral infection which can be significantly improved after treatments of the cells with γ radiations (7) or coinfection with other insect viruses as Chilo iridescent (CIV) (8). Mechanisms of the spreading of viral infection in cell culture are still under investigation. However, we recently showed the multiplication of infected cells and the propagation of the viral infection during at least ten infected cell passages (4).

We conducted studies in our laboratory on the behaviour of CPV in infected cells, its genomic stability and variability and on the recombination process between CPV types. These research goals were hampered in their achievement by the lack of efficient method for the release of viral particles from infected cells and for the cloning of the CPV. Since the viral particles are not released in the culture medium and the infected cells are not lysed (1), no plaque assay is available for the titration and purification, for serial passage of CPV in cell lines and for selection of viruses lacking polyhedrin gene. Several methods as freeze-thawing and sonication of infected cells could be used successfully to release virus particles (9). However, none of these techniques yielded high titers of infectious viruses for subsequent serial passages in cell cultures. To overcome this problem a solution of Trypsin-EDTA was used for lysis of cells and for liberation of cell associated CPV particles from infected monolayer cells as well as from one individually infected cell. A treatment at 37°C (Fig. 1) for two hours by a solution of Trypsin (10%) + EDTA (10%) in PBS yielded a maximal recovery of viruses. Extraction of viruses was also increased approximately 10^3 times by this treatment (Fig. 1). Sonication of the infected cells prior the Trypsin-EDTA treatment additionally increased a little more than 10 times the recovery of viruses (Fig. 1).

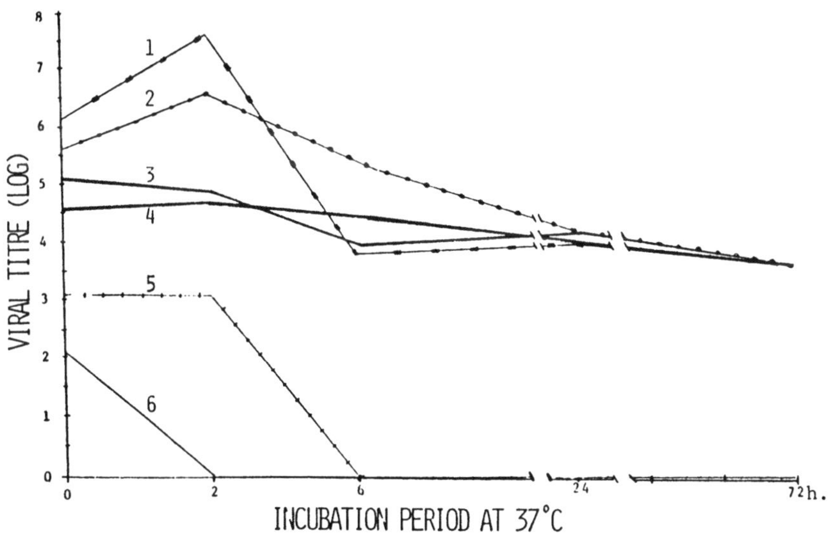

Fig. 1 Virus titer recovered following incubation of:
Sonicated infected cells in Trypsin-EDTA (1) or Grace medium (4)
Infected cell monolayer in Trypsin-EDTA (2) or Grace medium (5).
CPV suspension in Trypsin-EDTA (3) or Grace medium (6)

Since Trypsin has an effect on infectivity of other members of Reoviridae, such as rotaviruses, experiments were conducted in order to elucidate the mechanism of action on CPV. Treatments of CPV suspension and cells prior to infection by Trypsin, as well as the addition of enzyme in the cell culture medium, showed a lack of effect on the viral infectivity. It was concluded that the dissociating effect of this enzyme on the cells was mainly responsible for the increase of the titer of the recovered viruses.

Extraction and amplification of viral particles from infected cell monolayers

The cells were suspended in one ml of fresh Grace culture medium. Cell suspension was, thereafter, sonicated at 75 W for 1 minute and incubated in a solution of Trypsin (10%) and EDTA (10%) in PBS for two hours at 37°C. The viral extract was further diluted 10 times and was used for reinfection of other flasks.

Cloning of infected cells:

Cloning of infected cells was performed according to two different procedures:

1. CPV infected cells were suspended in fresh culture medium and the suspension was serially diluted in a 96 well plate to obtain one cell per well. Wells having a single cell containing or not polyhedra, were treated with 0.025 ml of Trypsin-EDTA solution for 2 hours at 37°C. Thereafter, 0.125 ml of Grace's medium and 0.15 ml of uninfected cell suspension (10^5/ml) were added. Infected monolayers were submitted to the same treatment seven days later with the Trypsin-EDTA solution for virus extraction. Thereafter, the virus titre obtained with such extraction was amplified either in cell culture flask or in E. scandens larvae.

2. Suspension of CPV infected cells at a concentration of 1 000 cells per ml was mixed with agar and distributed in 6 well plates (35 mm diameter). Isolated cells containing or not polyhedra were selected under inverted microscope and extracted from the agar using a Pasteur pipet. The mixture of agar (2%) and cells was thereafter treated with the solution of Trypsin (10%) and EDTA (10%) for two hours at 37°C and the suspension was diluted 10 times in Grace medium and used for infection of cell monolayer according to the standard protocol.

The above mentioned treatments of cells by EDTA Trypsin allowed us to extract high titers of virus from cells and to demonstrate genomic stability after serial passages of CPV in L. dispar cell cultures. Furthermore, this methodology was for the first time efficiently used for cloning infected cells, as well as virus isolation, and amplification starting from one isolated infected cell. This technique should overcome some major difficulties in mass cultivation and type purification of CPV using cell culture. The present approach could also be used for the culture and amplification of other types of CPV replicating at a low level in insect cell lines.

References

1. Quiot, J.M. and Belloncik, S. Arch. Virol., 55:145-153, 1977.
2. Grancher-Barray, S., Boisvert, J. and Belloncik, S. Arch. Virol., 70:55-61, 1981.
3. Belloncik, S., Rocheleau, H., Su, D.M. and Arella, M. Proc. Third Intern. Cell Culture Congress, Sendai, Japan, 1985.
4. Belloncik, S. and Arella, M. Fundamental and Applied Aspects of Invertebrate Pathology (Samson, R.A., Vlak, J.M. and Peters, D., eds.), pp. 69-72, 1986.
5. Hukuhara, T. and Midorikawa, M. Double-stranded RNA Viruses (Compans, R.W. and Bishop, D.H.L., eds.), pp. 406-414, 1983.
6. Inowe, H. and Mitsuhashi, J. Proc. Satellite Symposium, Third Intern. Cell Culture Congress, Sendai, Japan, 1985.
7. Belloncik, S. and Arella, M. Arch. Virol. 68:303-308, 1981.
8. Arella, M., Devauchelle, G. and Belloncik, S. Ann. Virol. 134E:455-463, 1983.
9. Belloncik, S. and Chagnon, A. Intervirology 13:28-32, 1980.

EFFECT OF INHIBITORS ON THE REPLICATION OF A NUCLEAR POLYHEDROSIS VIRUS OF THE SILKWORM, BOMBYX MORI

M. Funakoshi and K. Aizawa

Institute of Biological Control, Faculty of Agriculture,
Kyushu University, Fukuoka 812, Japan

The inhibitory effects of metabolic and antimicrobial inhibitors on a nuclear polyhedrosis virus of the silkworm, Bombyx mori (BmNPV) were investigated by using silkworm larvae or pupae and by plaque assay method using insect cell lines. Some drugs decreased infectivity of BmNPV in vitro, however, the therapeutic effect by these drugs has not been demonstrated except for tannic acid and antimycin A.
Tunicamycin, phosphonoacetic acid, and trisodium phosphonoformate remarkably inhibited the replication of BmNPV in insect cell lines.

Nuclear polyhedrosis viruses are members of Baculoviridae and are potential agents for the control of insect pests. On the other hand, in beneficial insects, such as the silkworm, Bombyx mori, and the honey bee, Apis mellifera, it is necessary to develop procedures for the control of virus diseases.

Investigations on the control of virus diseases of the silkworm have been done from the following three aspects: 1) disinfectants and methods of disinfection, 2) breeding of silkworm strains resistant to virus diseases, 3) chemical therapy. However, among all three procedures, studies on chemical therapy have been poor.

We have investigated the effect of metabolic and antimicrobial inhibitors on a nuclear polyhedrosis virus of the silkworm, B. mori (BmNPV) using silkworm larvae and pupae by feeding and by injection of drugs. Some drugs decreased the infectivity of BmNPV in vitro, however, therapeutic effects of these drugs have not been demonstrated except for tannic acid and antimycin A.

This paper describes the inhibitory effect of some drugs on BmNPV by injection of drugs into silkworm pupae or larvae and by plaque assay method using insect cell lines.

Inactivation of Silkworm Nuclear Polyhedrosis Virus by Inhibitors

Serial ten-fold dilutions of BmNPV infected silkworm hemolymph were prepared in 0.05M phosphate buffer, pH 7.2. The diluted virus was mixed with an equal volume of drug solution at 27°C for 60 min. and the mixture was injected into silkworm pupae. In the control, phosphate buffer was mixed with the virus. LD_{50} was determined according to the method of Reed and Muench (11). The inactivation index was calculated by the difference of -log LD_{50} values between the control and the drug treatment.

1. Antiviral effects of antibiotics, metabolic inhibitors, and mutagens

The antiviral effects of mytomycin C, actinomycin C_1, nalidixic acid, Nhydroxyurea, chloroquine-2-phosphate, daunomycin, rifamycin SV, formycin, rifampicin, 8-azaserine, tubercidin, 5-fluorouracil, BUDR, IUDR, cycloheximide, neomycin, kanamycin, amphotericin B, bacitracin, D-cycloserine, gramicidin D, polymyxin B, valinomycin, and antimycin A were tested. Among these drugs, mytomycin C, actinomycin C_1, daunomycin, and antimycin A showed the antiviral activity at the concentration of 1 mg/ml, 0.1 mg/ml, 1 mg/ml, and 0.5 mg/ml, respectively. The inactivation index of mytomycin C, actinomycin C_1, daunomycin, and antimycin A was 1.15, 1.54, 1.57, and 1.54, respectively.

2. Antiviral effects of proteolytic enzymes

BmNPV is strongly inactivated in vitro with the gut juice of silkworm larvae (1). We found that one of the antiviral principles in the silkworm gut juice was an alkaline protease. In this connection, we have examined the antiviral effect of proteolytic enzymes on BmNPV.

The antiviral effects of trypsin, nagarse, pronase E, pronase P, protease I, protease III, protease IV, peptidase, and acromopeptidase were tested. Nagarse, pronase E, pronase P, protease IV, and acromopeptidase showed the antiviral activity. The inactivation index of nagarse, pronase E, pronase P, protease IV, and acromopeptidase at the concentration of 1 mg/ml was $\geq$2.13, $\geq$3.80, $\geq$3.80, $\geq$2.13, and 1.80, respectively. Moreover, acromopeptidase inactivated alkali-liberated occluded virus by the treatment for 6 hrs at 27°C.

3. Antiviral effects of phenol compounds

The antiviral effects of tannic acid, pyrogallol, and gallic acid were tested. Among these drugs, tannic acid and pyrogallol showed the antiviral activity since the inactivation index of tannic acid and pyrogallol at the concentration of 500 µg/ml was 3.50 and 1.83, respectively.

Preventive and Therapeutic Effects of Drugs by In Vivo Assay

1. Preventive and therapeutic effects of mytomycin C, daunomycin, and antimycin A

Drug solution was injected into pupae 3 hrs and 1 hr before the virus injection to test the preventive effect, and 1 hr and 3 hrs after the virus injection to test the therapeutic effect. In the control, phosphate buffer was injected 1 hr before the virus injection. The defense index was calculated by the difference of -log LD_{50} values between the control and the drug treatment.

The defensive effect was observed when 1.25 µg of antimycin A was injected. When antimycin A was injected 3 hrs and 1 hr before the virus injection and 1 hr and 3 hrs after the virus injection, the defense index was 0.92, 2.34, 2.17, and 1.67, respectively. However, the defensive effect was not observed when 25 µg of mytomycin and 12.5 µg of daunomycin was injected, respectively.

2. Preventive and therapeutic effect of tannic acid

One hundred and twenty five µg of tannic acid was injected into silkworm larvae 2 hrs and 1 hr before the virus injection to test the preventive effect and 1 hr and 2 hrs after the virus injection to test the therapeutic effect. Defense indexes greater than 0.7 were obtained in repeated experiments and the results suggest that tannic acid has the defensive effect in vivo, however, its effect was not strong.

A Viral Inhibitory Factor Extracted from Virus Infected Hemolymph

A viral inhibitory factor (VIF) was extracted from the hemolymph of silkworm larvae infected with a nuclear polyhedrosis virus (2). The VIF inactivated the virus *in vitro* very strongly but did not show a therapeutic effect by injection since the VIF activity is inhibited by an inhibitor existing in insect hemolymph (4, 5). However, the VIF exhibits a defense reaction *in vivo* by the injection of phospholipase C (Funakoshi and Aizawa, unpublished).

Antiviral Activity of Drugs by Plaque Assay

We have employed a plaque assay method using insect cell lines for the assay of antiviral activity of drugs.

The effects of tunicamycin, phosphonoacetic acid, and trisodium phosphonoformate on the replication of BmNPV were examined using insect cell lines, BM-N and S.P.C. Bm 36. Tunicamycin inhibits glycosylation of protein and the replication of herpes simplex virus (10). Phosphonoacetic acid is an effective inhibitor for the replication of herpes simplex virus in cell culture (9) and *in vivo* (12). Biochemical and genetical investigations have clarified that phosphonoacetic acid inhibits the DNA polymerase activity at the pyrophosphate binding site (8).

The cell lines S.P.C. Bm 36 and BM-N were obtained from Dr. J.-M. Quiot and Dr. S. Maeda, respectively. These cells were grown and maintained in IPL-41 medium supplemented with 10 % fetal bovine serum. The supernatant at 3,000 rpm for 15 min. of BmNPV infected hemolymph of silkworm larvae was used as virus source.

Approximately 2×10^6 S.P.C. Bm 36 or BM-N cells was seeded in the tissue culture flasks and BmNPV of 1.8×10^6 PFU was inoculated to the tissue culture flasks. The flasks were maintained at 27°C for 60 min. and the inoculum was replaced by 4 ml of the medium containing a drug and then the cultures were incubated at 27°C for 72 hrs. The extracellular virus titers in the medium were determined by plaque assay method. In the tests on the effect of drugs on the formation of polyhedra, the percentage of cells forming polyhedra in approximately 100 BM-N cells observed was calculated.

1. Effect of tunicamycin on the replication of BmNPV in S.P.C. Bm 36 cells

BmNPV in S.P.C. Bm 36 cell culture was treated with 0.5 µg/ml and 10 µg/ml of tunicamycin at 27°C for 72 hrs. The virus titers (PFU/ml) were 8.2×10^4 and 3.2×10^4 at the concentration of 0.5 µg/ml and 10 µg/ml, respectively, while the virus titer (PFU/ml) in the control was 1.5×10^6. The replication of extracellular virus was markedly inhibited by 0.5 µg/ml of tunicamycin. However, the formation of polyhedra was not affected by 10 µg/ml of tunicamycin.

2. Effect of phosphonoacetic acid (PAA)

The replication of BmNPV was not inhibited by the treatment with 0.7 mM PAA. At the concentration of 3.5 mM PAA, the virus titer (PFU/ml) was 100 fold lower than that in the control in S.P.C Bm 36 cells and was 10 fold lower than that in the control in BM-N cells (Table 1).

3. Effect of trisodium phosphonoformate (PFA)

The replication of BmNPV was not inhibited by the treatment with 0.7 mM PFA. At the concentration of 3.5 mM PFA, the virus titer (PFU/ml) was 100 fold lower than that in the control in S.P.C. Bm 36 cells and was 10 fold lower than that in the control in BM-N cells (Table 2).

The inhibitory effect of PAA and PFA for the replication of virus was clearly shown in S.P.C. Bm 36 cells but not in BM-N cells.

4. Effect of phosphonoacetic acid (PAA) and trisodium phosphonoformate (PFA) on the formation of polyhedra in BM-N cells

The percentage of cells forming polyhedra was reduced to 57.7 and 20.0 by the treatment with 0.7 mM and 3.5 mM PAA, respectively, while the percentage in the control was 73.0. Similarly, the percentage of cells forming polyhedra was reduced to 34.7 and 12.1 by the treatment with 0.7 mM and 3.5 mM PFA, respectively, while the percentage in the control was 64.7. The formation of polyhedra of BmNPV was remarkably inhibited by PFA and PAA.

Table 1.
Effect of phosphonoacetic acid (PAA) on the replication of BmNPV in cell lines, S.P.C. Bm 36 and BM-N

Treatment	PFU/ml	
	S.P.C. Bm 36	BM-N
Control	3.6 X 10^5	5.7 X 10^7
PAA 0.7 mM	8.5 X 10^4	4.3 X 10^7
PAA 3.5 mM	2.4 X 10^3	5.4 X 10^6

BM-N cells were used for plaque assay.

Table 2.
Effect of trisodium phosphonoformate (PFA) on the replication of BmNPV in cell lines, S.P.C. Bm 36 and BM-N

Treatment	PFU/ml	
	S.P.C. Bm 36	BM-N
Control	5.0 X 10^6	1.0 X 10^7
PFA 0.7 mM	5.5 X 10^5	5.5 X 10^6
PFA 3.5 mM	1.5 X 10^4	9.0 X 10^5

BM-N cells were used for plaque assay.

Kelly and Lescott reported that tunicamycin had a marked inhibitory effect on the replication of Tricoplusia ni MNPV in Spodoptera frugiperda cells, however, the formation of polyhedra which contained no nucleocapsids occurred in the presence of tunicamycin (7).

Charlton and Volkman reported that glycosylation was completely inhibited at the concentration of 10 μg/ml of tunicamycin and the infectivity of budded virus of Autographa californica MNPV was reduced to 2 % of the control (3). In our study, the replication of BmNPV in S.P.C. Bm 36 cells was reduced to less than 1 % of the control by the treatment with 0.5 μg/ml of tunicamycin. However, the formation of polyhedra was not affected by 10 μg/ml of tunicamycin.

Kelly reported that 20 μg/ml of PAA did not inhibit DNA polymerase activity in S. frugiperda cells infected with T. ni NPV and in uninfected cells although the effect of PAA on the replication of T. ni NPV was not reported (6). In our study, PFA and PAA remarkably inhibited the replication of the extracellular virus of BmNPV in S.P.C. Bm 36 cells and inhibited the formation of polyhedra of BmNPV in BM-N cells.

References

1. Aizawa, K. J. Insect Pathol., 4: 72-76, 1962.
2. Aizawa, K. Proc. IVth Inter. Colloq. Insect Pathol., pp. 352-356, 1970.
3. Charlton, C. A. and Volkman, L. E. Virology, 154: 214-218, 1986.
4. Funakoshi, M. and Aizawa, K. Proc. Sericult. Sci. Kyushu, 2: 65, 1971.
5. Funakoshi, M. and Aizawa, K. Proc. Sericult. Sci. Kyushu, 5: 51, 1974.
6. Kelly, D. C. J. gen. Virol., 52: 313-319, 1981.
7. Kelly, D. C. and Lescott, T. J. gen. Virol., 64: 1915-1926, 1983.
8. Leinbach, S. S., Reno, J. M., Lee, L. F., Isbell, A. F., and Boezi, J. A. Biochemistry, 15: 426-430, 1976.
9. Overby, L. R., Robishaw, E. E., Schleicher, J. B., Rueter, A., Shipkowitz, N. L., and Mao, J. C.-H. Antimicrob. Agents Chemother., 6: 360-365, 1974.
10. Pizer, L. I., Cohen, G. H., and Eisenberg, R. J. J. Virol., 34: 142-153, 1980
11. Reed, L. J. and Muench, H. A. Amer. J. Hyg., 27: 493-501, 1938.
12. Shipkowitz, N. L., Bower, R. R., Apell, R. N., Nordeen, C. W., Overby, L. R., Roderick, W. R., Shleicher, J. B., and von Esch, A. M. Appl. Microbiol., 26: 264-267, 1973.

THE EFFECT OF THE SYNERGISTIC FACTOR ON THE IN VITRO INFECTIVITY OF NUCLEAR POLYHEDROSIS VIRUSES

T. Hukuhara and Y. Zhu

Faculty of Agriculture, Tokyo University of Agriculture and Technology, Fuchu, Tokyo 183, Japan

A synergistic factor (SyF) purified from the capsules of the Hawaiian strain of Pseudaletia unipuncta granulosis virus enhanced the infection of Spodoptera frugiperda cells in culture with plasma-membrane budded and polyhedra-derived virions of Trichoplusia ni multiple nucleocapsid nuclear polyhedrosis virus. Specific binding of the SyF on the cell surface and on the envelope of the two virus forms was visualized by immunofluorescent microscopy and electron microscopy using colloidal immuno-gold markers, respectively. The SyF had no specific affinity to cell lines from Bombyx mori, T. ni and Leucania separata nor to polyhedra-derived virions of B. mori nuclear polyhedrosis virus. Anti-SyF antibodies and chloroquine inhibited the activity of the SyF.

In studies on nuclear polyhedrosis virus (NPV) infection in the armyworm, Pseudaletia unipuncta, Tanada and Hukuhara (1) noted that the NPV infectivity was enhanced by a factor present in the capsule of an armyworm granulosis virus (GV). Of the two strains of the armyworm GV, Hawaiian and Oregonian strains, only the former had this factor, which was designated as the synergistic factor (2, 3). It had a molecular weight of approximately 126,000, and contained polypeptides and phospholipids (4, 5). Ohba and Tanada (6, 7) reported that the SyF also enhanced the NPV infection of several established cell lines of lepidopterous insects.

We studied the initial events involved in the enhancement both under in vivo and in vitro conditions. The SyF was purified from the capsules of the Hawaiian strain of the armyworm GV by anti-SyF antibody-conjugated affinity column chromatography as described previously (3). When purified SyF was combined with polyhedra of an armyworm NPV and perorally administered to P. separata larvae, the LD_{50} was reduced several thousand times, while the SyF did not enhance the virus infectivity in Bombyx mori larvae (8). These results indicated that the SyF displayed specificity in enhancing NPV infections, depending on the host insect and the virus.

We are currently employing immunolabeling methods to establish the basis for the specificity. The epithelial cells of armyworm larval midguts were dispersed by trypsinization. The cells were incubated with the SyF, and then processed for indirect fluorescent antibody staining using rabbit anti-SyF antibody and FITC-labeled goat antibody against the rabbit IgG. Examination of the final preparations by epifluorescence microscopy revealed the localization of the characteristic fluorescence in one end of columnar cells, where microvilli were present (9). Thus, the SyF appeared to be specifically adsorbed to the microvilli.

In the next set of experiments, polyhedra of an armyworm NPV were placed on electron microscope grids, and treated with weak alkali to liberate virions. The grids were incubated with the SyF and then processed for immunolabeling using rabbit anti-SyF antibody, and gold-labeled goat antibody against rabbit IgG. The gold particles were regularly distributed on enveloped nucleocapsids (ENCs) released from the polyhedra (Fig. 1). In some virions, the envelope was partially or completely removed by the alkali treatment. Gold particles were bound to the shed envelope but not

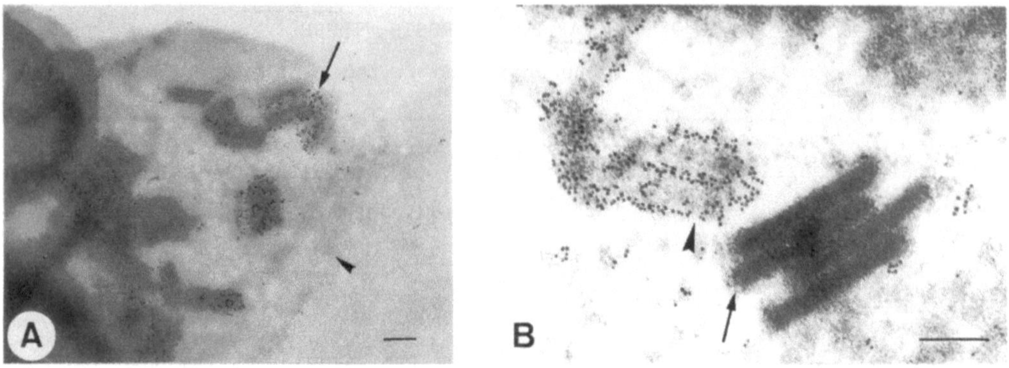

Figure 1. Virions of P. unipuncta NPV immunolabeled with 5-nm diameter gold particles. (A) ENCs (arrow) within a membrane (arrowhead) of a partially dissolved polyhedron. Bar=200 nm. (B) Free NCs (arrow) and a shed envelope (arrowhead). Bar=100 nm.

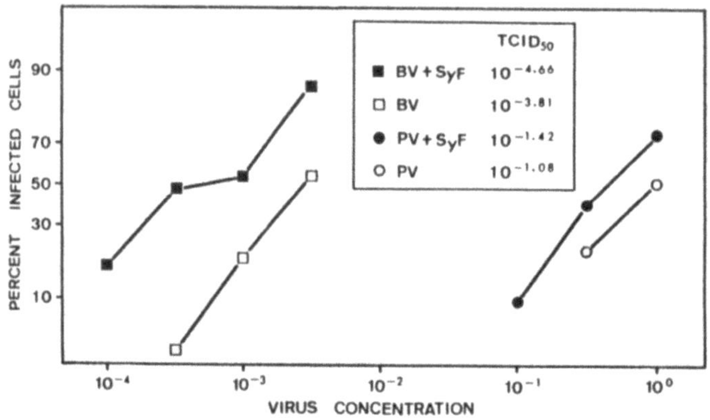

Figure 2. Infection rate of IPLB-SF cells inoculated with BV or polyhedra-derived virus (PV) of T. ni NPV in the presence or absence of SyF (0.025 mg/ml).

to the free NCs. These results indicated that the SyF had specific affinity to microvilli and viral envelopes. We presume that the initial event involved in the enhancement is the specific binding of SyF to columnar cells and ENCs. Incidentally, the SyF was bound neither to B. mori columnar cells nor to B. mori ENCs.

We have extended the above approach to in vitro condition with IPLB-SF-21AE cells and T. ni multiple nucleocapsid NPV. The effect of SyF was determined by the end-point dilution method using tissue culture microtiter plates (7). Two types of virus inocula were used. One was the supernatant of cell cultures 2 days post-inoculation, which presumably contained only the plasma-membrane budded virus (BV). The other was prepared by dissolving polyhedra in alkali and washing the released virions (10). Incorporation of SyF increased the proportion of infected cells and reduced the infectivity titer ($TCID_{50}$) irrespective of whether the cell culture- or polyhedra-derived virus was used as an inoculum (Fig. 2). The enhancement index which is defined as the increase in $-\log TCID_{50}$ value over the SyF-omitted assay, was greater at higher SyF concentrations.

When the SyF was incubated with anti-SyF antibodies before the incorporation to a BV preparation, the enhancement index was reduced (Table 1). The results indicated that the anti-SyF antibodies neutralized the enhancing activity of SyF. The enhancement of the BV infectivity was partially inhibited by the addition of chloroquine (Table 2). This chemical is known to diffuse across lipid membrane in its uncharged form and become trapped in acidic endosomes and lysosomes, thus raising

Table 1. Neutralization of the enhancing activity of the SyF by anti-SyF antibodies

Treatment	$-\log TCID_{50}$	Enhancement index
SyF (0.125 mg/ml)	8.64	2.39
SyF + anti-SyF	7.00	0.75
Anti-SyF (0.075 mg/ml)	6.33	0.08
Medium alone	6.25	—

Table 2. Effect of the SyF and chloroquine on virus infectivity

Treatment	$-\log TCID_{50}$	Enhancement index
SyF (0.150 mg/ml)	6.59	1.18
SyF + chloroquine	6.00	0.59
Chloroquine (1 mM)	4.88	-0.53
Medium alone	5.41	—

the pH,and acts as an inhibitor of viral uncoating (11). The results suggest that the SyF enhances the uptake of BV by endocytosis.

To determine the affinity of the SyF to IPLB-SF cells, the cells were incubated with the SyF and then processed by indirect fluorescent antibody staining. A surface location of the SyF was apparent. The results indicated that the cells had receptors for SyF. The SyF had no specific affinity to cell lines from Bombyx mori (BM-N), T. ni (TN368) and Leucania separata (NIAS-LeSe-11), in which it had no enhancing effect on NPV infections.

To determine the affinity of the SyF to T. ni NPV, ultrathin sections of infected IPLB-SF cells were incubated with the SyF and then processed for immunolabeling. Electron microscopy revealed the presence of gold particles on the envelope of virions outside the cells as well as on virions budding from the plasma membrane (Fig. 3). They were provided with peplomers on one end of the envelope. Gold par-

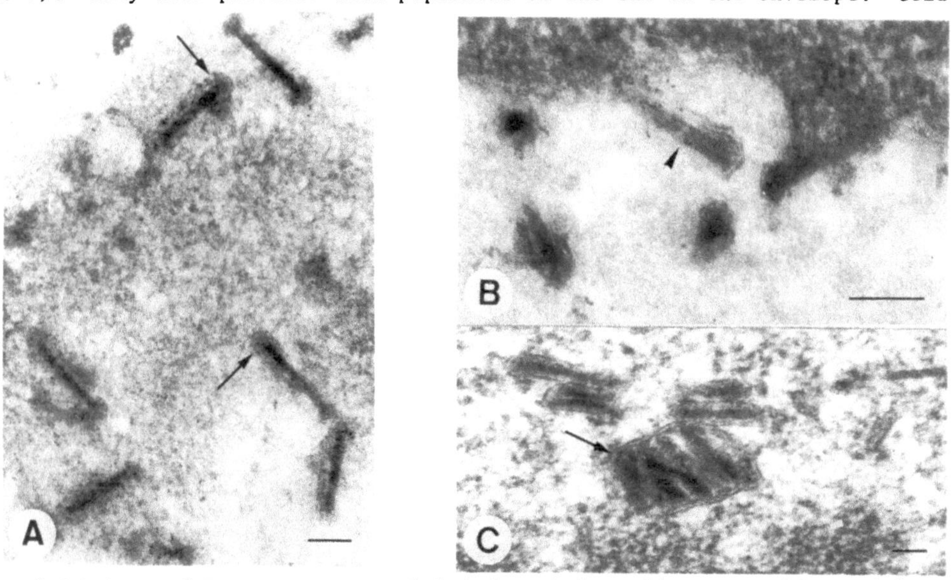

Figure 3. Virions of T. ni NPV immunolabeled with 5-nm diameter gold particles. (A) Extracellular ENCs having peplomers (arrow). (B) ENCs (arrowhead) budding from the plasma membrane. (C) ENCs (arrow) in the nucleus. Bar=100 nm.

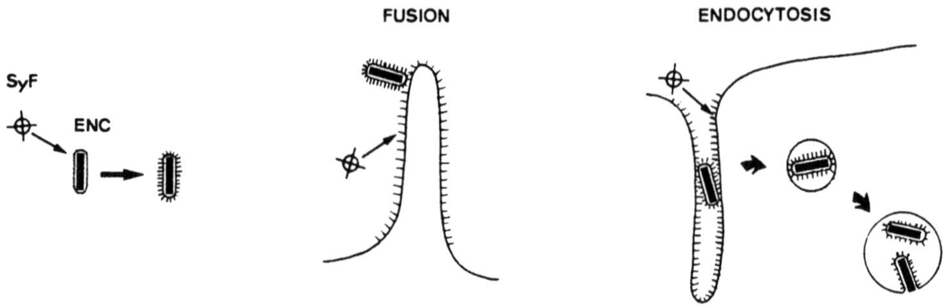

Figure 4. Postulated mechanisms of the interaction of the SyF with an ENC and a host cell. Structural alterations in the viral envelope and the plasma membrane are represented by spikes.

ticles were also observed on ENCs in the nucleus of infected cells. Thus, the SyF appears to have specific affinity to the envelope of the two forms of *T. ni* NPV.

We have demonstrated that the SyF is adsorbed to viral envelopes and to the cell surface in virus-cell combinations showing enhancement. Specific ligand-receptor interaction would seem to be a prerequisite for the enhancement. The adsorbed SyF may interact with the bi-layer membranes in such a way as to increase the binding of ENCs to cell surface (Fig. 4). An ENC can either be taken up by a cell by way of endocytosis or it can fuse with the plasma membrane. In either case the process is preceded by virus binding to the cell surface. Increased binding will result in the enhancement of virus infectivity. Inhibition of the enhancement by chloroquine can best be explained on the assumption that BV enters the host cell by endocytosis under *in vitro* condition.

We have made some advancements in understanding the early events of enhancement. This approach may allow us to understand baculovirus infection mechanism and to increase the effectiveness of viral insecticides currently being used for insect pest control.

Acknowledgements

A portion of this study was supported by the U.S. National Science Foundation under Grant PCM 8201247 provided to Dr. Y. Tanada, University of California.

References

1. Tanada, Y. and Hukuhara, T. J. Invertebr. Pathol., 17:116-126, 1971.
2. Tanada, Y. and Hukuhara, T. J. Invertebr. Pathol., 12:263-268, 1968.
3. Yamamoto, T., Kita, H. and Tanada, Y. J. gen. Virol., 45:371-381, 1979.
4. Hara, S., Tanada, Y. and Omi, E. M. J. Invertebr. Pathol., 27:115-124, 1976.
5. Yamamoto, T. and T, Y. J. Invertebr. Pathol., 31:48-56, 1978.
6. Ohba, M. and Tanada, Y. Naturwissenschaften, 70:613-614, 1983.
7. Ohba, M. and Tanada, Y. Ann. Virol. (Inst. Pasteur) E, 135:167-176, 1984.
8. Hukuhara, T., Tamura, K., Zhu, Y., Abe, H. and Tanada, Y. Appl. Ent. Zool., 22: 235-236, 1987.
9. Tamura, K., Zhu, Y. and Hukuhara, T. (unpublished).
10. Witt, D. J. and Janus, C. A. J. Invertebr. Pathol., 29:222-226, 1977.
11. Lenard, J. and Miller, D. K. Cell, 28:5-6, 1982.

INFECTION OF SILKWORM CELL LINE WITH AN INSECT VIRUS

H. Inoue, K. Taniai, T. Tamura, and T. Kanda

Laboratory of Silkworm Genetics, The Sericultural Experiment Station, Yatabe, Ibaraki 305, Japan

The susceptibility of the infection of the Bombyx cell line (SES-BoMo-15A) with B. mori NPV was examined after the cells were transferred from the MGM-448 medium to the MGM-443 and MM media. Polyhedra formation was observed in the cells cultured in the MGM-448 and the MGM-443 media, while the cells cultured in the less-expensive MM medium showed no polyhedra, though fluorescent antibody staining revealed the presence of viral and polyhedral proteins.

The development of a low-cost medium is essential for large-scale culture of insect cells in order to produce baculoviruses for control use. However, the occurrence of cell infection with baculoviruses *in vitro* seems to be influenced by the culture media. In the present study, attempts were made to evaluate differences in the infection with a *Bombyx mori* nuclear polyhedrosis virus (BmNPV) when various media were used for culturing the cells.

Materials and Methods

Media: Three media, MGM-448 medium containing 10 % FBS, MGM-443 medium containing 10 % FBS and MM medium containing 3 % FBS, were used (1). The cost of the MGM-443 and MM media was about 1/2 and 1/6 that of the MGM-448 medium, respectively.

Cell line: The SES-BoMo-15A cell line (2) established from embryos of the silkworm, *Bombyx mori*, was used. The cells were passaged in the MGM-448 medium, and parts of the cells were adapted to the MGM-443 medium and the MM medium.

Virus inoculation: The hemolymph of diseased silkworm larvae infected with a nuclear polyhedrosis virus was diluted 30 times with Carlson's fluid and filtered through a 450 nm pore size filter. The cells maintained in the three different media were subcultured and on the second day after the subculture, 2 drops of BmNPV inoculum were added with a Pasteur pipette into a culture Falcon type 50 ml flask containing 4 ml of medium. Then, the culture flasks were placed at 25 °C.

Results

I. Adaptation of cells cultured in MGM-448 medium to different media

The cells cultured in the MGM-448 medium were transferred to the MM medium. At first, the growth of the cells which was slow gradually improved and after nearly 18 passages cells could be subcultured in the medium in a 1:1 ratio every week. On the other hand, the adaptation of the cells cultured in the MGM-448 medium to the MGM-443 medium was very good and they were able to be subcultured in the medium in a 1:2 ratio every week. The population doubling time of these cells was 2.5 days in the MGM-443 medium and the MGM-448 medium, and 7 days in the MM medium.

II. Infection of cells with B. mori nuclear polyhedrosis virus

As shown in Table 1, the number of cells in which polyhedra were formed in the MGM-448 medium was the same as that in the MGM-443 medium. But, no cells with polyhedra formation were observed in the MM medium. When these cells were tested with fluorescent antibodies against the BmNPV, almost all of the cells in the MGM-443 medium and 20 % of the cells in the MM medium showed a greenish yellow specific fluorescence. When these cells were also tested with fluorescent antibodies against the polyhedral protein, a few cells in the MM medium showed specific fluorescence, suggesting the presence of polyhedral protein in these cells, though no polyhedra were observed.

Table 1. Number of cells with polyhedra formation in the nuclei

Media	No. of cells observed	No. of cells in which polyhedra were formed	Percentage
MGM-448	1376	180	13.1 %
MGM-443	1242	176	14.2
MM	1606	0	0.0

Observation at 5 days after inoculation

Table 2. Percentage of cells in which polyhedra were formed after transfer of the cells from the MM medium to the MM medium containing 10 % FBS and the MGM-443 medium

First medium	Second media	Percentage of cells in which polyhedra were formed
	MM (3 % FBS)	0.0 %
MM (3 % FBS)	MM (10 % FBS)	0.0
	MGM-443 (10 % FBS)	6.6

Observation at 5 days after inoculation

In order to analyse the effect of fetal bovine serum (FBS) as well as the effect of the medium on polyhedra formation, the cells cultured in the MM medium which contained 3 % FBS were retransferred to the MM medium containing 10 % FBS and the MGM-443 medium containing 10 % FBS. After several passages in those media, the occurrence of BmNPV infection was examined. A few cells containing polyhedra were observed in the MGM-443 medium, whereas no polyhedra were detected in the cells cultured in the MM media containing either 10 % FBS or 3 % FBS (Table 2).

Discussion

The MM medium (1) is one of the less-expensive media and the possibility of using it in insect cell culture has been considered. Recently, successful growth of several insect cell lines has been reported in MM medium that did not contain FBS (1). The results obtained in the present study showed that the MM medium containing 3 % or 10 % FBS was suitable for the growth of *Bombyx* cell line (SES-BoMo-15A). However, polyhedra of BmNPV were not formed though the polyhedral protein was produced. On the contrary, polyhedra were produced in the cells cultured in the MGM-448 medium and the MGM-443 medium.

Goodwin and Adams (3) reported that the initial formulation of the medium they developed promoted the growth of a *L. dispar* cell line but not the replication of *L. dispar* NPV. The supplementation of a-tocopherol acetate, cholesterol,methyl oleate and Tween 80 induced normal *L. dispar* NPV replication. These results suggest that the MM medium needs to be supplemented with nutrients for normal formation of polyhedra of BmNPV.

References

1. Mitsuhashi, J. In:Advances in Cell Culture, Vol.2 (Maramorosch, K. ed.), pp. 133-196, Academic Press Inc., 1982.
2. Inoue, H. and Mitsuhashi, J. J. Seric. Sci. Japan, 53:108-113, 1984.
3. Goodwin, R. H.and Adams, J. R. In:Invertebrate Systems *In Vitro* (Kurstak, E., Maramorosch, K. and Dubendorfer, A. eds.), pp.493-509, Elsevier/ North-Holland Biomed. Press, 1980.

USE OF INSECT CELL CULTURE FOR MULTIPLICATION AND STUDY OF PATHOGENIC AGENTS AND PARTICULARLY VIRUS CANDIDATES FOR BIOLOGICAL CONTROL OF PESTS : GENETIC RECOMBINATION STUDY WITH BACULOVIRUSES

J. Quiot and G. Croizer

Station de Recherches de Pathologie Comparée INRA-CNRS
30380 - St-Christol-Lez-Ales, France

Using Lepidopteran cell lines and restriction enzymes, the problem of the genetic recombination of baculoviruses has been investigated. These studies have permitted to show on the one hand the heterogeneity of the natural viral populations and on the other hand the possibility of obtaining genetic recombinants from baculoviral clones from parental populations by plaque assay.

Since the establishment of the first invertebrate cell cultures (1, 2, 3, 4) this field has considerably developed. Established cell lines from various tissues or organs (ovaries, dorsal vessels, embryos, hemocytes and fat bodies) of different orders (Lepidoptera, Diptera, Dictyoptera, Orthoptera, Homoptera...) (5, 6, 7) are now employed in many different disciplines.

This evolution has promoted the in vitro studies of pathogens (viruses, rickettsiella, mollicutes or protozoans) by the way of this multiplication in vitro : the study of their pathogenesis and morphogenesis at the cellular level and the relation between pathogenous agents and the host-cells. Now, these in vitro infections are an integral part of numerous programs in molecular biology, genetics and biotechnology, such as the study of the genetic recombination of viruses.

Baculoviruses have focused our interest : as candidates for microbiological control of pests, the structure of their populations and the characteristics of their evolution have to be elucidated. Moreover, such studies would be to facilitate the analysis of problems caused by the epizooties among invertebrates of economical importance. At last, in the invertebrates, the possibility to make virus infection in vivo and in vitro by turns assists greatly these studies and designates them as patterns of particularly attractive study.

Successive viral reproductions may significantly modify the characteristics of the initial viral population. Therefore, these populations are to be analysed through the different stages of their evolution from their isolation in the nature up to in vivo or in vitro multiplication of the viruses contained in the microbiological insecticides. Such studies are necessary to manage the biological control and to normalize the production of the virus preparations.

In a first time, it was neccessary to evaluate the genetic heterogeneity of the natural populations of baculoviruses. This approach is based on the individualization of the genomes by plaque assay by the rightly codified use of restriction enzymes which give many physical markers of the genomes.

The nuclear polyhedrosis baculoviruses obtained from Mamestra brassicae larvae in U.S.S.R. and in France are very different viruses. The infrequency of DNA segments migrating together permits to postulate that they are two really distinct viruses. While the preparations of viruses from France seems homogenous, these from U.S.S.R. show a great heterogeneity on account of the presence of many DNA segments in small quantities in the electrophoregrams.

Baculoviruses from Spodoptera littoralis from Madagascar, Egypt, Marocco and France (Lyon) have been easily differentiated by the examination of the restriction segments. In fact, these viruses are variants in so far as restricted by HinD III, they all have homologous segments. A part of the differences between the geographic isolates which mark the evolutionary character of the baculoviruses found in the insects of the same species, the heterogeneity within a given population provides essential reports about the evolutionary potentialities of the viral populations in the different geographic areas.

For example, cloning of the Sp. littoralis nuclear polyhedrosis virus shows homogenous populations containing very varied genomes (from Lyon). The demonstration of the heterogeneity of some viral populations offers the problem of the existence and of the role of the genetic recombination among the baculoviruses. Recombinants have been showed by Croizier et al. (9) and Smith and Summers (10) from supposed parental baculoviruses not cloned.

To take up this problem more accurately, it seemed to be necessary to first show the possibility to obtain genetic recombinants from clones from parental populations. We purposed to analyse them, then to detect the frequency of the genetic recombination by testing the destiny of some couples of clones.

The isolation of clones implicates of course the use of cell lines. Within these works (11) we have used two baculoviruses from Lepidoptera, the genomes of which are related : from Autographa californica (AcMNPV) and from Galleria mellonella (GmMNPV) and the Bombyx mori cell line (SPCBm36) (12).

We extracted many clones from the GmMNPV and AcMNPV populations. Among these we kept the GmMNPV 1-1 and AcMNPV 1-2 clones and we established their genetic maps after digestion of their DNA by the restriction enzymes, Eco RI, Bam HI, HinD III and Sma I. Once genetically characterized by the choice and the localisation of seven markers, these two parental clones were mixed in equal parts and injected into the general cavity of G. mellonella larvae. Five transfers from larva to larva were then successively carried out. The DNAs of the five viral populations thus obtained, were compared to the DNAs of the parental clones by electrophoresis after cleavage by Eco RI enzyme.

These analysis shows that immediately the first passage on larvae of Galleria, the parental clone GmMNPV 1-1, has almost disapeared and the clone AcMNPV 1-2 is also little abundant. At the fifth passage, we observed an almost complete disappearance of the parental clones and a concomitant appearance of new types.

The study of these viral forms has been made after cloning on cell lines. Among the many clones thus obtained, two (Ac 389 and Ac 390) were chosen to be analysed after digestion by Eco RI, Bam HI and HinD III enzymes.

They respectively carry the genome of the GmMNPV 1-1 clone in the 5 to 59.2% and 5 to 18.9 % regions of the restriction map and the genome of the AcMNPV 1-2 clone in the 70.2 to 99.2 and 43.9 to 99.2 % regions (Table 1).

In an ulterior study with the same methods we compared the destiny of 7 couples of clones (Ac2 - Gm1 ; Ac1 - Gm2 ; Ac3 - Gm1 ; Ac4 - Gm1 ; Ac3 - Gm2 and Ac4 - Gm2). The Ac2 - Gm1 couple from which the Ac 389 and the 390 clones derive gave the highest degree of recombination as it produced recombinants which progressively substituted it. On the contrary, with the couples of Ac1 - Gm1 and Ac1 - Gm2 clones there was no production of recombinants, only the viral population of Ac1 type was left as the Gm1 and Gm2 were eliminated.

Table 1. Repartition of the genetic markers to the AG 389 and AG 390 recombinants

Markers	BamHI	EcoRI.L	EcoRI.H	BamHI	EcoRI.A	BamHI	HinDIII
Position in the genome	5	6-7	18.9-25.7	43.9	59.2-70.2	74.4	99.2
Parental origin to AG 389	AcNPV 1-2	AcNPV 1-2	AcNPV 1-2	AcNPV 1-2	AcNPV 1-2	GmNPV 1-1	GmNPV 1-1
Parental origin to AG 390	AcNPV 1-2	AcNPV 1-2	AcNPV 1-2	GmNPV 1-1	GmNPV 1-1	GmNPV 1-1	GmNPV 1-1

The Ac3 - Gm1 ; Ac4 - Gm1 ; Ac3 - Gm2 and Ac4 - Gm2 couples, after 4 passages in Galleria larvae, gave intermediate results ; a low percentage of recombinants with the viral populations where the two parental clones still remained.

These works show that it is possible to select genetic recombinants from 2 different viral populations but that it is not a general principle. However, the genetic recombination likely has to be taken into account within the epidemiologic studies with respect to the baculovirosis. Moreover, it is of a great interest for the microbiological control of pests, as it could enable to select the recombinants more virulent than the viruses present in the parental populations. Of course, the results described just before are subjected to the fact that the baculovirus populations which are genetically related were used. Then, it was attractive to know whether the genetic recombination happens so often when the parental clones are much less genetically related.

With this end in view, we tried to find other natural situations generative of recombinants. When mixed, A. californica virus eliminates Rachiplusia ou virus without production of recombinants. The absence of recombinants cannot be ascribed to the incapacity of 2 viruses to exchange their genetic materials (10). The non-production of recombinants is directly in relation with the great selective value of the A. californica virus in the G. mellonella larvae. In the mixture very much unbalanced (1 L.D.50 Ac MNPV and 500,000 L.D.50 ROMNPV) in favour of the RO.MNPV, the AcMNPV supplants the RO.MNPV and the eventual recombinants. We ascertained by cotransfection of the genome of the ROMNPV in its whole and of the genome of AcMNPV segmented by Sma I and Bam HI that nothing is opposed to the production and to the natural selection of recombinants AcMNPV - ROMNPV.

It follows that what limits the advent of natural recombinants is more connected on the one hand to the small selective value of the recombinants with regard of one of their parents and on the other hand, to the limits of our observing facilities than to the capacities of these viruses to exchange their genetic material. In other words, in the nature, the counter-selection towards the recombinant viruses must be particularly effective.

Two other patterns of recombination between variants of baculoviruses from M. brassicae on the one hand and from Sp. littoralis on the other hand tested in laboratory did not enable us to prove the presence of recombinant viruses. We have analyzed the Ac-MNPV è GmMNPV recombinants previously obtained with regard to their biological value as for the control of various Lepidoptera. The L.D. 50% ascribed to these 2 viruses was always of the same order of importance when recorded with G. mellonella, M. brassicae, Sp. littoralis, Sp. exigua, Scotia segetum, Plutella xylostella. Four recombinants analyzed represent the properties of these of the parents. It follows that in the actual state of the knowledges the practical possibilities of recourse to the natural recombination as for the quest of viruses useful to the biological control are likely restricted.

On the contrary, the blind induced recombination and the molecular recombination must be explored for that purpose. They look primissing in so far as they would allow the association in the same virus of information more distant than these reassociated by the natural recombination. With regard to the genetics of populations, the works show that the natural recombination must be involved within the mechanisms which give some plasticity to the viral populations. They permit also to judge the way which remains to be done in order to practise this phenomenon for the biological control and the difficulties to state precisely the extent of the recombination within the evolution of the natural populations.

References

1. Grace, T.D.C. J. Gen. Physiol., 41:1027-1034, 1958.
2. Vago, C. and Chastang, S. Experientia, 14:110-113, 1958.
3. Aizawa, K. and Vago, C. C. R. Acad. Sci. Paris, 249:928-930, 1959.
4. Grace, T.D.C. Nature, 195:788-789, 1962.
5. Hink, W.F. Invertebrate Tissue Culture, Research Applications (Maramorosch, K., ed.) pp. 363-387, Academic Press. 1972.
6. Hink, W.F. Invertebrate Tissue Culture, Research Applications. (Maramorosch, L., ed.) pp. 319-369, Academic Press, 1976.
7. Hink, W.F. Invertebrate Systems In Vitro (Kurstak, E., Maramorosch, K. and Dübendorfer, A., eds.) pp. 553-578, Elsevier Biomedical, 1980.
8. Croizier, G., Quiot, J.M., Paradis, S. and Boukhoudmi-Amiri, K. Entomophaga, 31:385-392, 1986.
9. Croizier, G., Godse, D. and Vlak, J. C. R. Acad. Sci. Paris, 290:579-582, 1980.
10. Smith, G.E. and Summers, M.D. J. Virol., 33:311-319, 1980.
11. Croizier, G. and Quiot, J.M. Ann. Virol. Inst. Pasteur, 132:3-18, 1981.
12. Quiot, J.M. Serologia, 22:25-31, 1982.

REPLICATION OF PIERIS RAPAE GRANULOSIS VIRUS IN AN INSECT CELL LINE

Chen Q. and Dai X.

Insect Virus Research Laboratory, Department of Biology, Huazhong Normal University, Wuhan, China

This paper reports the replication and development of Pieris rapae granulosis virus (PrGV) in a cell line of Plutella xylostella(PX). Under light microscope, 24 hours postinfection some granules existed in the cytoplasm of cells. Six days postinfection, many tiny particles moved Brownianly within the cells. Under electron microscope, the ultrathin sections of 48 hours postinfection showed that tens of enveloped virions aligned in lines in the high electron density blots. Some virions randomly distributed in the nucleus, on the nuclear membrane or outside the nuclear membrane pores. The infected PX cells were sonicated and centrifuged. When the supernatant was used to infect PX cells, the same results appeared as described above.

No reproducible replication of granulosis virus (GV) in vitro had been reported until Naser et al.(1) multiplied Cydia pomonella GV in the primary culture of C. pomonella eggs in 1984. And then in 1986, Granados and his co-workers (2) reported the replication of Trichoplusia ni GV in several egg-derived Trichoplusia ni cells. In the same year in our laboratory, Chen et al. (3) also replicated Plutella xylostella GV in an established cell line of Bombyx mori, Bm21E-HNU5, which originated from B. mori embryos. This paper reports replication of another GV, PrGV, in vitro.

Materials and Methods

The cell line, BCIRL-RX2-HNU3 (PX), which derived from the pupae of Plutella xylostella (4), was cultured in TC-199-MK medium supplemented with 10% bovine serum. The floating, round-shaped cells were split at 1:3 ratio at a three-day interval.

For multiplication of PrGV, the fourth instar larvae of P. rapae which were collected from the field, were inoculated with 10^{-1}mg PrGV capsules/ml. Four or five days post inoculation, the diseased larvae were scissored up and sonicated at 70 W for 60 sec in an ice bath. After centrifugation (2500g, 20 min) two times, the supernatant was filtrated through a 0.45um millipore filter. The filtrate was used as viral inoculum to infect the PX cells in logarithmic growth stage. After two-hour adsorption, the viral inoculum was washed off with medium and fresh medium was added to the vessel to culture the cells. At the third and the eighth days, the old medium was replaced with fresh medium.

At the intervals of 48 hr, 64 hr, 96 hr and 110 hr postinfection (PI), the cells were fixed with glutaraldehyde and osmium tetroxide, dehydrated through a series of ethanol and embedded in Epon 812. The ultrathin sections were then made and examined in a Hitachi H-600 electron microscope.

Results and Discussion

Under light microscope, the suspended cells began to attach to the bottom of the flask 10 min PI. 24 hours PI, the nuclei of the cells became hypertrophied and dense, and granules existed in the cytoplasm . 48 hours PI, the cells were loosely attached to the bottom and all cells became detached and scattered in the medium 110 hours PI. 6 days PI, many tiny particles moved Brownianly within the cells that were large, completely round and full of dense materials.

Under electron microscope, the ultrathin sections of 48 hours PI showed that tens of enveloped virions aligned in lines in the high electron density blots, i.e. virogenic stroma. But randomly some virions were distributed in the nucleus, on the nuclear membrane or outside the nuclear membrane pores (Figure 1). 64 hours PI, the nuclear membrane began to disintegrate. The virogenic stroma and the virions with a size of 200 x 60 nm spread to cytoplasm. 110 hours PI, electron microscopy revealed that virions were being occluded from one side or one end (Figure 2, 3). The longitudinal section of the mature capsule was oval, containing a virion with a size of 370 x 200 nm (Figure 4).

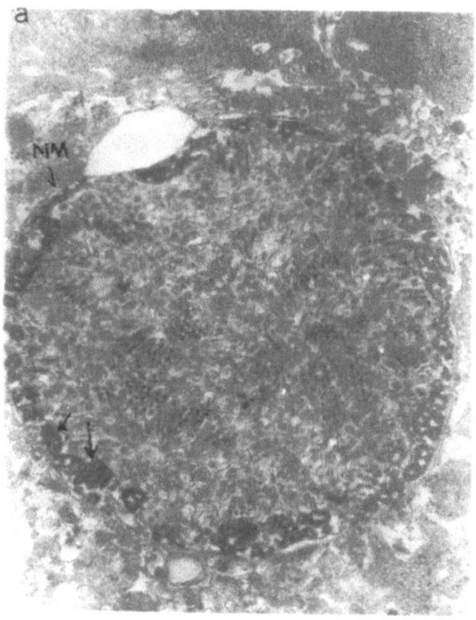

Fig. 1. 48 hours PI, the virogenic stroma and enveloped virions existed in the nucleus.
NM: nuclear membrane;

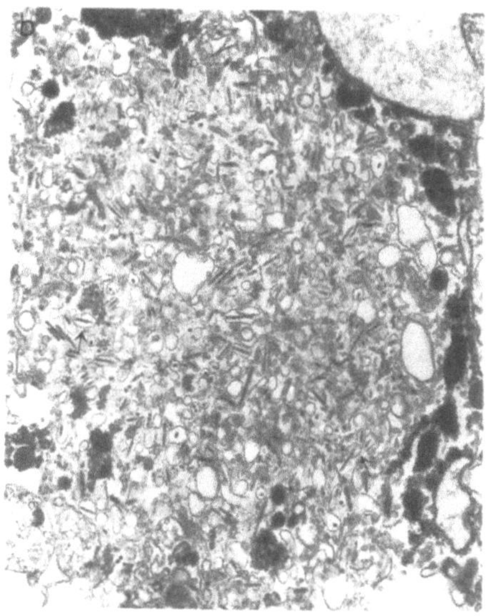

Fig. 2. 110 hours PI, virions were being occluded from one side or one end. Capsules also appeared in the cytoplasm.

In order to test the infectivity of the in vitro replicated PrGV, the infected PX cells were sonicated and centrifuged. Then, the supernatant was employed to infect PX cells. Electron microscopic examination testified that the same results as described above appeared. When P. rapae larvae were fed with the capsules produced in vitro, the typical symptoms of GV infection were observed three days later. The color of the larvae became yellow-green, yellow-white and white, progressively. Their body contents were liquefied and they died with their heads hung downwards, frequently in the shape of an inverted "V".

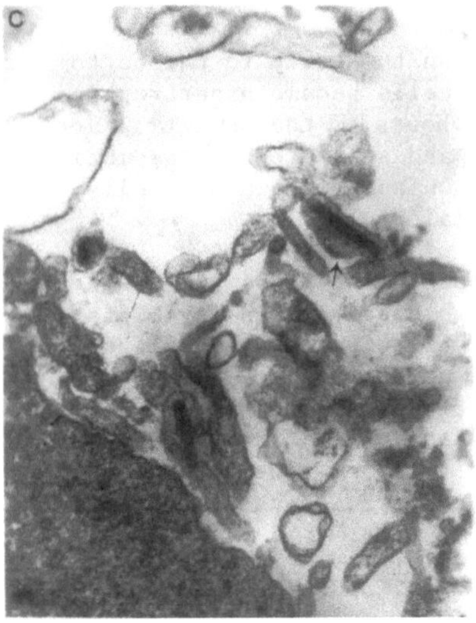

Fig. 3. 110 hours PI, virions were being occluded from one side or one end.

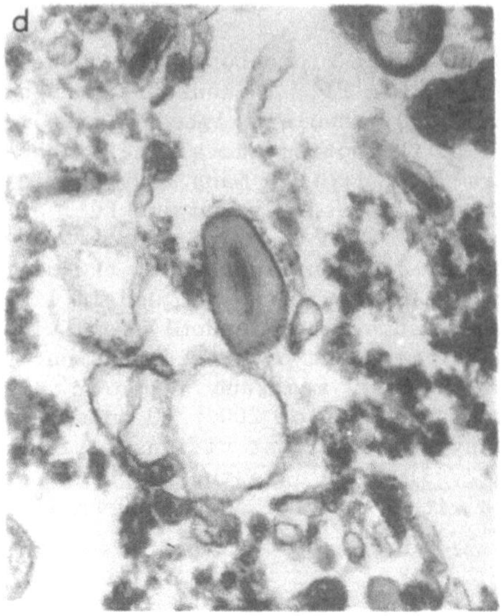

Fig. 4. 110 hours PI, the longitudinal section of the mature capsule.

Probably insufficient infectious viral units in the viral inoculum caused the unsuccessful replication of GV in vitro in the past. Usually hemolymph or GV infected larvae were used to infect insect tissue cultures. However, the main tissues that are attacked by GV in vivo are fatbodies, midgut epithelium and epidermal cells, not hemocytes (5). In these tissues exist many infectious virions outside the cells and a lot of virions, infectious viral units, enzymes and precursors that are required for GV replication inside the cells. That is why we prepared the viral inoculum with tissue extract.

No cell line of P. rapae in our laboratory is available at the present time. But in P. xylostella cells there may be the factors that are needed for PrGV replication because P. xylostella larvae can be infected with PrGV. The success of the replication of PrGV in PX cells showed that a low host specificity appeared in the in vitro replication system.

References

1. Naser, W. L., et al., FEMS Microbiol. Lett. 24:117-121, 1984.
2. Granados, R. R., Virology. 152:472-476.
3. Chen Quhou, et al., Submitted to the IVth International Conference on Invertebrate Pathology. In press.
4. Chen Quhou, et al., J. Central China Teacher's College (Natural Science Edition) No. 3:99-107, 1983.
5. Liu Lancui, et al., J. Wuhan Univ. (Natural Science Edition) No. 2:49-60, 1981.

STUDIES ON RECEPTOR OF DELTA-ENDOTOXIN FROM BACILLUS THURINGIENSIS VAR. AIZAWAI ON CULTURED INSECT CELLS

M. Himeno

Department of Polymer Engineering, Faculty of Technology,
Tokyo University of Agriculture and Technology,
Koganei, Tokyo 184, Japan.

Cultured insect line cells, TN-368 cells, swelled and burst by treatment with the delta-endotoxin from Bacillus thuringiensis var. aizawai. The delta-toxin bound to cellular receptor, but did not induce the cell swelling at 0°C. When the cells were treated by sialidase before treatment with the toxin, the cell swelling induced by the toxin was inhibited. The cell swelling was also inhibited by presence of NANA, colominic acid, GalNAc, crude ganglioside, ManNH, ManNAc or soybean agglutinin, but not GluNH. These results suggest that a derivative of NANA, GalNAc or ManNH is a constituent of the delta-endotoxin receptor.

Crystalline protein bodies (CPB) produced by Bacillus thuringiensis (Bt) are strong toxin against Lepidopteran, Dipteran or Coleopteran insect larvae. When CPBs were ingested by susceptible insects, they were dissolved in their midgut juice and changed to delta-endotoxin. The delta-endotoxin attacks epithelium cells in the midgut, especially, columnar cells. The columnar cells swelled and were released from the basement membrane and then finally burst by treatment with the toxin. The delta-endotoxin induces the swelling of cultured insect cells but not mammalian cells (1). The cell swelling was observed in the isotonic solution with NaCl or KCl but not in sucrose isotonic solution (2, 3). Knowles and Ellar proposed that a glycoprotein containing N-acetyl-galactosamine (GalNAc) was a candidate for the cellular receptor for the lepidopteran specific delta-endotoxin (4). Also, Thomas and Ellar proposed lipid components in cellular membrane for a dipteran specific delta-endotoxin (5).

In this presentation we report on the sugar compounds related with the receptor for lepidopteran specific delta-endotoxin obtained by dissolution of CPB from Bt var. aizawai with silkworm gut juice on cultured insect line cell, TN-368 cells, derived from cabbage looper.

When the cell was previously chilled at 0°C, the cell swelling was not observed. But after incubation in the presence of the toxin at 0°C, the toxin was washed out and the cells were subsequently incubated in PBS without the toxin at 27°C. The cell swelling was induced without additional delta-endotoxin. These results suggest that the cellular receptor interacted with the toxin at 0°C, but the toxin did not induce the swelling at 0°C (Fig. 1). The membrane fractions of disrupted TN cells were separated on a sucrose density column (6). The separated components were assayed on the TN cells and delta-endotoxin. Fractions included plasma membranes or broken cell fragments and nuclei inhibited the cell swelling. When cell surface of the TN cell was treated with sialidase (neuraminidase) and washed with PBS, the cell swelling by delta-endotoxin did not occur on these treated cells (Fig. 2). These results suggest that the sugar components relating to the receptor on the cell surface were removed by the enzyme treatment.

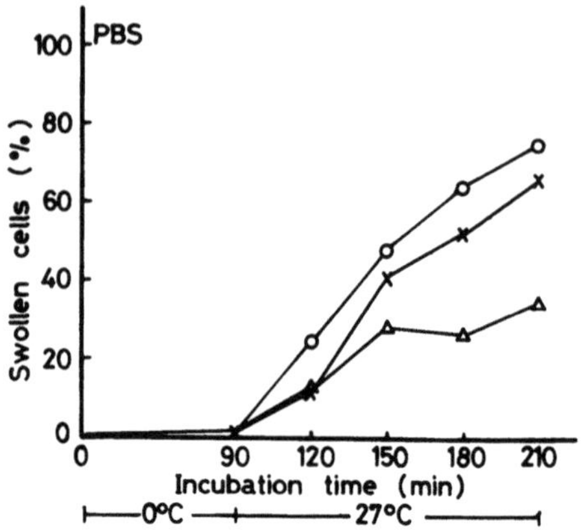

Fig. 1. The TN-368 cells were previously chilled at 0°C and were treated with previously chilled delta-endotoxin (13.5 µg/ml) for 90 min at 0°C. The cells removed from the toxin solution at 90 min were washed with PBS and were incubated in PBS at 27°C (-Δ-). After washing with PBS the cells were again treated with the toxin at 27°C (-o-). After incubation at 0°C the cells were directly incubated without washing at 27°C (-x-).

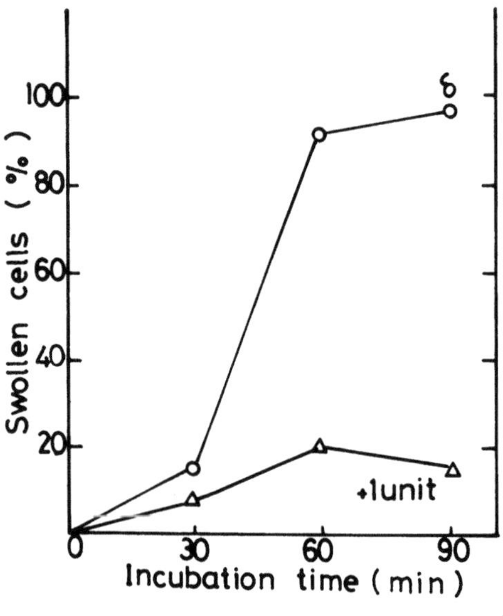

Fig. 2. After the TN-368 cells were treated with sialidase (1 unit) for 2.5 hr at 27°C, the cells were treated with the delta-toxin (3.7 µg/ml) for 90 min at 27°C (-Δ-). As control the toxin was added to the cell sheet without treatment of the enzyme (-o-).

In order to search for a receptor compound of delta-endotoxin in this cell line system, the inhibition of several sugar compounds against the delta-endotoxin activity was determined in the TN 368-cell system. When the TN-368 cells were treated with delta-endotoxin in the presence of GalNAc, the cell swelling was not observed. 50 mM concentration of GalNAc completely inhibited the cell swelling in the NaCl or KCl isotonic solution (Fig. 3). The 0.5 mM concentration of N-acetylneuraminic acid (NANA) inhibited the activity of delta-endotoxin (Fig. 4). NANA most effectively inhibited the activity in tested compounds. Also, colominic acid (2.5 µg/ml) which is a polymer of NANA inhibited the delta-endotoxin activity. Crude ganglioside had weak inhibition effect for the toxin activity. However, purified gangliosides GM_1 or GM_2 did not inhibit the delta-endotoxin activity. NANA was not detected in the TN cells and midgut tissues of silkworm by resorcinol-HCl method. This evidence suggests a possibility that the cellular receptor does not contain NANA but contains a derivative of NANA. Glucose or galactose also did not inhibit the activity. But galactosamine (GalNH) slightly inhibited the delta-endotoxin activity.

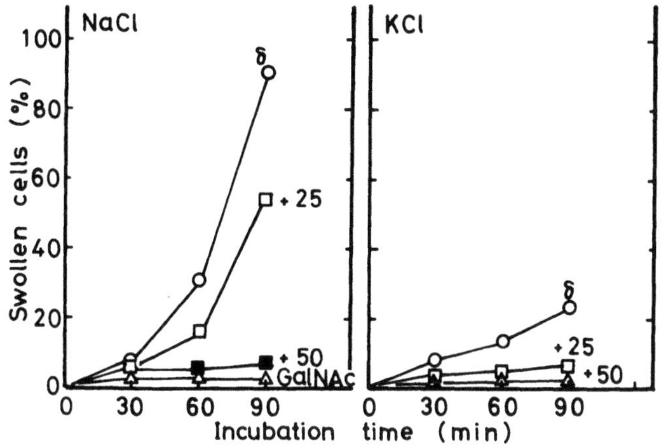

Fig. 3. The TN-368 cells were treated with delta-endotoxin (13.8 μg/ml) in the presence of GalNAc (50 mM: -■-, 25 mM: -□-) at 27°C. As control the cells were treated only with toxin (-o-) or GalNAc (-△-). GalNAc itself did not swell the cells; the cells were incubated in NaCl (a) or KCl (b) isotonic solution.

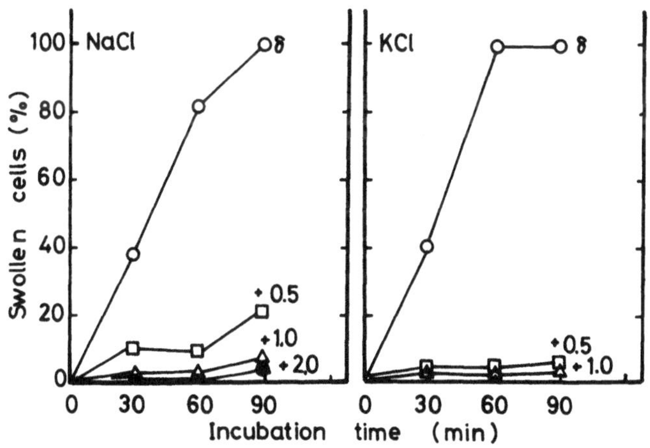

Fig. 4. The TN-368 cells were treated with delta-endotoxin (13.8 μg/ml) in the presence of NANA (2 mM: -•-, 1.0 mM: -△-, 0.5 mM: -□-) in NaCl (a) or KCl (b) isotonic solution. As control cells were treated only with toxin (-o-).

Mannosamine (ManNH) or N-acetylmannosamine (ManNAc) also have the inhibition activity (Fig. 5). However, mannose has weak inhibition activity. These compounds were compared (Fig. 6). The inhibition activities of tested monosaccharides were in the order of NANA, ManNH, GalNAc, ManNAc, and mannose.

These results suggest that the cellular receptor of lepidopteran specific delta-endotoxin (Bacillus thuringiensis var. aizawai) on the TN-368 cell surface was a compound including a derivative of NANA which is susceptible to sialidase, GalNAc or ManNH.

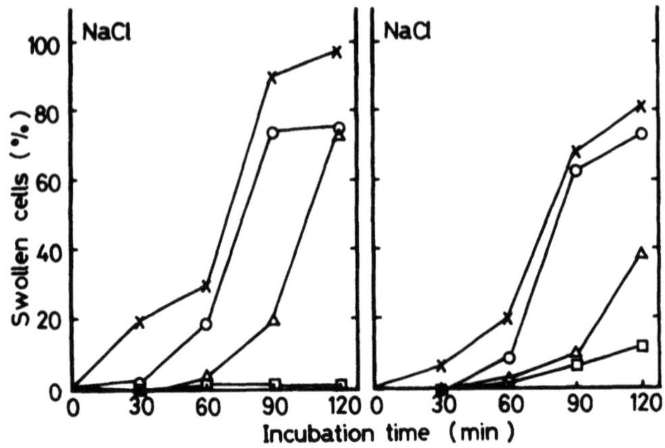

Fig. 5. The TN-368 cells were treated with delta-endotoxin (11.3 µg/ml) in the presence of 50 mM (-o-) or 100 mM (-△-) of ManNH (left), and also in the presence of 50 mM (-o-) or 100mM (-△-) of ManNAc (right) at 27°C. As control, ManNH (-□-) ManNAc (-□-) or the toxin (-x-) alone in NaCl isotonic solution was added to the cells.

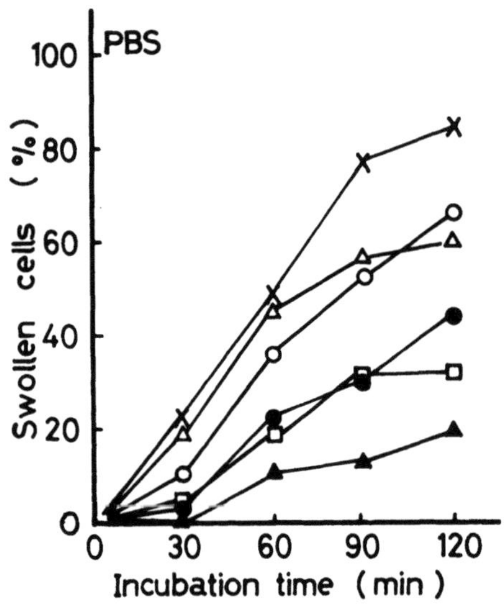

Fig. 6. The TN-368 cells were treated with delta-endotoxin (13.5 µg/ml) in the presence of 5 mM NANA (-▲-), 50 mM ManNH (-□-), 50 mM GalNAc (-●-), 50 mM ManNAc (-△-), or mannose (-o-); as control the cells were treated only with toxin (-x-) at 27°C. These monosaccharides alone in PBS did not affect the cell swelling without the toxin.

References

1. Nishiitsutsuji-Uwo, J., Endo, Y. and Himeno, M. Appl. Ent. Zool., 15:133-139, 1980
2. Nishiitsutsuji-Uwo,J., Endo, Y. and Himeno, M. J. Invertebr. Pathol., 34: 267-275, 1979
3. Himeno, M., Koyama, N., Funato, T., and Komano, T. Agric. Biol Chem., 49: 1461-1468, 1985
4. Knowles, B. H. and Ellar, D. J. J. Cell Sci., 83:89-101, 1986
5. Thomas, W. E. and Ellar, D. J. FEBS Lett., 154:362-368, 1983
6. Atkinson, P. H. Methods in Cell Biology. VII (Prescott, D. M. ed.), pp. 157 - 188, Academic Press. 1973

INOCULATION OF INSECT CELL CULTURES WITH HYDROGEN PEROXIDE PRIMED SPORES OF MICROSPORIDIA

T. Kawarabata[1)] and S. Hayasaka[2)]

1) Institute of Biological Control, Faculty of Agriculture, Kyushu University, Fukuoka 812, Japan
2) Sericultural Experiment Station, Tsukuba, Ibaraki 305, Japan

Spores of microsporidians(Nosema bombycis, Nosema sp. M11, and Nosema sp. M12) pathogenic to lepidopterous insects were primed by the treatment with a 0.1 M potassium chloride solution containing 3-12% hydrogen peroxide at 0°C for 5 min. Primed spores were mixed with the Antheraea eucalypti cell cultures at 27°C to induce the germination of the spores. Sporoplasms were observed in the cytoplasm of A. eucalypti cells 1 hr postinoculation. However, no development of the sporoplasms injected into the host cells was observed afterwards. Disintegration of the sporoplasms in the cytoplasm of the host cells was evident 24 hr post inoculation.

Artificial germination methods of microsporidian spores have been studied by several investigators (3,10,11,14), and later, those methods have been applied to infect insect cell cultures (2,5,6,7,8,9,12,13,14). Potassium hydroxide (2,4,5,7,12,14) or potassium chloride (13) was used to induce the germination of microsporidian spores and subsequent infection of insect cell cultures. Also, spores of N. bombycis were able to germinate after hydrogen peroxide treatment (6,11). However, there are few reports on the infection of insect cell cultures with spores of microsporidians primed by treatment with hydrogen peroxide.

In the present report, we describe inoculation of the A. eucalypti cell cultures with hydrogen peroxide primed spores of microsporidians pathogenic to lepidopterous insects.

Materials and Methods

Microsporidians used were derived from the laboratory culture of the Division of Sericulture, the Sericultural Experiment Station, Tsukuba, Japan. They consisted of Nosema bombycis NIS 001 (standard strain of N. bombycis in Japan), Nosema sp. NIS M11, and Nosema sp. NIS M12. Microsporidians were grown in larvae of the silkworm, Bombyx mori (strain: C 137 X J 137). Spores used for the inoculation of the insect cell cultures were prepared by single step gradient centrifugation on percoll (Pharmacia, Sweden). Usually, 1 volume of the spore suspension was layered onto 9 volumes of percoll in a small glass test tube (18 x 125 mm). The tubes were centrifuged at 3,000 rpm for 20 min at 4°C and the sedimented spores were resuspended in distilled water. The purification cycle of spores on single step percoll gradient was repeated 3 times. Purified spores were finally suspended in distilled water and spore concentrations were determined with a hemacytometer. The purified spores were stored for not more than two days at 4°C.

A cell line of Antheraea eucalypti (1) was obtained from by Dr. J. L. Vaughn (U. S. Department of Agriculture, Beltsville, Md.). The A. eucalypti cells were propagated in Grace's medium (1) supplemented with 5% fetal bovine serum (Gibco) and 1% heat-inactivated (60°C, 30 min) silkworm hemolymph. The A. eucalypti cells were subcultured every 7 days. The cultures were maintained at 27°C. The purified microsporidian spores were primed with a 0.1 M KCl solution containing 3-12% (w/v) hydrogen peroxide (6) at 0°C for 5 min. The spore suspension was mixed with an equal volume of ice-cold priming solution containing hydrogen peroxide at 0°C for 5 min. After the treatment, the spores were washed repeatedly with an ice-cold 0.1 M KCl solution to remove hydrogen peroxide. Finally, the primed spores were suspended in an ice-cold 0.1 N KCl solution, and 1.0 ml of the primed spore suspension (8 X 10^8 spores/ml) was added to 40 ml of the A. eucalypti cell suspension (10^6 cells/ml) by mixing for a few seconds. The spore-cell suspension in a plastic culture flask (Falcon:3024) was gently and continuously shaken for 10 min and each inoculated cell suspension (2 X 10^6 cells in 2 ml) was placed in a plastic culture flask (Falcon:3013). The flasks were kept at 27°C for 1 hr and cells were allowed to settle at the bottom of the flask. The supernatant was removed carefully by aspiration and replaced with 4 ml of the complete culture medium supplemented with 200 units penicillin/ml and 200 ug streptomycin/ml. The spores of N. bombycis primed with 0.1 N KOH at 27°C for 30 min were also used to infect the A. eucalypti cells (7) as a control of spore viability.

One infected flask and one control culture flask were used for each sample collection. The proportion of the germinated spores was determined in 1,000-spore samples stained with Giemsa at various time intervals. The percentage of infected cells was determined in 1,000-cell samples.

Results

1. Priming of microsporidian spores with hydrogen peroxide

Purified microsporidian spores were suspended in the solution containing various concentrations of hydrogen peroxide in 0.1 M KCl solution and the suspensions were kept at 27°C for 30 min. After the incubation, the spore suspensions were observed by phase-contrast microscopy to determine the optimal concentration of hydrogen peroxide for the germination of the microsporidian spores. The optimal concentrations of the hydrogen peroxide solution for the germination of the spores of N. bombycis, N. sp. M11 and N. sp. M12 were 3%, 6% and 6%, respectively.

Priming of microsporidian spores was carried out at 0°C using the optimal concentration of the hydrogen peroxide solutions for the germination of the spores of each microsporidian species. Priming effect of hydrogen peroxide was prompt and vigorous. It was found that the microsporidian spores were primed sufficiently by the hydrogen peroxide treatment within 5 min at 0°C under the conditions used.

Before the spores were washed with an ice-cold 0.1 M KCl solution to remove hydrogen peroxide, the primed spores of N. bombycis showed a relatively high germination rate (ca. 90%). However, the germination rate of the primed N. bombycis spores rapidly decreased after repeated washing with the ice-cold 0.1 M KCl solution. After seven washings with the ice-cold 0.1 M KCl solution, the germination rate of the primed N. bombycis spores decreased to less than 10%.

2. Infection of A. eucalypti cells with hydrogen peroxide primed spores

The infection rate of N. bombycis in the A. eucalypti cell line with sporoplasms emerging from the hydrogen peroxide primed spores is shown in Figure 1A. A. eucalypti cells suspended in 20 ml of the complete medium (5 X 10^5 cells/ml) at 27°C were mixed with 2.0 ml of the primed spore suspension (2.5 X 10^9 spores/ml) at 0°C. The initial level of infection was relatively high and 40% of the A. eucalypti cells were infected with sporoplasms 1 hr postinoculation. The percentage of spores that germinated in the cultures 1 hr postinoculation was 8 % and the proportion of hatched spores to host cells was 40 spores to a cell. Up to 24 hr post inoculation, about 25% of the spores had germinated and more than 50% of the A. eucalypti cells were infected with sporoplasms. The levels of the host cell infection and spore germination rate did not change from 3 to 48 hr postinoculation. However, no development of sporoplasms invading into the host cells was observed even 48 hr postinoculation. Instead, the partial disintegration of the sporoplasms in the cytoplasm of the host cells was evident 24 hr post inoculation.

Figure 1 B shows the infection rate of *N.* sp. NIS M11 in the *A. eucalypti* cell line with sporoplasms derived from hydrogen peroxide primed spores. *A. eucalypti* cells suspended in 10 ml of the complete medium (5 X 10^5 cells/ml) at 27°C were mixed with 1.0 ml of the primed spore suspension (2.5 X 10^9 spores/ml) at 0°C. About 74% of inoculated spores germinated and 43% of the *A. eucalypti* cells were infected 1 hr postinoculation. Changes in the cell infection level and spore germination rate in the inoculated cultures were not clearly demonstrated 3 hr postinoculation and afterwards. Also, initiation of the sporoplasm development in the host cells was not observed during 48 hr post inoculation. Gradual degradation of the sporoplasms parasitizing the cytoplasm of the host cells started 24 hr post inoculation. Infection of *N.* sp. NIS M12 in the *A. eucalypti* cell line with sporoplasms hatched from hydrogen peroxide primed spores is shown in Figure 1C. *A. eucalypti* cells suspended in 5 ml of the complete medium (5 X 10^5 cells/ml) at 27°C were mixed with 0.5 ml of primed spore suspension (1.3 X 10^9 spores/ml) at 0°C. The initial level of infection and the germination rate of the spores in the inoculated cultures were similar to those of *N.* sp. NIS M11. Inoculation of *A. eucalypti* cells with hydrogen peroxide primed spores of *N.* sp. NIS M12 failed to produce schizonts of the parasite.

It was apparent that the priming of microsporidian spores by treatment with hydrogen peroxide produced non-viable sporoplasms.

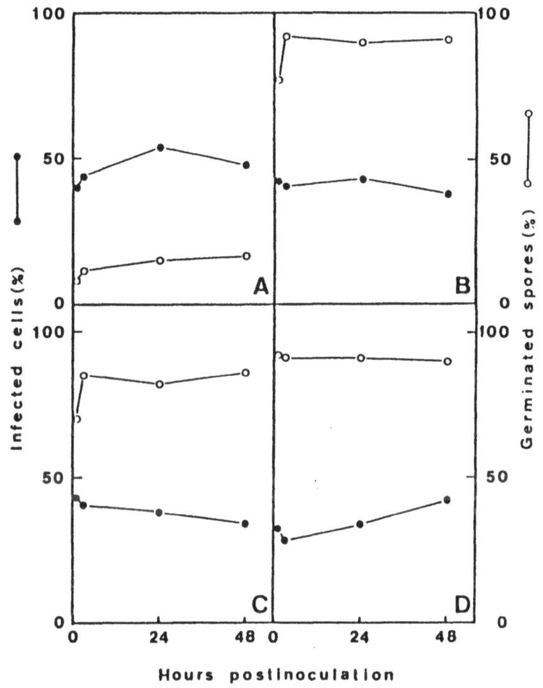

Fig. 1. Inoculation of *A. eucalypti* cell cultures with primed spores of microsporidians.

Figure 4 D shows the infection rate of *N. bombycis* in *A. eucalypti* cells with sporoplasms derived from potassium hydroxide primed spores. *A. eucalypti* cells suspended in 5 ml of the complete medium (5 X 10^5 cells/ml) were mixed with 0.25 ml of the primed spore suspension (1.3 X 10^8 spores/ml) at 27°C. Normal development of the sporoplasms parasitizing the host cells was observed, and the sporoplasms with diffused nuclei and increased size clearly started to grow 3 hr postinoculation.

Discussion

Spores of microsporidians have been induced to germinate artificially with various chemical treatments, and the sporoplasm of a microsporidian spore emerges from the spore (3,10,11,14). Also, some of these treatments have been shown to have a priming effect on spores of microsporidians under certain conditions (6,10,11). It is well known that potassium hydroxide (10) or hydrogen peroxide (11) treatment is one of the most effective ways of inducing the germination of spores of microsporidians parasitic to lepidopterous insects. The hydrogen peroxide germination method has frequently been applied to determine the hatchability of the spore preparations (6), as it was the only method available to assay the hatchability of some microsporidian spores when the ordinary potassium hydroxide germination method was not effective. Iwano and Ishihara (6) reported that spores of *N. bombycis* could be primed by hydrogen peroxide treatment under low temperature conditions and a relatively long incubation period. We applied a modified hydrogen peroxide method for the determination of the hatchability of primed microsporidian spores, and for the inoculation of insect cell cultures with spores of *N.* sp. M11 and *N.* sp. M12 which do not germinate readily.

However, the viability of the sporoplasms derived from microsporidian spores which had artificially germinated has not been well studied. Several investigators have reported on the development of microsporidians in insect cell cultures inoculated with spores germinating artificially, and apparently sporoplasms derived from microsporidian spores primed by treatment with potassium chloride or potassium hydroxide were viable (2,5,7,8,9,12,13,14).

In the present report, we demonstrated that the sporoplasms emerging from the spores primed by treatment with hydrogen peroxide could be injected normally into the cytoplasm of *A. eucalypti* cells. However, the development of sporoplasms to schizonts did not occur, and our results strongly suggest that the hatchability of the microsporidian spores was not affected by the treatment with cytotoxic chemicals which stimulate spore germination (3,11). On the other hand, the sporoplasms were easily inactivated by the treatment applied for the induction of the germination mechanism of microsporidian spores. We also found that the priming with hydrogen peroxide partially impaired the invasion mechanism of *N.* sp. M12 spores. Although the polar tube invaded into the cytoplasm of the host cells, the transfer of the sporoplasm from germinated spores into the host cells was incomplete.

Further development of priming techniques is required for the efficient germination of microsporidian spores and production of viable sporoplasms *in vitro*.

References

1. Grace, T. D. C. (1962). Nature (London), 195, 788-789.
2. Hayasaka, S. (1983). Sanshi-Kenkyu, 127, 22-29.
3. Ishihara, R. (1967). Can. J. Microbiol., 13, 1321-1332.
4. Ishihara, R. (1969). J. Invertebr. Pathol., 14, 316-320.
5. Ishihara, R. and Sohi, S. S. (1966). J. Invertebr. Pathol., 8, 538-540.
6. Iwano, H. and Ishihara, R. (1979). J. Sericult. Sci. Japan, 48, 142-146.
7. Kawarabata, T. and Ishihara, R. (1984). J. Invertebr. Pathol., 44, 52-62.
8. Kurtti, T. J. and Brooks, M. A. (1971). Curr. Top. Microbiol. Immunol., 55, 204-208.
9. Kurtti, T. J. and Brooks, M. A. (1977). J. Invertebr. Pathol., 29, 220-224.
10. Oshima, K. (1964a). Japan. J. Zool., 14, 209-229.
11. Oshima, K. (1966). Japan. J. Zool., 15, 203-220.
12. Streett, D. A., and Lynn, D. E. (1984). J. Parasitol., 70, 452-454.
13. Streett, D. A., Ralph, D., and Hink, F. (1980). J. Protozool., 27, 113-117.
14. Undeen, A. H. (1978). Misc. Publ. Entomol. Soc. Am., 11, 29-63.

DEVELOPMENT IN CULTURE CELLS OF A MICROSPORIDIUM FROM THE LAWN GRASS CUTWORM

H. Iwano and R. Ishihara

Laboratory of Applied Entomology, Department of Agriculture, Nihon University, Kanagawa 252, Japan

Life cycle of a Nosema sp. from the lawn grass cutworm, Spodoptera depravata in Antheraea eucalypti cell was studied. Spores were produced 30 hr after inoculation. Distribution of this Nosema sp. also started to increase around that period. Spores in culture cells discharged sporoplasm in the medium spontaneously. Those observations support the claim that sporoplasm discharged from spores distributes the microsporidia from cell to cell.

Although Ishihara claimed that an infectious agent (secondary infective form=SIF) distributes Nosema bombycis from culture cell to cell (1), the origin of SIF was left to be identified. Sato and Watanabe showed sporoplasm discharged from spores spontaneously has a role in distributing the microsporidia within a host (2). Kawarabata and Ishihara, however, pointed out that the period of spore formation comes later than that of appearance of SIF and also that of the increase in the rate of cells infected with the parasite (3). A microsporidium obtained from the lawn grass cutworm, Spodoptera depravata was successfully inoculated into cells of Antheraea eucalypti. The development of this microsporidium was studied and an attempt was made to identify the origin of SIF.

Materials and Methods

Microsporidium: A microsporidium isolated from the lawn grass cutworm, Spodoptera depravata in Kanagawa, Japan was used. The microsporidium was propagated in larvae of the silkworm, Bombyx mori which was found to be susceptible to the parasite. Spores were collected from infected silkworm larvae. Spore suspension after partial purification by repeated centrifugation at 1,500g for 5 min was purified through Percoll® (Pharmacia, Sweden) at 73,000g for 30 min.

Cell line and medium: Antheraea eucalypti cell line was received from Dr. T. Kawarabata, Kyushu University, Fukuoka, Japan. The A. eucalypti cells were propagated at 26° C in Grace's medium supplemented with 5% FBS (Gibco), 5% heat inactivated (65° C, 5 min.) silkworm (B. mori) haemolymph, 200 units penicillin/ml and 200 µg streptomycin/ml.

Infection of cell line: The purified spores of the microsporidium

($2.4X\ 10^9$/ml) were primed with 0.1N KOH at 25° C for 40 min (4). Subsequently, spores were introduced into culture cell suspension. Cell suspension, after inoculation, was partitioned in small lots (0.8ml). 1 hr after inoculation, the supernatant of suspension was replaced with 2 ml of fresh medium and incubated at 26° C.

Staining, counting and microscopy: Of the lots above, one lot was sacrificed to prepare a smear slide every six hours after inoculation. Slides were, after staining with Giemsa, used for determining the rate of cells infected with the microsporidium and its developmental stage. A fraction of the cell culture was maintained beyond 72 hr after inoculation by replacing the medium once a week. Samples were also extracted for electron microscopy. Cells were fixed with 2.5% glutaraldehyde at 4° C for 40 min, followed by postfixation with 1% osmium tetraoxide at 4° C for 2 hours. After embedding the specimen into Epok 812, a section was made to stain with uranium acetate and lead. Observation was made under an electron microscope (JEM 1200EX).

Results

Extent of distribution: Of culture cells found in the microscopic field of view, 44% were invaded by the microsporidium just after inoculation. The extent of distribution remained at around the same level until 24 hr after inoculation. From 30 hr after inoculation, the incidence of infected cells increased, reaching as high as 81% 72 hr after inoculation. Thereafter, the distribution showed a decline (Fig. 1).

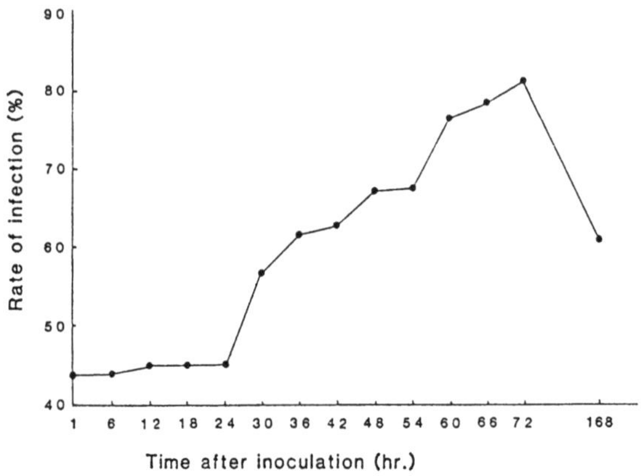

Figure 1. The prevalence of the Nosema sp. infection in A. eucalypti cells.

Development of the microsporidium: There were only sporoplasm in culture cells just after inoculation. The microsporidia grew up to schizont 12 hr after inoculation. The number of schizonts increased 24 hr after inoculation. At the same time, some of the parasites were found to enter into sporogony. Sporonts divided to produce two sporoblasts indicating this microsporidium belongs to the genus Nosema. Electron microscopic observation of the specimen for the same period showed the presence of polar filament in the protozoa and also of its thicker membrane (Fig. 2). Spores began to appear 30 hr after inoculation. They had two compact nuclei, a cytoplasm of high electron density, polar filament and a thick shell (Fig. 3). SIF was also recognized from this time on (Fig. 4). From culture cells which were maintained for many weeks after inoculation were observed the polar filaments of the Nosema sp. piercing through the cell wall of the host cell to the medium and the sporoplasm at the tip of the filaments (Fig. 5). This observation showed that spores of this Nosema sp. extrude polar filaments spontaneously and sporoplasm as well. 36 hr after inoculation, another form of developmental stage of this Nosema sp. appeared. It was bigger than the spore mentioned above and had a smooth but thinner shell, inside of which very poor contents were found (Fig. 3). 72 hr or more after inoculation, spores were found with features somewhat different from those found in the earlier period (Fig. 6). They were smaller in size and the core of the spore was stained with Giemsa (Fig. 7), but it was hard to identify nuclei. Electron microscopically, spores had a thicker shell and its nuclei were also hard to differentiate. We often failed to get a good section of spores at this period, owing to the poor infiltration of resin.

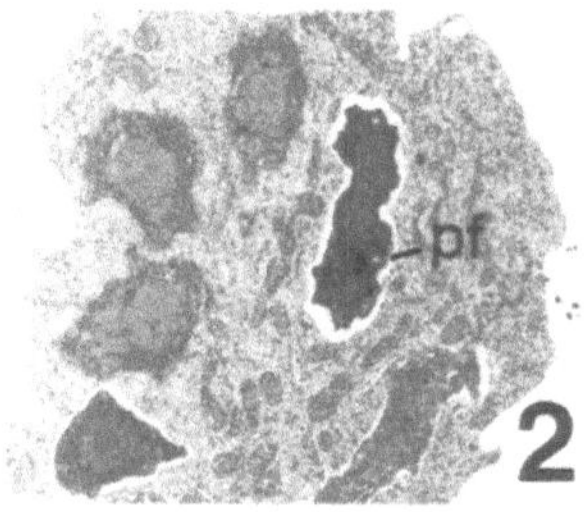

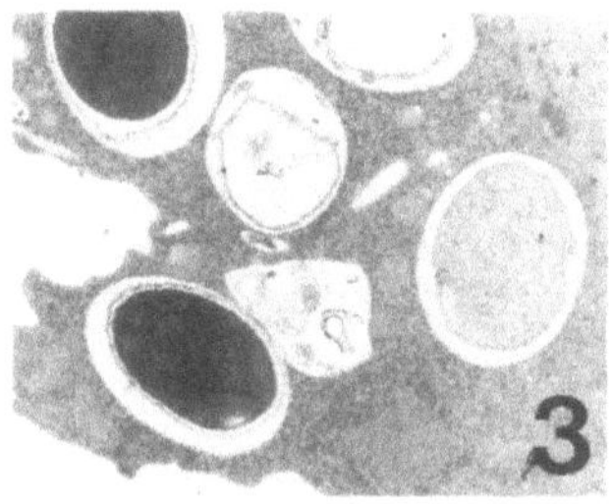

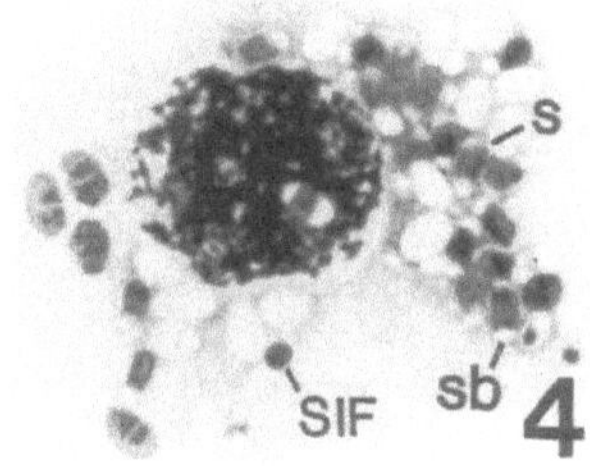

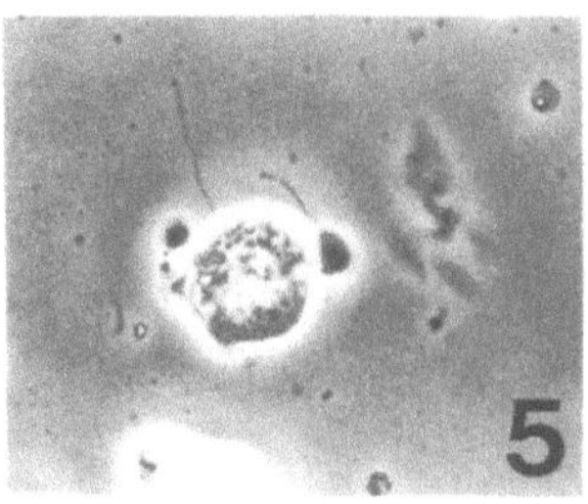

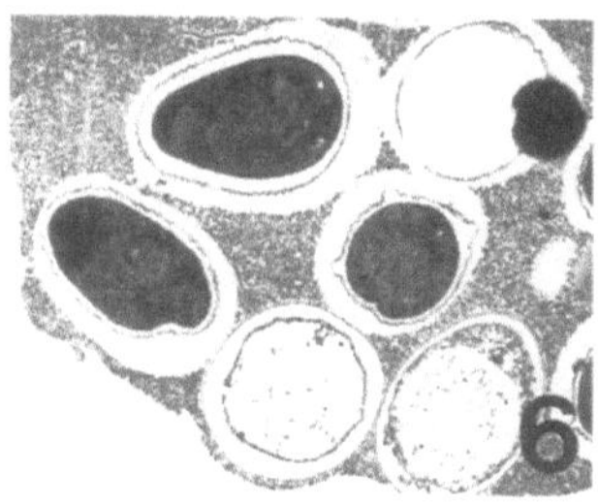

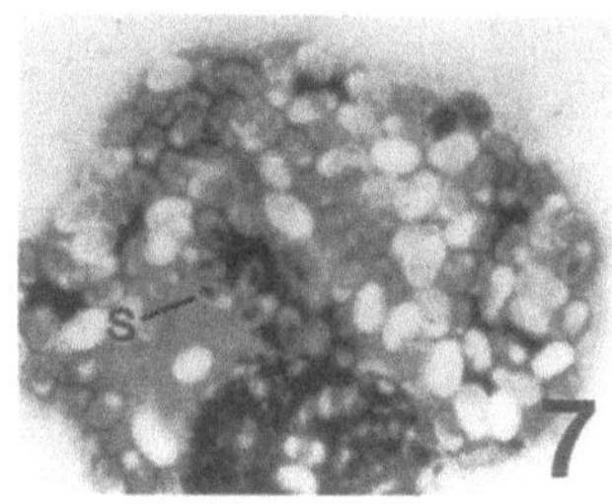

Figure 2. Sporogonial stages of the Nosema sp. in A. eucalypti cell 30 hr after inoculation, as revealed by electron microscope. pf:polar filament.

Figure 3. Spores of a Nosema sp. 36 hr after inoculation. An empty spore-like stage is also visible.

Figure 4. Sporogony as revealed by light microscope. Sporoblast(sb), spore(s), and sporoplasm(SIF=secondary infective form) were seen.

Figure 5. Polar filaments coming out of culture cells from maintained cultures. Portion of sporoplasm appeared at the tip of polar filament.

Figure 6. Mature spores 72 hr after inoculation. Plasm of spores was higher in electron density compared to that of spores formed earlier. Lower infiltration of resin was recognized too.

Figure 7. Spores as revealed by light microscope 72 hr after inoculation. Giemsa stain failed to differentiate the two nuclei of spores formed at this period.

Discussion

As is true of Nosema bombycis, sporoplasm of this Nosema sp.,if spores are primed to discharge their sporoplasm, can develop in culture cells which are derived from insects not related to its host insects, Spodoptera depravata. This fact suggests that the microsporidia, though they are obligate parasites, can develop beyond in a wide difference of cell types.

The proportion of cells infected with this Nosema sp. increased markedly around 30 hr after inoculation. At the same period, newly formed spores were detected and SIF also appeared from this period on. On the other hand, spores were found to discharge their polar filament spontaneously. So, it is possible that spores, when formed, send out sporoplasms to host cells nearby. The marked increase of culture cells infected with the microsporidia may be caused by those sporoplasms. The fact that the period when new spores formed coincides with the period when the rate of infection went up considerably, suggests that the invasion of the Nosema sp. 30 hr after inoculation was done by sporoplasms from spores formed during this period, as reported by Sato and Watanabe (2).

SIF is different from sporoplasm in its stainability by Giemsa stain, however. We recognized that spores formed during the later period of infection were somewhat different from these spores formed earlier. For the time being, we speculate that the Nosema sp. may produce two types of spore: one formed earlier which easily discharges its sporoplasm which takes Giemsa stain more and distributes the Nosema sp. within a host, while the other discharges its sporoplasm only when primed by specific agents and transmits the Nosema sp. from host to host. The latter type discharges sporoplasm which stains less with Giemsa.

Acknowledgements

We are grateful to Dr. T. Kawarabata, Kyushu University for his kind sharing of A. eucalypti cell with us and also thank Dr. J. Mitsuhashi, Forestry and Forest Product Research Institute for his kind instruction in tissue culture technique.

References

1. Ishihara, R. J. Invertebr. Pathol., 14:316-320, 1969.
2. Sato, R. and Watanabe, H. J. Seric. Sci. Jpn., 55:28-32, 1986.
3. Kawarabata, T. and Ishihara, R. J. Invertebr. Pathol., 44:52-62, 1984.
4. Ohshima, K. Parasitology, 29:220-224, 1937.

THE ROLE OF PHENOLOXIDASE IN THE CELLULAR DEFENSE REACTIONS TO FOREIGN MATERIALS BY THE GRANULAR CELLS OF THE SILKWORM, BOMBYX MORI

H.Wago and Y.Ichikawa

Department of Bacteriology, Saitama Medical School, Moroyama, Iruma, Saitama 350-04, Japan

Prophenoloxidase-activating system (proPO-AS) is well known to play an essential role of nonself recognition in insect host defense. So, this study was undertaken to elucidate further the real role of proPO-AS, particularly of the presence of PO, in the initial cellular reaction of the granular cells against foreign materials in Bombyx mori. The results obtained here clearly indicated that (1) the attachment of the granular cells would be promoted by the formed PO in plasma, (2) even the attached cells secreted PO to the exterior of the cells, thus resulting in the strong melanin formation, (3) PO secretion from the cells depended on the presence of calcium ion , and (4) the zymosan-induced secretion of PO was largely due to β-1,3-glucan.

The granular cells of the silkworm Bombyx mori are responsible for the cellular defense reactions such as phagocytosis, nodule formation and encapsulation (1,2). Moreover, melanin deposition is seen to be associated with nodule formation or encapsulation by the granular cells. Recently, it has been shown that the prophenoloxidase-activating system (proPO-AS) plays a key role in nonself recognition in insects (3,4). However, the relationship between proPO-AS and the initial cellular reaction by the granular cells is poorly understood, although the proPO-AS has an opsonic effect in crustaceans (5). The components of proPO-AS are present in plasma of B.mori and are stimulated by a serine protease enzyme (6,7). Therefore,the present study was done to elucidate the real role of proPO-AS in the phagocytic processes including both attachment and ingestion by the granular cells.

Materials and Methods

Final 5th instar larvae of B.mori reared on an artificial diet were used as an experimental animal. They were raised under a photoperiod of 9 hr light and 15 hr dark.

To examine whether or not phenoloxidase (PO) will be needed in one of the phagocytic processes such as attachment by the granular cells, 50 μl of 0.5 mM p-nitrophenyl-p'-guanidobenzoate (pNPGB), 50 μl of 0.1 % phenylthiourea (PTU) and 50 μl of 50 μg/ml zymosan suspensions were first injected into the larval hemocoel with a fine needle inserted into the second leg tarsus, respectively. Each drug was dissolved in insect physiological saline (IPS) consisting of 150 mM NaCl and 5 mM KCl. Zymosan (1 mg/ml in IPS) was boiled for 2 hr and washed three times by centrifugation at 300 g for 10 min before use, and then a concentration of 50 μg/ml zymosan suspension was prepared by dilution with IPS. One hour later postinjection, the blood was collected from each group of larvae into an ice-cold small tube by cutting the cordal horn. Subsequently, PO activity was measured spectrophotometrically at 25°C by directly mixing 25 μl of each hemolymph sample containing hemocytes with 1 ml of 0.01 M L-dopa dissolved in 0.1 M phosphate buffer containing 150 mM NaCl at pH 6.0, and the change in absorbance at 490 nm where dopa-chrome was produced was measured. Next, 25 μl of 0.25 % sheep erythrocytes (SRBC) in Dulbecco's phosphate buffered saline (PBS,pH 7.2) was injected into the drug-preinjected larvae ; 30 min after SRBC injection, the phagocytic capacity by the granular cells was compared by the examination of collected hemolymph with the aid of neutral red stain.

Then, on experiment was designed to verify whether the produced PO in plasma had a certain role in hemocyte adhesion phase. First, 1 ml of the blood was directly collected into 2 ml of 0.5 mM pNPGB in IPS (PO^-hemolymph) or into 2 ml of 0.1 % PTU in IPS (PO^+hemolymph). In each hemolymph sample, hemocytes were totally contained at a concentration of 6 ~ 8 X 10^5 cells/ml where the granular cells made up 62 % and plasmatocytes 21 %. Each was then incubated immediately in a 35 mm-plastic petri dish (Corning), and the numbers of unattached hemocyte types were counted in a hemocytemeter at a 10 min interval after the onset of incubation. The granular cells could be easily distinguished from the plasmatocytes on the basis of cell morphology.

Since the presence of PO in the incubation medium was demonstrated to promote the adhesion of the granular cells and plasmatocytes, 1 ml of the blood was collected into 2 ml of 0.1 % PTU in IPS and 1 ml of it was placed on a 35 mm-dish to allow both the granular cells and plasmatocytes to attach for 60 min at 25°C. Then, the unattached hemocytes were removed and rinsed twice with IPS. Adhered cells were made up of 15.3 % plasmatocytes and 84.7 % granular cells. Next, 1 ml of 0.01 M L-dopa was added to the attached cells, and the absorbance at 490 nm in the supernatant was measured to know the enzyme activity. On the other hand, the granular cells failed to attach in the presence of pNPGB during a 60 min incubation after the onset of incubation in contrast to the plasmatocytes. So, by employing this characteristic difference, the granular cells proved to be separable from the plasmatocytes by violent pipetting procedures within 60 min incubation. Detached granular cells as well as the other hemocytes were recovered and spun at 40 g for 10 min. The supernatant was discarded and 1 ml of IPS without pNPGB was added to it. The incubation was continued for 60 min to allow only the granular cells to attach. Then, unattached hemocytes were washed out to obtain a monolayer of only the granular cells. One ml of 0.01 M L-dopa was provided with plasmatocytes or with the granular cells, thus the PO activity being measured to demonstrate the PO production from the attached hemocytes.

The frequency of melanized hemocytes and speed of melanization were examined after the addition of L-dopa under the light microscope by the counting of the number of melanin-formed cells with the incubation time.

The monolayers of the granular cells or plasmatocytes were prepared as above, and 1 ml of 2.5 mM EGTA or EDTA, 0.5 mM pNPGB and 50 µg/ml zymosan particles, each of which was dissolved or suspended in IPS containing 0.01 M L-dopa, was added to the hemocyte monolayers to know that (1) calcium ion would be required for the hemocyte melanization and (2) PO production would be affected by zymosan or pNPGB as observed in plasma. After the addition of test substances, the absorbance at 490 nm was measured in the same way with the incubation time.

Next, to examine the effects of zymosan or pNPGB on the *in vitro* phagocytosis by the granular cells, the granular cells and plasmatocytes were allowed to attach to the dish as mentioned above. Then, 1 ml of 0.25 % SRBC suspension in IPS with or without 50 µg/ml zymosan or 0.5 mM pNPGB was replaced by the supernatant and further incubation was done for 60 min at 25°C, followed by the examination of attachment of SRBC to the granular cells or its ingestion.

Finally, the experiment was performed to know the active components in zymosan to induce the PO production resulting in melanin formation by the granular cells. As mentioned above, the attached granular cells were challenged with 50 µg/ml β-1,3-glucan, 50 µg/ml mannan or 50 µg/ml β-1,3-glucan plus 50 µg/ml mannan in IPS with 0.01 M L-dopa, respectively. β-1,3-glucan was stocked as a 1 mg/ml suspension in IPS after 2hr boiling and ultrasonic treatments and a concentration of 50 µg/ml was prepared with 0.01 M L-dopa. After the addition of stimulus, PO activity was measured.

Results and Discussion

To investigate the role of PO in the *in vivo* phagocytosis, zymosan or pNPGB was preinjected to produce or to inhibit PO activity in plasma, and then SRBC was injected into the larval hemocoel. From this experiment, it was shown that though zymosan-preinjection enhanced PO activity in plasma and erythrophagocytosis by the granular cells, pNPGB-preinjection suppressed them. This strongly suggested that PO production played a certain role in the phagocytic processes. Subsequently, the granular cells and plasmatocytes were incubated with PTU or pNPGB in vitro to know the effect of PO presence on the attachment phase. The result indicated that though plasmatocytes were able to attach and to spread extensively in the presence of pNPGB, the granular cells failed to attach during a 60 min incubation. PTU didn't affect the attachment phase of either hemocyte. The plasmatocytes incubated with pNPGB, however, were rounded up after 4 hr of incubation and began to detach from the dish. Thus, PO was thought to be an important enzyme for the attachment phase by the granular cells and for spreading behaviour of plasmatocytes. In addition, it was observed that the granular cells became attached to the dish surface with the incubation time even in the presence of pNPGB. Namely, attachment was increased after 60 min of incubation, and over 90 % of granular cells could adhere 5 hr after the onset of incubation. As the granular cells could not adhere during a 60 min incubation in the presence of pNPGB, it was possible to separate plasmatocytes from the granular cells by the repeated pipetting procedures. Then, the replacement of the incubation medium with pNPGB-free IPS after centrifugation of detached hemocytes produced the adhesive granular cells alone. By incubating the hemocytes in the petri dish, the granular cells were separated from other hemocytes except for the plasmatocytes. Hence, the monolayers of plasmatocytes or the granular cells were utilized.

PO activity has also been found to be located in the hemocytes of Locusta migratoria and Galleria mellonella (8,9). Furthermore, the granular cells are reported to react by violently discharging their cytoplasmic contents over nonself materials in contact with foreign surfaces (10). On the other hand, the plasma of B.mori contained the proPO-AS and this enzyme activation is initiated by elicitors such as glucans with β-1,3-glycosidic linkages (zymosan) (7), and melanin-formation has been observed in encapsulation or nodule formation. Accordingly, it was examined whether the adhered granular cells of the silkworm would also discharge PO or proPO after adhesion to the foreign materials to the exterior of the cells and then whether zymosan and its main subcomponent such as β-1,3-glucan or mannan would augment the immediate-onset secretion from the cells. Moreover, the activation of proPO by β-1,3-glucan is reported to significantly increase the percentage of phagocytosis (11). So, the effect of zymosan on in vitro phagocytosis by once-attached granular cells was also examined.

It was clearly shown that the separated granular cells after attachment could release or discharge PO but not proPO from the cells, resulting in the formation of melanin pigment, because PO activity could be measured even in the presence of pNPGB after the onset of incubation with L-dopa. This means that PO was not formed at the outside of the granular cells by the splitting of proPO by a serine protease. Interestingly, the aggregated granular cells showed much stronger melanization than a single cell, and the neuron-like processes were observed amongst the aggregated cells, which were just like in vivo early encapsulation reaction (Fig.1). The addition of EGTA or EDTA to the incubation medium completely inhibited PO release from the attached granular cells, indicating the calcium requirement for PO discharge. Conversely, it was noted that there was an initial rapid secretion of PO following zymosan exposure, which was then followed by a prolonged stimulation of secretion (Fig.2). Moreover, zymosan augmented erythrophagocytosis by the granular cells. On the other hand, PO activity from the plasmatocytes was hardly detected, although it was found that only about 2 % of the plasmatocytes, mainly large size, showed melanization 20 hr after the onset of incubation. The function of this type is unknown. There is a possibility that plasmatocytes may consist of a certain type of subpopulation in view of the degree of melanization.

Additionally, β-1,3-glucan was found to induce the secretion of PO from the granular cells, while the mannan, the second major subcomponent of zymosan, had no effect on PO secretion. Therefore, the active component in zymosan in inducing PO secretion from once-attached granular cells was considered to be β-1,3-glucan. This aspect was similar to proPO-AS provoked in plasma.

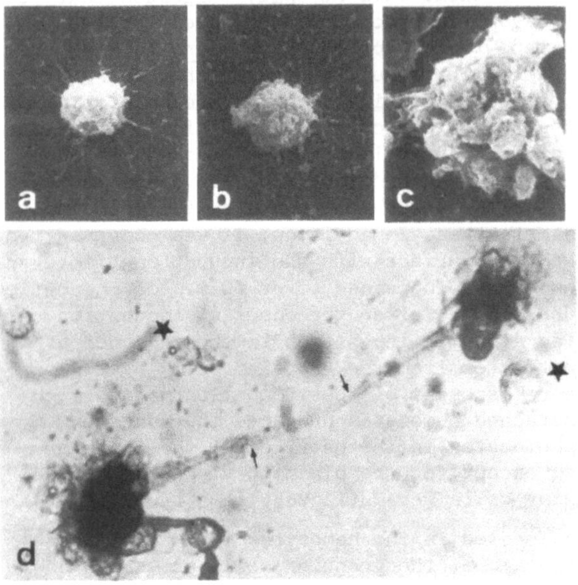

Fig.1 SEM of an attached granular cell of B.mori (a). In the presence of L-dopa, melanin deposit was formed over the cell (b), and the degree of melanization by the aggregated cells was not only more notable than a single cell (c,d:stars), but also the neuron-like processes were seen between the aggregated cells (d:arrows)

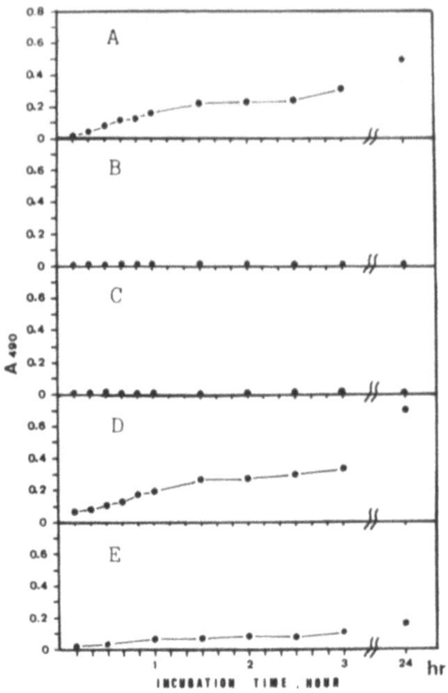

Fig.2 The changes in absorbance at 490nm after the addition of EDTA (B), EGTA (C),zymosan (D) or pNPGB (E) to once-attached granular cells.

These results obtained in the present study apparently suggest that (1) attachment phase of the granular cells to foreign materials would be promoted by the produced PO in plasma, thus PO serves as opsonin, (2) even once-attached granular cells secrete PO enzyme so that melanin pigment can be formed more strongly around foreign surfaces to depress any biotic invaders and (3) plasmatocytes probably have an important role in elimination of foreign substances without proPO-AS in their structure in place of the cellular defense reactions by the granular cells. In this manner, PO production triggered by invasion of foreignness and PO secretion from the reacted granular cells seem to cooperate in host defense reactions in insects, and these hemocyte types may play different roles in the initial defense reaction .

References

1.Wago,H. Dev.Comp.Immunol.,6:591-599,1982.
2 Wago,H. Dev.Comp.Immunol.,7:199-208,1983.
3.Ratcliffe,N.A.,Leonard,C.and Rowley,A.F. Science,226:557-559,1984.
4.Leonard,C.,Ratcliffe,N.A.and Rowley,A.F. J.Insect Physiol.,31:789-799,1985.
5.Söderhäll,K.,Vey,A.and Ramstedt,M. Dev.Comp.Immunol.,8:23-29,1984.
6.Ashida,M.,Ishizaki,Y.and Iwahana,H. Biochem.Biophys.Res.Commun.,113:562-568,1983.
7.Ashida,M.and Dohke,K. Insect Biochem.,10:37-47,1980.
8.Hoffmann,J.A.,Porte,A.and Joly,P. C.R.Acad.Sci.D.Paris, 270:629-631,1970.
9.Schmit,A.R.,Rowley,A.F.and Ratcliffe,N.A. J.Invertebr.Pathol.,29:232-234,1977.
10.Ratcliffe,N.A.and Rowley,A.F. Insect Hemocytes (Gupta,A.P.,ed.),pp.331-414, Cambridge University Press,1979.
11.Smith,V.J.and Söderhäll,K. Biol.Bull.,164:299-314,1983.

Part 6

FISH CELL CULTURES

FISH CELL CULTURES: An OVERVIEW

B. L. Nicholson

Department of Microbiology and Center for
Marine Studies, University of Maine,
Orono, Maine 04469, USA

Many fish cell cultures from a variety of species have been developed since the first continuous fish cell line was originated over 25 years ago. Most of these cell lines have been established for the isolation and study of viruses causing diseases in artificially propagated species. In recent years, awareness has increased about the importance of fish cell cultures as experimental and educational in vitro models for basic and comparative studies of cellular physiology, cytogenetics, toxicology, carcinogenesis, host-pathogen interactions, etc..

Many continuous fish cell cultures from a variety of species have been developed since the first fish cell line was originated from rainbow trout over 25 years ago. A comprehensive list of most fish cell lines developed before 1980 has been published (27) and several comprehensive reviews of maintenance procedures and applications of fish cell cultures are available (19,24,26-29). The purpose of this report is to provide a concise overview of the procedures for the preparation and maintenance of fish cell cultures as well as some current and potential applications of such cell cultures.

In general, the procedures for the preparation and maintenance of fish cell cultures do not differ from those applied to higher vertebrates. The first consideration in the preparation of a primary cell culture is the choice of tissue to be used. If a particular cell type or function is to be studied, the appropriate tissue must be used. However, for general purposes a variety of tissues provide good sources of cells. Gonads and embryonic tissue usually yield the greatest number of actively dividing cells; however, kidney, heart, liver, fin, muscle, spleen, and other tissues can also be used. Most fish cell cultures have been derived from normal tissues; however, several continuous cell lines have been initiated from neoplastic tissues.

In some ways, the preparations of fish tissues for culture are easier than with mammalian tissues. Unlike cells which must be maintained at or near 37°C, fish tissues can be taken at field sites, placed in growth medium, and transported at ambient temperature to the laboratory for preparation. Cell suspensions are usually prepared by enzymatic dissociation; however, in cases where this has failed initial success has been achieved with explant culture.

The growth media routinely employed in the culture of fish cells are generally the same used for homeotherm cell lines. The most widely used media for fish cells are Eagle's Minimum Essential Medium

(MEM) and Leibowitz Medium L-15 supplemented with 5-10% fetal bovine serum (FBS). In some cases with certain marine fish cell lines, it may be necessary to increase the NaCl concentration of standard media (6); however, this is not true for most marine fish cell lines. Although FBS is usually used as a supplement to the basal medium, many fish cell lines can be adapted to media supplemented with calf serum (17) and an undefined serum-free medium for the growth of 5 fish cell lines has been reported (23).

Fish cell cultures generally proliferate over a wide range of incubation temperatures (17). The optimal growth temperature and the temperature range over which a particular culture will grow usually reflect the fish species and its natural environment. Temperatures of 15-20°C are usually optimal for cells from "coldwater" species with growth ranging from 2 to 27°C. Cells from intermediate or "coolwater" species have a somewhat higher limit and an optimum between 20 and 28°C. Most "warmwater" fish cell cultures have an optimum growth temperature between 25-37°C.

Virtually all fish cell lines are anchorage dependent, growing as monolayer cultures. In addition to standard culture vessels a modified procedure has been developed for growing several fish cell lines on microcarrier beads (16). The efficiency of microcarriers in growth vessel and medium requirements provides significant advantages for the large-scale production of fish cell cultures, viruses and cellular products. In a few instances, fish cell lines have been adapted to growth in suspension although the ability of such cultures to support the replication of some viruses is either abolished or diminished (14).

In general, the routine maintenance of fish cell cultures is easier than mammalian lines. In contrast to most mammalian cultures, it is not necessary to change the growth medium between subcultures. This characteristic also provides some advantages in the storage of cultures. Although fish cell lines can be preserved by standard methods for freezing cells, most lines can be easily stored when such equipment is not available by keeping without media change at temperatures well below the optimum so that cells remain viable but little or no growth occurs. For example, salmonid cell lines can be kept at 4-6°C in a refrigerator for periods of 2-6 months.

Unlike the advances in vertebrate fish cell culture, there has been virtually no progress in developing cell cultures from marine or aquatic invertebrate shellfish. Indeed, a freshwater snail line is the only such cell line in existence (9). Despite numerous attempts utilizing a wide variety of growth media, supplements, and cultural conditions, no other shellfish cell cultures have been developed. Modified media and procedures for the maintenance of viable oyster hemacytes for periods of up to 6 months have been developed; however, these cells do not divide (3,25). Presumably, unique growth factors and/or cultural conditions are required by shellfish cells that have not been identified nor can be supplied by homologous hemolymph or tissue extracts. Increased efforts and new approaches are needed to develop much needed shellfish cell cultures comparable to those from vertebrate fish species.

Fish cell cultures have both fundamental and practical importance. The most widely employed application of fish cell cultures and, until recently, the rationale for developing most such cultures is the isolation of fish viruses that are agents of epizootics of commercially important aquaculture species. The first fish cell line was developed to facilitate the isolation of infectious pancreatic necrosis virus (IPNV) and other viruses of trout. Because of the importance of trout and other salmonid fish culture in North America, Europe, and Japan, this led to the development of other salmonid cell lines for similar purposes. Consequently, most early cell lines were derived from "coldwater" species. Only a few such lines were developed from "warmwater" species such as catfish which are farmed in the southern parts of the United States. Very few cell lines from marine fish species were developed, primarily because marine fish culture (other than salmonids) is not extensively practiced in these geographical areas. However, aquaculture of warmwater species, with the associated disease problems, is extensive in Asia and other parts of the world. Consequently, there has been a rapid increase in the number of cell lines developed from such species, including carp, loach, tilapia, perch, grouper, eels and others (4,5,18). These new cell lines are being used for the isolation and characterization of previously unknown viruses.

In addition to serving solely as a means of virus isolation, fish cell cultures are very useful as in vitro models for studying the replication and genetics of these viruses, the establishment and maintenance of virus carrier states, the effects of antiviral drugs, and the production of experimental vaccines. Also, in vitro culture techniques have been used to investigate unique viruses that do not replicate in standard fish cell lines but require highly differentiated cells. One example is viral erythrocytic necrosis (VEN), a viral infection of the red blood cells of several species of marine and anadromous fish. Although fish red blood cells will not divide, viable in vitro red blood cell cultures have been maintained for several weeks and used in investigations of the replication of these viruses and as in vitro models of viral induced anemia (20).

Fish cell cultures are also being developed as in vitro models for studying various biological processes. They have been used for determining karyotypes and other aspects of cytogenetics including chromosomal polymorphism and speciation, chromosome abnormalities, and evolution (7,19,22). They are also finding widespread use in studying cellular physiology. Examples include: organ cultures from tilapia, eel and trout pituitary glands for studying the production of the growth hormone prolactin (1,2,8); carp pituitary organ cultures to determine the efficacy of hypothalamus extracts on the production of gonadotropin and a continuous carp pituitary cell line producing gonadotropin (2); blastoderm explants from zebra fish for comparative studies of pigmented retinal epithelial cell morphology, function and differentiation (13); melanoma cell cultures for the study of the processes leading to proliferation and differentiation of melanoma cells (12); and explant cultures of fish kidney tubules for the investigation of active transport and correlation between structure and function at the ultrastructural level (10). Fish cell cultures are finding application in toxicology for evaluating effects of various chemicals, pesticides and industrial wastes and in the study of carcinogenesis as in vitro models for investigating cell transformation by fish viruses, chemical agents, and the interaction of viruses and chemical carcinogens. Also, fish cell cultures are now being used for studying the immune response in fish, both from the practical view of vaccine development and for the study of the comparative evolutionary aspects of the immune response. Examples include in vitro stimulation of antibody production for basic studies and testing the potency of vaccines, in vitro models of immunoglobulin synthesis, and the role of natural killer (NK) cell activities in disease resistance in fish (11,15).

And finally, because of their ability to grow over a wide

temperature range including most ambient or room temperatures, fish cell cultures should be very useful for high school and college classroom instruction.

References

1. Baker, B.I. and Ingleton, P.M. J. Comp. Physiol.,100 : 269-282, 1975.
2. Benjamin, M. and Baker, B.I. Cell Tissue Res.,174:547-564, 1976.
3. Brewster, F. and Nicholson, B.L. J. Fish Res. Bd. Can.,36: 461-467, 1979.
4. Chen, S.N., Chen, J.H. and Kow, G.H. Abs. Proc. ROC-Japan Sym. Fish Dis. Tungkang, Republic of China, p. 36, 1982.
5. Chen, S.N., Ureno, Y. and Kou, G.H. Proc. Natl. Sci. Counc., Repub. China,6: 993-100, 1982.
6. Clem, L.W. Sigel, M.M. and Friis, R.R. Ann. N.Y. Acad. Sci., 126: 343-361, 1965.
7. Etlinger, H.M. Dev. Comp. Immunol.,2:263-277, 1978.
8. Grau, E., Nishioka, R.S.,and Bern, H.A. Am. Zool.,19: 875, 1979.
9. Hansen, E.L. Tissue Cult. Assoc. Man.,5:1009. 1979.
10. Jones, R.T., Klaunig, J.E., Sanefuji, H. and Hinton, D.E. J. Tissue Cult. Met.,7: 151-154, 1982.
11. Kaattari, S. and Irwin, M.J. Dev. Comp. Immunol.,9:433-444, 1985.
12. Kuhn, C., Vielkind, U. and E. Anders, F. In Vitro, 15:537-543, 1979.
13. Laale, H.W. In Vitro 17: 701-705, 1981.
14. Lidgerding, B.C. in Developments in Biological Standards, Fish Biologics: Serodiagnosis and Vaccines (D.P. Anderson and W. Hennessen, eds.). Vol. 49, pp. 233-241, S. Karger. 1981.
15. Moody, C.E., Serreze, D.V., and Reno, P.W. Dev. Comp. Immunol., 9:51-64, 1985.
16. Nicholson, B.L. Appl. Environ. Microb.,39: 394-397, 1980.
17. Nicholson, B.L. in Techniques in the Life Sciences (E. Kurstak, ed.), Cl, 1-16, Elsevier Ltd. 1985.
18. Nicholson, B.L., Danner, D.J.,and Wu, J.L. In Vitro Cell. and Dev. Biol.,23: 199-204, 1987.
19. Regen, J.D., Sigel, M.M., Leg, W.H., Llamas, K.A. and Beasley, A.R. Can. J. Genet. Cytol.,10: 448-453, 1968.
20. Reno, P.W. and Nicholson, B.L. Can. J. Fish. Aq. Sci.,37: 2276-2281, 1980.
21. Ribeiro, L. and Ahne, W. In Vitro 18: 419-420, 1982.
22. Roberts, F.L. Trans. Am. Fish Soc., 99: 105-111, 1970.
23. Shea, T.B. and Berry, E.S. In Vitro,11: 818-824, 1983.
24. Sigel, M.M. and Beasley, A.R. Tissue Culture, Methods and Applications (P.F. Kruse, Jr. and M.K. Patterson, Jr., eds.), pp. 133-135, Academic Press, New York. 1973.
25. Stephens, E.B. and Hetrick, F.M. Tissue Cult. Assoc. Man.,5: 991, 1979.
26. Wolf, K. in Tissue Culture,Methods and Applications (P.F. Kruse, Jr. and M.K. Patterson, Jr., Eds.), pp. 138-143, Academic Press, New York. 1973.
27. Wolf, K. and Ahne, W. in Advances in Cell Culture (K. Maramorosch, ed.), Vol. 2, pp. 305-328, Academic Press, New York. 1982.
28. Wolf, K. and Mann. J.A. In Vitro,16: 168-179, 1980.
29. Wolf, K. and Quimby, M.C. Tissue Cult. Assoc. Man.,2: 453-456, 1976.

CELL CULTURES FROM HEMATOPOIETIC TISSUE AND OVARY OF PENAEID SHRIMP, PENAEUS MONODON

S. N. Chen, K. J. Jong, and G. H. Kou.

Department of Zoology, National Taiwan University, Taipei, Taiwan

Primary cell cultures have been established from hematopoietic tissue and ovary of grass prawn, *Penaeus monodon*, using 2X L-15 plus fetal calf serum, muscle extract of grass prawn and lobster hemolymph with an osmolarity of approximate 720 mmol/kg at 28±1°C. The optimal temperatures for the cell culture of hematopoietic tissue ranged from 21 to 32°C. Ovary at undeveloping or spent stage is superior to the other maturation stages for the cultivation of cells *in vitro*.

Although many investigations have been performed on the viruses of penaeid shrimp including *Penaeus duorarum, P. setiferus, P. aztecus, P. stylirostris, P. japonicus* and *P. monodon*, these studies are limited in histopathological observations using light and electron microscopes (Lightner, 1983). To date, little is known about the biochemical or biological properties of the viruses. Research on these problems could be faciliated if culture cells susceptible to shrimp viruses were available.

Previously, we have successfully established monolayer cell sheets for either hematopoietic tissue or ovary of *P. monodon* and *P. penicillatus* using culture medium consisting of Leibovitz's L-15 medium supplemented with fetal calf serum, shrimp muscle extract and lobster hemolymph (Chen et al., 1986; 1987). In this paper conditions for the cultivation of hematopoietic tissue and ovary of *P. monodon* were determined. In addition, ovaries of *P. monodon* at different developmental stages were tested for their capability to form a confluent cell sheet *in vitro*.

Materials and Methods

Adult female grass shrimp, *P. monodon*, weighing approximately 30-100 gm each was used for the present study. Prior to the experiment, the shrimps were sterilized as described previously (Chen et al., 1987). Subsequently, paired lobes of the hematopoietic tissue or Nakamura organ(Nakamura and Yonekura, 1980) located on dorso-lateral surface of cardiac stomach, weighing approximately 0.025-0.07 gm(in wet weight) each, were removed using sterile forceps. The tissue was washed in 2X Leibovitz's L-15 (L-15) containing 1,000 µg/ml streptomycin, 1000 units/ml penicillin and 100 µg/ml fungizone three times, then minced with sterile scissors. Subsequently, approximately 0.07 gm of the minced tissue was cultured in each culture medium as described in Table I,respectively.

The osmolarity of each culture medium was adjusted to 720±10 mmol/kg using a sterile salt solution containing 102.4 gm/l of NaCl, 1.8 gm/l of KCl, 5.1 gm/l of $CaCl_2$, 10.8 gm/l of $MgSO_4$ and 11.8 gm/l of $MgCl_2$. Lobster hemolyph (LHL) was prepared from spiny lobster, Panulirus homarus of approximately 300 gm as described previously (Chen et al., 1987). Leibovitz's L-15 medium, fetal calf serum (FCS) and antiseptic compounds were purchased from GIBCO Laboratories, USA. The cell cultures were observed daily under an Olympus IM inverted microscope.

To detect the effect of temperature on the adherence and cell outgrowth of hematopoietic tissue fragment, shrimp tissue culture medium (STCM) with an osmolarity of 720±10 mmol/kg was used. Following inoculation of approximately 0.07 mg tissue fragments, the flasks were then incubated at a temperature of 17±1°C, 22±1°C, 25±1°C, 28±1°C, 31±1°C, or 37±1°C, respectively. The STCM is 55% 2X L-15 plus 25% muscle extract of grass shrimp, 15% of LHL and 5% FCS. Using the similar medium, ovaries of grass shrimp at undeveloped, developing, ripe stages were cultured and the ability to forming a monolayer sheet for each tissue was observed under an Olympus IM inverted microscope. The maturation stage of ovary was categorized by ovum size and gonad color or expansion as described by Motoh (1981). After sterilization and cutting of tissue as described above, approximately 1-2 gm of minced ovary was cultured in STCM under an incubation temperature of 28±1°C. The flasks were also observed daily under inverted microscope.

Results and Discussion

When the hematopoietic tissue fragments were inoculated into a culture medium consisting of 2X L-15, 15% LHL, 5% FCS and 25% ME with an osmolarity of 720±10 mmol/kg approximately 80-90% of tissue fragments adhered on the flask surface within 4hrs after incubation at 28±1 °C. Twenty four hrs after incubation at the similar temperature, a confluent cell sheet was formed (Fig.1 and Table I). The similar result was also obtained in a culture system using 55% 2X L-15 plus 20% FCS and 25% ME. However, when 80% 2X L-15 and 20% FCS with an osmolarity of 720±10 mmol/kg was used, approximately 50-60% tissue fragment adhered on the flask surface. It is noted that a confluent cell sheet could be formed in this culture medium without muscle extract of shrimp and lobster hemolymph.

Table I. Primary cell culture of hematopoietic tissue of Penaeus monodon at 28±1°C

Culture medium	Adherence of the tissue fragment	Formation of confluent cell sheet
55% 2X L-15 + 15% LHL + 5% FCS + 25% ME	+++	Yes
55% 2X L-15 + 20% FCS + 25% ME	+++	Yes
80% 2X L-15 + 20% FCS	++	Yes

* Into each flask approximately 0.07 gm of tissue fragment was inoculated.
1 Osmolarity was adjusted to 720±10 mmol/kg using a salt solution as mentioned in the section of Materials and Methods.
2 The results were obtained by observations under inverted microscope
2X L-15 : Leibovitz's L-15 medium in double strength.
LHL : Lobster hemolymph FCS : fetal calf serum
ME : Muscle extract of grass shrimp
++ : Approximately 50-60% of tissue fragments adhered on the flask surface
+++: Approximately 80-90% of tissue fragments adhered on the flask surface
Each experiment was performed at least in duplicate.

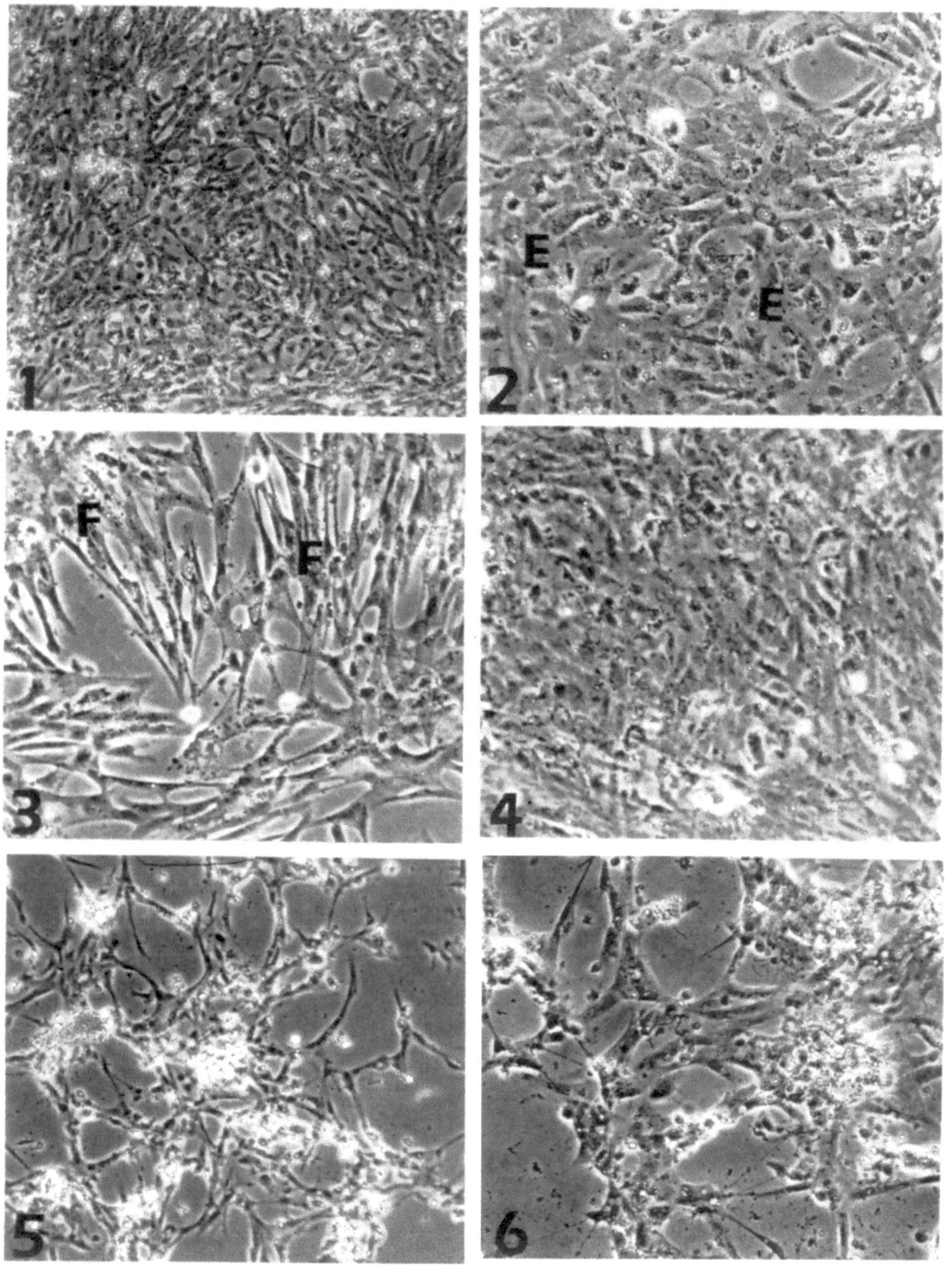

Figs. 1-6 Primary cell culture of hematopoietic tissue of grass prawn, Penaeus monodon.

Fig. 1 2 days after incubation at $28 \pm 1^{o}C$ (400X)
Note the formation of monolayer cell sheet

Fig. 2 Note the population of epithelioid cells (E) (800X)

Fig. 3 Note the population of fibroblast-like cells (F) (800X)

Fig. 4 3 days after incubation at $28 \pm 1^{o}C$.
Note the cell sheet was too dense to determine the cell morphology. (800X)

Figs.5-6 Primary cell culture of ovary of grass prawn, Penaeus monodon.
Fig. 5 (400X), Fig. 6 (800X).

The primary culture of grass shrimp hematopoietic tissue consists of fibroblastic-like(Fig. 2) and epithelioid (Fig. 3) cells. However, 3 days after incubation at 28±1°C, the cell masses become too dense to determine the cell morphology (Fig. 4). This observation is identical to those obtained from the cultivation *in vitro* of hematopoietic tissue of red-tail shrimp, *P. penicillatus* (Chen *et al.*, 1987). Experiments performed on the cultivation of hematopoietic tissue in STCM with various temperatures showed that monolayer cell sheet was formed in the flasks incubated at 22±1°C,25±1°C,28±1°C and 31±1°C. In contrast, limited tissue fragments adhered on the flask surface under an incubation temperature of 37±1°C or 17±1°C for 24 hrs . This result may suggest that the optimal temperature for cultivation of hematopoietic cells ranged from 21 to 32°C.

The result obtained from the cultivation *in vitro* of ovary fragment in STCM showed that the best adherence or cell outgrowth occurred when undeveloped or spent stages were used and a confluent cell sheet (Figs. 5 and 6) was formed 48-72 hrs after incubation at 28±1°C. No confluent cell sheet was formed when ovary at developing or ripe stage was used.

In this series of experiments we determined the conditions for the cultivation *in vitro* of primary cell cultures of hematopoietic tissue and ovary. These two confluent cell cultures could be established by using a simple culture medium including L-15 supplemented with lobster hemolymph, muscle extract of grass shrimp or with fetal calf serum alone. The successful establishment of these two monolayer cultures may provide a useful tool for the investigation of virology and cellular biology of penaeid shrimp. Studies are in progress on the isolation and identification of shrimp viruses using these cell cultures and the results will be published elsewhere.

References

1. Lightner, D.V., CRC Handbook of Mariculture Vol. 1. Crustacean Aquaculture (McVey, J.P. ed.), PP. 289-320, CRC Press, Inc., 1983.
2. Chen, S. N., K. J. Jong and G. H. Kou. Invertebrate Tissue Culture(J. Mitsuhashi,ed.) , CRC Press, Inc., 1987 (In press).
3. Chen, S. N., Chi, S.C., Kou, G.H. and Liao, I.C. Fish Patho., 21, 161-166. 1986.
4. Nakamura, K. and Yonekura, K. Mem. Fac. Fish., Kagoshima Univ., 29, 259-266, 1980.
5. Motoh, H., Studies on the Fisheries Biology of the Giant Tiger Prawn,*Penaeus monodon* in the Philippines. PP. 16-18, Aquaculture Department, Southeast Asian Fisheries Development Center, Philippines, 1981.

ESTABLISHMENT AND PROPERTIES OF A CELL LINE DERIVED FROM SALMONID FISH LIVER

T. Watanabe and T. Moritomo

Laboratory of Fish Pathology, Department of Veterinary Medicine,
Nihon University, Fujisawa, Kanagawa 252, Japan

The primary culture of embryonal liver cells was performed using the embryonated eggs of Yamame *Oncorhynchus masou*. The cells, in 5 flasks of 22 attempts cultivated in TC199 with 10% FBS at 20°C, consist of epithelial cells with a round nucleus and abundant granular intracytoplasmic organelles.One of them, designated YEL-13, grew in the manner of contact-inhibited monolayer until the formation of confluent cell sheets. The cells have been subcultured 90 times. YEL-13 cells were found to produce and release a protein which had the antigenicity similar to one of Yamame serum protein by fluorescent antibody method and immunoelectrophoresis.

Various fish cell lines have been established for the purposes of detecting and isolating fish viruses (1). However, there are few studies on the development of the fish-originated tissue cultures possessing the differentiated functions of derived tissue cells. In mammals, many studies on functional cell cultures have been performed for investigating cell genetics, cell differentiation, malignant transformation, or cell metabolisms (2,3). Recently, the importance of the functional cell cultures of fish in the study of fish science has been recognized. Hayashi and Ooshiro investigated gluconeogenesis by using the culture cells primarily derived from a Japanese eel *Anguilla japonica* liver and found that cultured eel liver cells had higher activities of glycogen and protein synthesis than those of freshly isolated liver cells (4).

We tried to establish a cell line possessing a differentiated liver function of embryonated eggs of Yamame, an indigenous and landlocked form of masu salmon *Oncorhynchus masou*. The salmonid fishes are the most important family for the North Pacific coastal countries in fish industry and fish science. Many scientific data on fish biology and biochemistry have been obtained from these fishes. The size of embryonated eggs plays an important and practical role in starting the primary culture because of the difficulty to separate liver tissue from a small egg under a binocular microscope. We chose the species Yamame for this study owing to three reasons; firstly, the culture of Yamame has increased in commercial fish farms in the last 15 years, and the embryonated eggs of this fish are easy to obtain; secondly, the egg size of Yamame is 6mm in diameter. The size is larger than those of *Salmo* and *Salvelinus* species and slightly smaller than that of chum salmon *O. keta*, being the largest egg size in salmonid fishes caught along Japanese coast and rivers; thirdly, the most important reason of three, we had already established a cell line derived from the kidney of Yamame, YNK (5), and obtained information on the tissue culture of this species.

In this study, some transferable cells which showed an epithelial cell monolayer morphology were obtained. This report describes the early cultivation of Yamame embryonated liver cells and some properties of the established cell line.

Materials and Methods

Primary Culture: The embryonated eggs of Yamame were obtained from Yamaguchi Fish Farm, Yamanashi. Upon confirmation that a few eggs had hatched (Fig. 1), eggs were dipped in a solution containing 100 I.U./ml of penicillin G potassium for 24 hours. After the surface of eggs was disinfected with 70% ethanol, egg membrane and yolk were removed under a binocular microscope. Liver tissues were transferred in a sterilized petri dish and minced into small fragments with scissors in a small amount of TC199 medium supplemented with 10% FBS, 200 I.U./ml of penicillin, 200μg/ml of streptomycin sulfate, and 30 I.U./ml of mycostatin. The cells and fragments were suspended in TC199 medium with 10% FBS (199-10) at a density of 100,000 cells per ml, transferred into plastic tissue culture flasks, and then cultivated at 20°C. One half of the medium in culturing was replaced by fresh one twice a week throughout the period of primary culture. Each flask contained 3-5 eggs for the primary culture and was designated YEL-.

Cell growth, electron microscopy, and method of virus infectivity assay were described previously (6).

Immunological methods: Rabbit antiserum against Yamame serum was prepared by intravenous injections 4 times at an interval of 7 days and collected at the 7th day after the last injection. Antiserum was heated at 63°C for 20min before use. For indirect fluorescent antibody method, YEL13 or YNK were grown on glass coverslips. The cells were fixed in cold acetone two days later, and stored at -20° C until use. In the staining procedures, diluted antiserum was applied to the cells for 30min at 37°C. After washing with phosphate buffered saline (PBS, pH7.4) the cells were overlaid for another 30min at 37°C with a solution of FITC-conjugated goat IgG against rabbit IgG. After washing with PBS, the cells were examined with a fluorescence microscope. Immunoelectrophoresis and immunodiffusion tests were performed using the methods described by Ouchterlony and Nilsson (7).

Results

Primary Culture: Flasks seeded with Yamame embryonated liver cells were observed daily with a phase contrast tissue culture microscope. Two days after the primary culture, epithelial and fibroblastic cells were observed migrating from tissue fragments in every flask. After 3-4 days, several colonies consisting of epithelial cells were observed in various parts of the flasks. These epithelial cells seen in 12 flasks multiplied to form colonies and terminated the growth in the manner of contact-inhibited monolayer in 27-38 days. These colonies were trypsinized and reseeded in new flasks. Cell growth was not observed in 8 flasks, and fibroblastic cells grew and were predominant in 2 other flasks (Table 1).

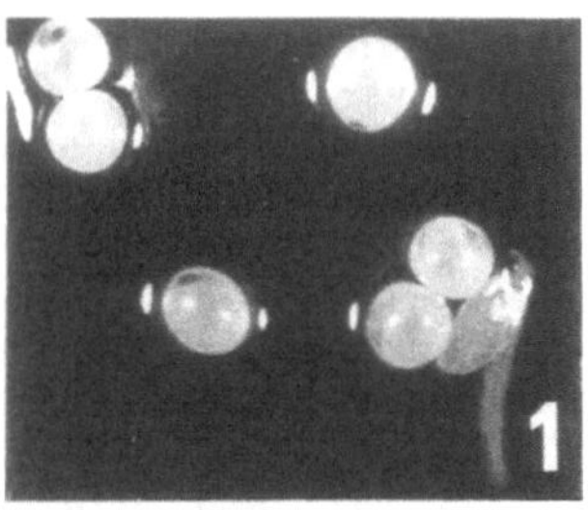

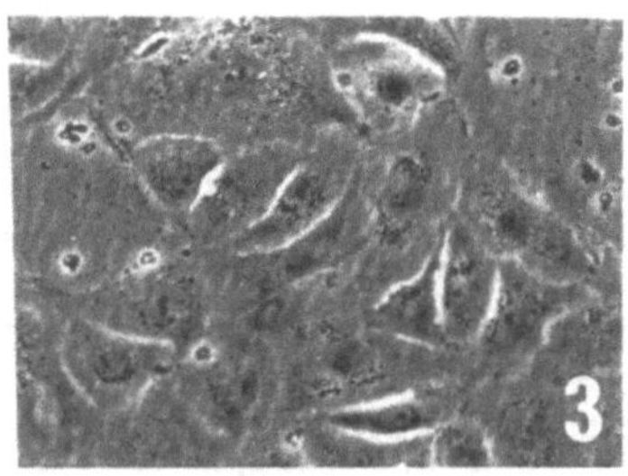

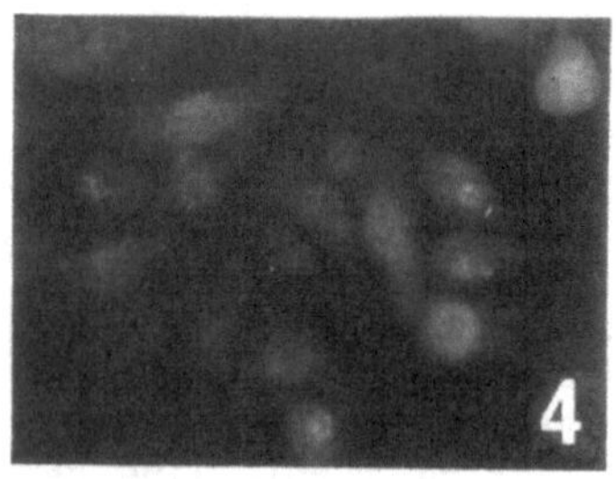

Fig. 1. Embryonated eggs of Yamame used in this work, slightly before hatching.

Fig. 3. Monolayer cell sheet of YEL-13 cells, consisting of large epithelial cells with a round nucleus and granular intracytoplasmic organelles around a nucleus, and a few fibroblastic spindle-shaped cells.

Fig. 4. Immunofluorescent photomicrograph of YEL-13 cells. The cytoplasms were stained intensely by immunofluorescence with antiserum to Yamame serum.

Passage of YEL-13: The secondarily passaged five cells, YEL-7, -8, -13, -15, and -17, showed an epithelial morphology and grew to form a cell sheet until the contact-inhibition restricted the growth. Fig. 3 shows the relationships between the doubling times and the numbers of passage of YEL-8, -12, and -13. From the fifth passage, YEL-13 shortened its doubling time gradually, and were able to be transferred at the intervals of 7-10 days after the tenth passage when seeded at the density of 100,000 cells per flask. YEL-13 has been subcultured for 90 times from its initiation in October, 1984 to June, 1987.

Table 1. Culture of Yamame embryonated liver.

Morphology	No. of flasks per 22 exps.	Flask No. of successive culture
Epithelial	12	YEL-7(14*),8(23), 13(90),15(16),17(20)
Fibroblastic	2	-
No growth	8	-

*Passage number.

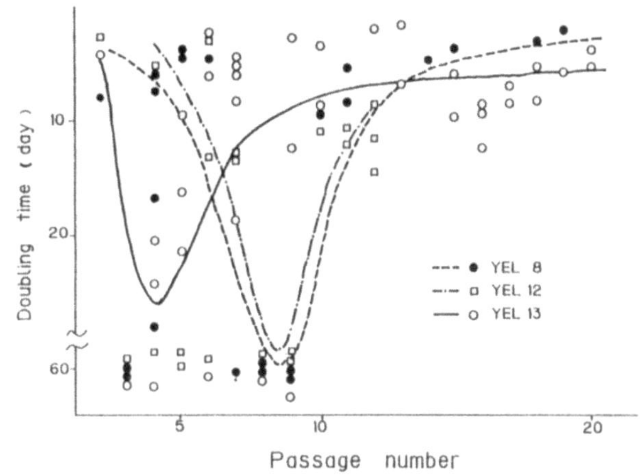

Fig. 2. Doubling times of YEL- cells from 2nd to 20th passage.

Morphology of YEL-13: YEL-13 consisted of epithelial cells with a round nucleus and abundant granular intracytoplasmic organelles suspected to be mitochondria (Fig. 3) around the nucleus. A few fibroblastic spindle-shaped cells with a small oval nucleus were observed. YEL-13 had no PAS positive granules even cultured in 199-10 medium supplemented with 0.1% glucose.

Electron microscopically, most of the YEL-13 were large and round with a round nucleus containing a nucleolus. The cells had a smooth surface with poorly developed microvilli and well-developed rough surfaced endoplasmic reticulum (rER) in cytoplasm. Some of these cells had expanded rER in which slightly electron-dense materials were observed. There were a few cells frequently seen with an elongated and markedly indented nucleus, microfibrils and numerous ribosomes. Some of them had extensive electron-densed granules in the center of the cytoplasm.

The infectivities of the three salmonid fish viruses, infectious hematopoietic necrosis virus (IHNV), infectious pancreatic necrosis virus (IPNV), and Oncorhynchus masou virus (OMV), were compared by using plaque assay method on two cells, YEL-13 and YNK. YEL-13 had susceptibilities similar to or slightly lower than YNK to IHNV and IPNV. YEL-13 had a slightly higher susceptibility than YNK to OMV. OMV did not form syncytium but necrotic plaques on the cells.

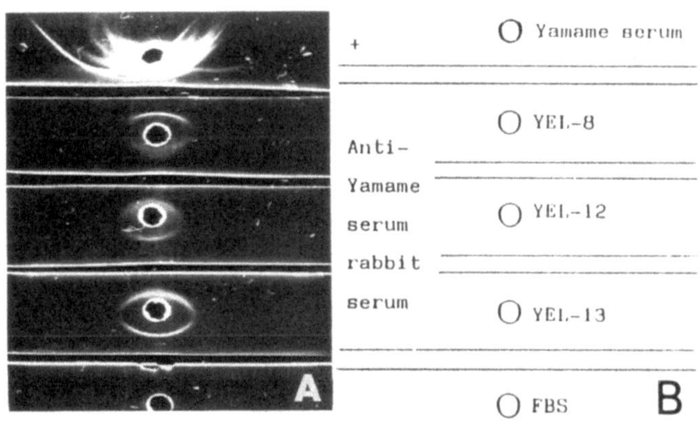

Fig. 5. Immunoelectrophoresis of the supernatants of YEL- cells (A) and diagram f(B).

Protein Synthesis by YEL-13: YEL-13 cells were examined under a fluorescent microscope, after reacting with rabbit serum against Yamame serum and with FITC-conjugated goat IgG against rabbit IgG. As shown in Fig. 4, the cytoplasm of YEL-13 was stained intensely by immunofluorescence with antiserum against Yamame serum. Immunoelectrophoresis study showed that only one precipitating line was formed between ten times concentrated culture supernatants of YEL-8, -12, -13 and antiserum against Yamame serum (Fig. 5). No precipitating line was observed between FBS (Fig. 5, bottom well) and antiserum. The double diffusion tests also showed that a protein, produced and released by YEL-8, 12, and 13, was one of the Yamame serum proteins.

Discussion

The liver cells transferred from Yamame embryo into tissue culture flasks multiplied with comparative rapidity and took 27-38 days for primary cultures. In 12 flasks from 22 experiments, epithelial cells had dominantly grown. Cells of YEL-7, -8, -13, -15, and -17 grew rapidly and were transferable. Of them, YEL-13 was subcultured more than 90 passages. Many tests by using mammalian hepatocytes (8,9) have been performed to culture liver cells with differentiated hepatic functions. There is no direct evidence that YEL- cells possess the properties of fish hepatocytes, but the morphology of the contact-inhibited epithelial monolayer cells with round nuclei resembles hepatic parenchymal cells. The findings of the electron microscopical study suggest that YEL-13 were derived from hepatocytes but apparently consist of two or more kinds of cells. The cloning of the cells has not yet been successful.

Immunological studies showed that YEL-13 produced and released a protein which reacted with antibodies against Yamame serum. The production of serum protein, especially albumin, is one of the properties of hepatocytes. We are examining the relationship of this protein with the component of serum proteins.

OMV, isolated from Yamame by Kimura (10), is an oncogenic herpesvirus of salmonid fish and is known to form syncytium when it infects salmonid fish cell lines, such as YNK. OMV did not fuse YEL-13 cells but formed necrotic plaques on the cell line. The results suggest that YEL-13 cells might offer a useful tool to investigate OMV replication and oncogenesis.

The authors are grateful to Dr. I. J. Pan of Nihon University for correcting of this manuscipt and Dr.T. Kimura of Hokkaido University for his kind supply of OMV. The authors also thank Mr. N. Hasegawa of Yamaguchi Trout Farm for his supply of embryonated eggs, and Miss Y. Nakayama for her excellent technical assistance. This work was supported in part by Grant-in-Aid for Scientific Research from the Ministry of Education.

References

1. Wolf, K. and Mann, J. A., In Vitro, 16:168-179, 1980.
2. Namba, M., Hyodo, F. and Kimoto, T. in Nutritional Requirements of Cultured Cells (KATSUTA, H., ed.), pp.277-291, Jap. Sci. Soc. Press, Tokyo,1978.
3. Morrow, J. in Eukaryotic Cell Genetics, pp.99-112, Academic Press,1983.
4. Hayashi, S. and Ooshiro, Z. Bull. Japan. Soc. Sci. Fish., 51:765-771, 1985.
5. Watanabe, T., Kobayashi, N., Sato, Y. and Ishizaki, Y. Bull. Japan. Soc. Sci. Fish., 44:415-418, 1976.
6. Watanabe, T., Nakano, M., Asakawa, H. and Moritomo, T. Bull. Nippon Suisan Gakkaishi, 53:537-542, 1987.
7. Ouchterlony, O. and Nilsson, L.-A. in Handbook of Experimental Immunology (Weir, D. M., ed.), pp. Vol. 1 32.1-50, Blackwell Sci. Pub.,1986.
8. Richman, R. A., Claus, T. H., Pilkis, S. J. and Friedman, D. L. Proc. Natl. Acad. Sci. USA, 73:3589-3593, 1976.
9. Guguen-Guillouzo, C. and Guillouzo, A. Mol. Cel. Biochem., 53/54:35-56, 1983.
10. Kimura, T., Yoshimizu, M. and Tanaka, M., Fish Health News, 9:2-3, 1980.

GROWTH AND MAINTENANCE OF ETHISTERONE-INDUCED ANAL-FIN PROCESSES OF A RICEFISH, ORYZIAS LATIPES, IN VITRO

H. Uwa and A. Iwata*

Department of Biology, Faculty of Science, Shinshu University, Matsumoto 390, Japan, and Department of Zoology, University of Washington, Seattle, Washington 98195, U.S.A*

The growth and maintenance of the horny processes of a ricefish, *Oryzias latipes*, were investigated *in vitro* with the organ-culture of the anal fin. Processes were induced in adult females with oral administration of ethisterone. The early process developed on the fifth day of the treatment. Anal fins were removed on this day and cultured with media containing androgen. Early processes elongated in medium 199 and CMRL-1066 during six to eight days of culture, whereas they degenerated to various extents in Eagle's MEM and Dulbecco's modified EM. Osteoclast-like multinuclear cells occasionally appeared during the culture, and started to take up the elongated processes and pieces of the joint plates into their cytoplasm phagocytotically.

In a ricefish, *Oryzias latipes*, the horny processes on the posterior rays of the anal fin are the most prominent secondary sexual character of the male. They are formed by the mesenchymal cell mass (scleroblast) which has appeared at the postero-external margin of the joint plate (1).

Anal-fin processes are also induced in adult females by the transplantation of testis (2) or administration of androgens (3, 4). Using a radioautographic technique with ^{3}H-thymidine, Uwa and Nagata (5) showed that the scleroblast mass is formed by the proliferation and migration of the precursor cells which have appeared around the sites of process formation after the treatment with androgen. The scleroblast mass begins to secrete horny substance and the early process is formed within a few days of the androgen treatment. Hormonal antagonism has been noticed by several investigators during androgen-induced process formation (3, 4, 6). Thus, the process formation seems to be a simple model for the studies of cell differentiation and organogenesis induced with sex hormones.

In higher vertebrates, a number of investigations have been carried out to reveal the role of sex hormones on the development and maintenance of the secondary sex organs *in vitro* (7). On the other hand, little attention has been directed to this field in fish. In the present study, we examined the growth and maintenance of the early process *in vitro*, to study the mechanism of the process formation with sex hormones.

Materials and Methods

Adult females of the orange-red variety of *Oryzias latipes* were used. Anal-fin processes were induced with oral administration of ethisterone-diet (12 μg/g diet) at 25 °C (6). After 4 days of the hormone pretreatment, anal fins were removed for culture (7). They were sterilized with 1 % sodium hypochlorite and 70 % ethanol, and washed three times with phosphate-buffered saline. Two or three sheets of fins were placed on a membrane filter (millipore HA type, 13 mm in diameter) and incubated in Falcon organ culture dishes in a humidified incubator at 25 °C.

Eagle's minimum essential medium (MEM), Dulbecco's modified Eagle medium (DEM), medium 199 and CMRL-1066 were used as culture media. L-15 Leibovitz medium (L-15) was also tested. Each medium was supplemented with 10 - 20 % horse or fetal calf serum, 100 U/ml of penicillin, 100 μg/ml of streptomycin and 60 μg/ml of kanamycin. Ethisterone and testosterone were suspended in media at final concentrations of 1 μg/ml and 1 IU/ml, respectively. The pH of the media was adjusted to 7.2 with 5 % CO_2-air or HEPES. One half volume of the medium was changed every other day.

Fins were fixed in Bouin's fluid 0, 2, 4, 6 and 8 days after culture. They were embedded in paraffin, sectioned at 4 μm and stained with Azocarmine-Aniline blue. The number of processes was counted in serial sections of the anal-fin. At least ten fish were examined in each group of each experimental series.

Results

On the initial day of culture, most of the anal-fin processes were at the early process stage; the apex of a young process is covered by the scleroblast mass. They elongated during the culture and some scleroblasts were embedded into the formed processes. Comparisons of the media for the growth and maintenance of the early process are summarized in Table 1.

Table 1. Total number of anal-fin processes and percentage of the elongated ones among total number (in parentheses) cultured with media containing testosterone or ethisterone[a]. *Most processes were degenerated.

Medium	Days after culture						
	0	2		4		6	
		Cont.	Exptl.	Cont.	Exptl.	Cont.	Exptl.
MEM	28(0%)	28(0%)	46(15%)	* 2(0%)	36(44%)	*15(0%)	*38(53%)
DEM	53(0%)	*32(0%)	39(23%)	17(0%)	*26(19%)	* 2(0%)	23(22%)
199	61(5%)	42(0%)	62(85%)	*28(0%)	38(82%)	*28(0%)	49(78%)
CMRL	35(14%)	-	42(7%)	-	31(39%)	-	36(56%)
L-15[a]	21(5%)	-	33(31%)	-	27(41%)	-	18(44%)

During culture, elongation of the process became obvious in each medium containing androgen, whereas no appreciable elongation was detected in androgen-free control media. In medium 199 and CMRL, horny processes elongated to nearly the same extent as in intact fins and their apex was covered by the scleroblast mass. The scleroblasts were also embedded in the elongated process. L-15 also gave similar results. In MEM and DEM, on the other hand, scleroblast masses frequently disappeared and elongated processes degenerated variously, even in media containing androgen. No notable difference was detected in the action of testosterone and ethisterone, nor horse and fetal calf serum on the growth and maintenance of the horny process _in vitro_.

In some fins, especially those cultured in MEM, DEM or androgen-free control media, joint plates frequently became broken and fin rays became dispersed during culture. In these fins, osteoclast-like multinuclear cells occasionally appeared. Frequency of the appearance of the osteoclast-like cells during culture is shown in Table 2. These cells appeared after 2 to 4 days of culture, and then increased in number, size and nuclear number. They started to take up pieces of joint plates into their cytoplasm phagocytotically.

Table. 2. Percentage of anal-fin in which osteoclast-like cells appeared. The fins were cultured with medium 199 containing testosterone.

Medium	Days after culture				
	0	2	4	6	8
androgen-free	0 %	0 %	43 %	64 %	74 %
androgen	0 %	0 %	6 %	21 %	27 %

Discussion

Since the initial work by Wolf et al. (9), unmodified mammalian-type media, such as MEM, DEM and medium 199, have been considered appropriate for culture of tissues and cells of fresh-water fishes. In the present study, we found that medium 199 and CMRL-1066 supplemented with 10 - 20 % sera were more suitable than MEM and DEM for the growth and maintenance of anal-fin processes induced in adult females with androgen. L-15 was also useful for culture. Rapid and sequential increases in RNA-, DNA-, and protein-syntheses occur in the scleroblast and its precursor cells during the process formation(10). Scleroblasts also incorporated much ^{3}H-proline for the synthesis of collagen which is subsequently secreted to form the horny process (11). Medium 199, CMRL-1066 and L-15 contain qualitatively and quantitatively more amino acids, vitamins, nucleic acid components and intermediary metabolites than MEM and DEM. The scleroblast seems to require nutritionally rich conditions for the rapid growth of the horny processes.

However, the present nutritional conditions were insufficient for the maintenance of the process. The elongated process degenerated gradually after 6 to 8 days of culture even cultured with these nutritionally rich media. Osteoclast-like multinuclear cells occasionally appeared in the anal fin, with disappearance of the scleroblast mass. These cells increased in size and number, and began to resorp joint plates. The appearance of the osteoclast and/or resorption of the bone occasionally occurs when culture conditions, such as media, sera and oxygen tension, are inadequate for the growth and maintenance of the bone *in vitro* (12). Moreover, the resorption of the bone is promoted with parathyloid hormone and vitamin A *in vitro*, whereas it is suppressed with calcitonin (13). Presently, we are trying to optimize the culture conditions for the maintenance of the anal-fin process *in vitro*.

It has been suggested by Oka (1) that the anal-fin process is formed by a osteoblast-like mesenchyme cell mass (scleroblast) appearing at the postero-external margin of the joint plate. From electron microscopic (14) and radioautographic (10, 11) evidences during androgen-induced anal-fin process formation, we have inferred that the precursor cell of the scleroblast arises from mesenchymal cells detected along the postero-external margin of the joint plate in intact fish. They proliferate and migrate to the site of process formation, form a mass, and begin to secrete a horny substance. The scleroblast mass is located at the apex of the early process, then some of them are embedded in the elongated process. The differentiation of the scleroblast seems to be synchronized with that of the osteogenic cells: preosteoblast, osteoblast and osteocyte.

(+) Androgen ------→ Precursor cells
↓ Proliferation and migration
Scleroblast mass
↓ Secretion of horny substance
Elongation of the process
Scleroblasts embedded in the process
Maintenance of the process
:
:
(-) Androgen ------→ Precursor cells (?)
↓ Proliferation and fusion
Osteoclast-like multinuclear cells
Resorption of the process

Fig. 1. A hypothetical scheme for the development and degeneration of the anal-fin process of a ricefish, *Oryzias latipes*.

Osteoclast-like multinuclear cells also arise during culture. Several investigators have pointed out that osteoclasts arise by fusion of cells of osteoblast series (15, 16, 17, 18), though osteoclasts in general have a hematogenous origin (19). It is clear that the osteoclast-like multinuclear cells arise during culture of the anal-fin. They may be derived from the scleroblast series, though little is known about the relation between the disappearance of the scleroblast and the appearance of the osteoclast-like multinuclear cells.

Androgen is essential not only for the differentiation of the scleroblast but also for the maintenance of the elongated processes. This possibility is supported by the fact that the anal-fin processes degenerated in castrated males (2). From the evidences so far studied, we summarized the development and degeneration of the anal-fin processes as shown in Fig. 1. Vaughan (20) has suggested that the differentiation and dedifferentiation of osteogenic cells occur reversibly between osteoblast and osteoclast series through the pluripotent osteogenic mesenchyme cell. Development and degeneration of the anal-fin processes may be controlled with androgen through differentiation and dedifferentiation of the scleroblast and osteoclast-like multinuclear cell.

References

1. Oka, T. B. Jour. Fac. Sci. Univ. Tokyo, 2:209-218, 1931.
2. Nagata, Y. Zool. Mag. (Japan), 48:102-108, 1936 (in Japanese).
3. Okada, Y. K. and Yamashita, H. Jour. Fac. Sci. Univ. Tokyo, 6:383-437, 1944.
4. Egami, N. Annot. Zool. Japon., 27:122-127, 1954.
5. Uwa, H. and Nagata, T. Develop., Growth and Differ., 18:279-288, 1976.
6. Uwa, H. Embryologia, 10:173-180, 1968.
7. van Tienhoven, A. Reproductive Physiology of Vertebrates, pp. 8-28, W. B. Saunders, 1968.
8. Iwata, A. and Uwa, H. Develop., Growth and Differ., 23:245-248. 1981.
9. Wolf, K., Quimby, C., Pyle, E. A. and Dexter, R. P. Science, 132:1890-1891, 1971.
10. Uwa, H. Develop., Growth and Differ., 11:77-87, 1969.
11. Uwa, H. Develop., Growth and Differ., 13:119-124, 1971.
12. Sato, A. Tissue Culture (Nakai, J. et al., eds.), pp. 276-291, Asakura, 1976 (in Japanese).
13. Reynold, J. J. and Dingle, J. T. Calc. Tiss. Res., 4:339-349, 1970.
14. Uwa, H. Develop., Growth and Differ., 16:41-53, 1974.
15. Tonna, E. A. Anat. Rec., 137:251-270, 1960.
16. Messier, B. and Leblond, C. P. Am. J. Anat., 106:247-285, 1960.
17. Kember, N. F. J. Bone Joint Surg., 42:824-839, 1960.
18. Young, R. W. J. Cell Biol., 14:357-370, 1962.
19. Jotereau, F. V. and Le Douarin, N. M. Develop. Biol., 63:253-265, 1978.
20. Vaughan, J. M. The Physiology of Bone, pp. 23-60, Clarendon Press, 1970.

FISH CELL LINES : SUSCEPTIBILITY TO SALMONID VIRUSES

M. Yoshimizu, M. Kamei, S. Dirakbusarakom, and T. Kimura

Laboratory of Microbiology, Faculty of Fisheries,
Hokkaido University, Hakodate 041, Japan

Optimum growth temperature and viral sensitivity of 32 cell lines derived from salmonid and non-salmonid fresh or sea water fish were studied using the salmonid viruses : IPNV, IHNV, OMV, *H. salmonis* and CSV, and non-salmonid rhabdovirus HRV. Viral replication in 19 cell lines was examined by employing the IPNV, IHNV, OMV and *H. salmonis* and chromosome numbers were observed in 8 cell lines. Optimum growth temperature of salmonid cell lines was 20 °C and non-salmonid warm water fish cell lines were 25 or 30 °C. Most of the cell lines had sensitivity to IPNV and IHNV, but OMV and *H. salmonis* showed the CPE only on the salmonid cell lines. On CHH-1, SE, RTG-2 and CHSE-214 cell lines, salmonid viruses showed a high rate of replication.

For the study of viral fish disease, the selection of fish cell line to determine the high susceptibility and high response to the viruses is very important. To isolate the salmonid viruses, RTG-2 and CHSE-214 cell lines are now widely used. Fendrick et al.(4) and Kelly et al.(5) reported that on EPC and FHM cell lines, infectious hematopoietic necrosis virus; IHNV (9) and infectious pancreatic necrosis virus; IPNV (9) respectively show the high titer and are better than RTG-2 or CHSE-214 cell lines for the isolation of those viruses.

In our laboratory, we are currently maintaining more than 40 cell lines derived from the salmonid and non-salmonid fresh or sea water fish. In this report we describe the optimum temperature for the growth of 32 representative fish cell lines, and then compare the susceptibility to the following salmonid viruses, IPNV, IHNV, *Oncorhynchus masou* virus; OMV (6), *Herpesvirus salmonis* (9), and chum salmon virus; CSV (12), and non-salmonid hirame rhabdovirus; HRV (7). For the comparative study of virus replication, 19 cell lines were selected, and 8 cell lines were employed for the study of chromosome analysis.

Cell line employed :

The cell lines derived from salmonids were CHSE-214 (chinook salmon; *Oncorhynchus tschawytscha*), KO-6 and HIME (kokanee salmon; *O. nerka*), CHH-1, SE, SET and SEH (chum salmon; *O. keta*), YNK and MSE (masu salmon; *O. masou*), RTG-2, RTT, RTE, RTE-2, RTH-149 (rainbow trout; *Salmo gairdneri*), STE (steelhead trout; *S. gairdneri*), AS-6 and ASE (Atlantic salmon; *S. sala*). Non-salmonid fish cell lines derived from fresh and sea water fish were AF-29 (ayu; *Plecoglossus altivelis*), BB (brown bullhead; *Ictalurus nebulosus*), FHM (fathead minnow; *Pimephales promelas*), EPC (carp; *Cyprinus carpio*), EK-1 and EO-2 (eel; *Anguilla japonica*), GCF (grass carp; *Ctenopharyngodon idella*), SCF (silver carp; *Hypophthalmichthy molitrix*), SHH (snakehead fish; *Channa striatus*), WF-1 and WF-2 (pond smelt; *Hypomesus olidus*) those are fresh water fish, and SBF (sea bass; *Latus calcarifer*), RSF (red snapper; *Lutjanus argentimaculatus*), HF-1 (Japanese flounder; *Paralichthys olivaceus*) and SBK (red sea bream; *Chrysophrys major*). These are sea water fish. Species and original tissue of these cell lines are shown in Tables I and II. Cell lines were

Table 1. Optimum growth temperature and virus susceptibility of fish cell lines derived from salmonid fish

Cell line	Species	(Tissue)	Reference	Optimum temperature	Cytopathic effect IPNV	IHNV	OMV	*H.s*	CSV	HRV
CHSE-214	chinook salmon	(embryo)	(8)	20	+	+	+	+	+	-
KO-6	kokanee salmon	(ovary)	(8)	20	+	+	+	+	+	-
HIME	kokanee salmon	(embryo)	by author	20	+	+	+	+	-	-
CHH-1	chum salmon	(heart)	(8)	20	+	+	+	+	+	-
SE	chum salmon	(embryo)	(10)	20	+	+	+	+	+	+
SEH	chum salmon	(embryo)	by author	20	+	+	+	+	+	-
SET	chum salmon	(embryo)	by author	20	+	+	+	+	+	-
MSE	masu salmon	(embryo)	by author	20	-	+	+	+	-	-
YNK	yamame	(kidney)	(11)	20	+	+	+	+	-	+
RTE	rainbow trout	(embryo)	(*)	20	+	+	+	+	+	+
RTE-2	rainbow trout	(embryo)	by author	20	+	+	-	+	+	+
RTH-149	rainbow trout	(hepatoma)	(8)	20	+	+	+	+	+	+
RTG-2	rainbow trout	(gonad)	(13)	20	+	+	+	+	+	+
RTT	rainbow trout	(tail)	(*)	20	+	+	+	+	+	+
STE-137	steelhead trout	(embryo)	(8)	20	+	+	+	+	+	+
AS-6	Atlantic salmon	(embryo)	(*)	20	+	+	+	+	-	-
ASE	Atlantic salmon	(embryo)	(*)	20	+	+	+	+	-	-

(*): provided by Dr. B.J.Hill

subcultured with Eagle's MEM or Leibovitz L-15 containing 10 % FBS (Gibco), 100 I.U. penicillin and 100 μg streptomycine (Sigma) /ml.

Optimum growth temperature :

For each cell line, a set of 8 25-cm² flasks was prepared. Each flask was seeded with 1.5×10^5 cells and replicate flasks were incubated at 15, 20, 25, and 30 °C. After 7 days of incubation the cells from each pair of flasks were trypsinized, and counted in a hemocytometer.

Cell lines derived from salmonids showed the highest growth at 20°C and could not grow at 30 °C. The optimum growth temperature of those salmonid cell lines and HF-1 cell line from hirame were 20 °C (Tables 1 and 2), and cell lines derived from ayu (AF-29), grass carp (GCF), silver carp (SCF), smelt (WF-1 and 2) and red sea bream (SBK) were 25 °C, and brown bullhead (BB), fathead minnow (FHM), eel(EK-1 and EO-2), carp (EPC), snakehead fish (SHH), red snapper (RSF) and sea bass (SBF) were 30 °C or higher (Table 2).

Table 2. Optimum growth temperature and virus susceptibility of fish cell lines derived from non-salmonid fresh and sea water fish

Cell line	Species	(Tissue)	Reference	Optimum temperature	Cytopathic effect IPNV	IHNV	OMV	*H.s*	CSV	HRV
AF-29	ayu	(fin)	(*)	25	+	-	-	-	+	+
BB	brown bullhead	(caudal	(13)	30	+	+	-	-	+	+
FHM	fathead minnow	trank)	(13)	≥ 30	+	+	-	-	-	+
EK-1	eel	(kidney)	(1)	≥ 30	+	-	-	-	+	+
EO-2	eel	(ovary)	(2)	≥ 30	+	+	-	-	+	+
EPC	carp	(epithelioma)	(13)	≥ 30	+	+	-	-	+	+
GCF	grass carp	(fin)	by author	25	-	+	-	-	+	-
SCF	silver carp	(fin)	by author	25	-	+	-	-	+	-
SHH	snakehead fish	(heart)	by author	30	+	+	NT	NT	-	-
WF-1	smelt	(fin)	by author	25	-	+	-	-	-	+
WF-2	smelt	(fin)	by author	25	-	+	-	-	-	+
HF-1	hirame	(fin)	by author	20	-	-	-	-	-	+
RSF	red snapper	(fin)	by author	≥ 30	+	+	NT	NT	-	NT
SBF	sea bass	(fin)	by author	≥ 30	+	+	NT	NT	-	NT
SBK	red sea bream	(kidney)	(**)	25	+	+	-	-	-	+

(*): provided by Dr. M. Tanaka, (**): provided by Dr. T. Nagabayashi

NT : not tested

Viral replication and sensitivity :

The cell lines were tested for their susceptibility to the following salmonid viruses: IPNV (Buhl), IHNV (ChAb), OMV (00-7812), *H. salmonis* and CSV and also non-salmonid virus; HRV. Stocks of virus prepared in RTG-2 or CHSE-214 cells were inoculated at a multiplicity of infection (MOI) of 0.1 on each of the cell lines, and examined daily to monitor progress of the cytopathic effect (CPE). These cell lines were incubated at 15 °C except *H. salmonis* inoculated cells. *H. salmonis* inoculated cell lines were cultured at 10 °C. In addition, the cell lines were tested for their ability to replicate the salmonid viruses (IPNV, IHNV, OMV and *H. salmonis*). Pairs of 25 cm^2 flasks of each cell line were inoculated at MOI of 0.0005 of each virus. After 7 days of incubation at 15 °C, culture media from replicate flasks were pooled and assayed for virus titer on RTG-2 cells using the 50 % tissue culture infectious dose assay ($TCID_{50}$/ml).

IPNV and IHNV showed the CPE on almost all of the cell lines derived from salmonid and non-salmonid fish but OMV and *H. salmonis* showed the CPE only on the salmonid cell lines. CSV did not have a tendency to show the CPE. HRV also did not show the CPE on the epithelial cells derived from genus *Oncorhynchus* (Tables 1 and 2). The representative growth curves of IPNV, IHNV, OMV and *H. salmonis* on 19 cell lines at 15 °C are shown in Fig. 1. AS-6, CHH-1, CHSE-214, SE and YNK cell lines showed the high response and high yield of the IPNV, while CHSE-214, CHH-1, EPC and RTE cell lines showed the high response and high yield of IHNV. In RTG-2, CHH-1, SE and KO-6 cell lines, OMV replicated rapidly and showed high yield and SE, RTG-2, YNK and CHH-1 cell lines supported the replication of the *H. salmonis* rapidly. From the result of this study we selected the 4 cell lines, CHH-1, SE, CHSE-214 and RTG-2, to show the high response to the salmonid viruses. Currently most of the laboratories are using the CHSE-214 and RTG-2 cell lines, and the results presented here suggest that the use of CHH-1 and SE cell lines along with RTG-2 and CHSE-214 cell lines is better for the routine virological examination of salmonids.

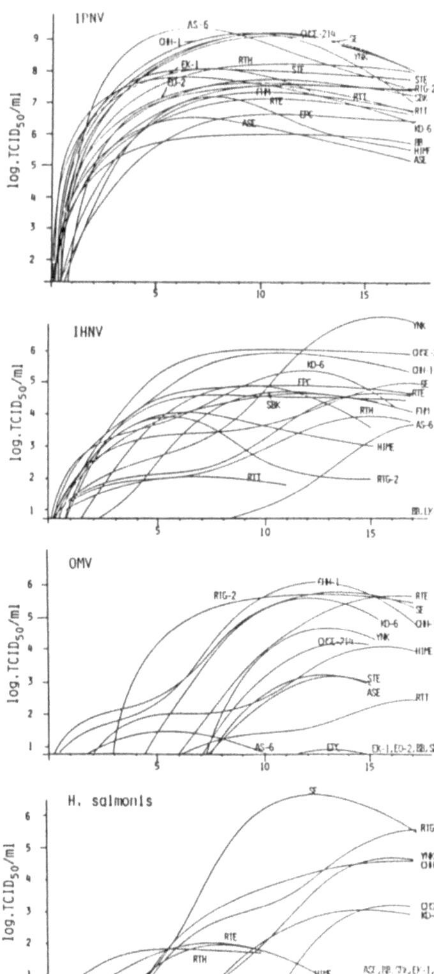

Fig. 1. Representative growth curve of IPNV, IHNV, OMV and *H. salmonis* in 19 selected fish cell lines.

Chromosome counts :

Chromosome preparation was made according to the method of Evans et al. (3). The chromosome spreads of 60 cells were counted for each cell line.

Distributions of chromosome number for 8 cell lines, CHH-1, SE, CHSE-214, RTG-2, AS-6, EPC, RTE and YNK are given in Fig. 2. Chum salmon has a diploid chromosome number of 74, but the CHH-1 cell line had a modal chromosome number of 64 and SE cell line had 58 to 60. The diploid number of chinook salmon is 68 but the modal number of CHSE-214 cell line was 64, and the usual diploid number of rainbow trout is 60 whereas modal numbers determined for RTG-2 and RTE cell lines were 90 and 54 respectively. While the diploid number of Atlantic salmon, yamame and carp are 53 to 60, 60 and 100, respectively, the AS-6, YNK and EPC cell lines derived from those fishes had the modal numbers 45, 68 and 52, respectively. The CHH-1, SE, AS-6, RTE and YNK cell lines consisted of 2n cells, but RTG-2, CHSE-214 and EPC cell lines consisted of 3n, 2n and 3n, and n and 2n cells, respectively.

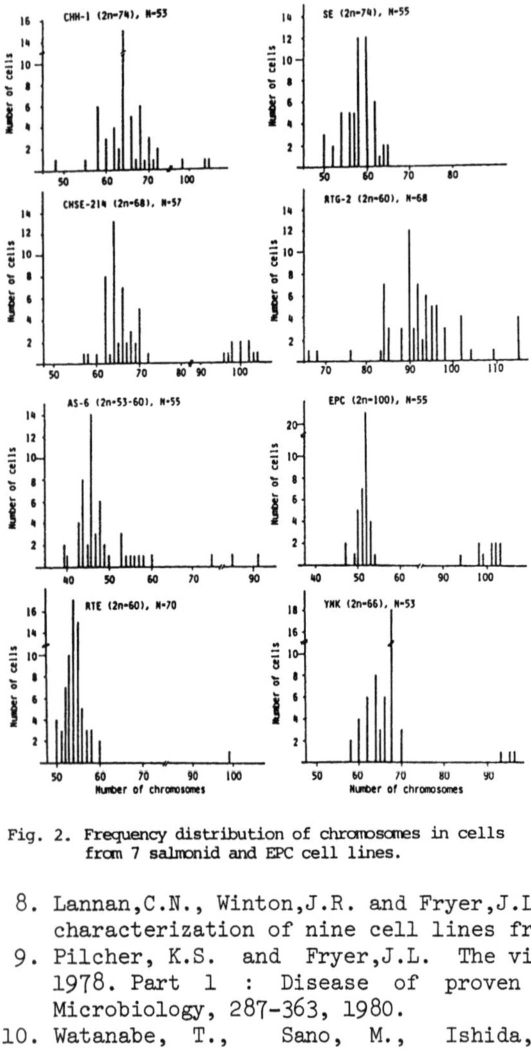

Fig. 2. Frequency distribution of chromosomes in cells from 7 salmonid and EPC cell lines.

References

1. Chen,S.N.,Ueno,Y.and Kou,G.H. A cell line derived from Japanese eel (*Anguilla japonica*) kidney. Proc. Natl.Sci.Coun.,ROC, 6, 93-100, 1982.
2. Chen,S.N. and Kou,G.H. A cell line derived from Japanese eel (*Anguilla japonica*) ovary. Fish Pathology, 16, 129-138, 1981.
3. Evans, E.P.,Breckon,G. and Ford, C.E. An air-drying method for meiotic preparations from mammalian testes. Cytogenetics, 3, 289-294, 1964.
4. Fendrick,J.L., Groberg,W.J. and Leong, J.C. Comparative sensitivity of five fish cell lines to wild type infectious haematopoietic necrosis virus from two Oregon sources. J. Fish Dis., 5, 87-95, 1982.
5. Kelly,R.K.,Souter,B.W. and Miller,H. R. Fish cell lines : comparisons of CHSE-214, FHM and RTG-2 in assaying IHN and IPN virus. J.Fish.Res.Board Can., 35, 1009-1011, 1978.
6. Kimura,T.,Yoshimizu,M. and Tanaka,M. Studies on a new virus (OMV) from *Oncorhynchus masou*-I. Characteristics and pathogenicity. Fish Pathology,15, 143-147, 1981.
7. Kimura,T., Yoshimizu,M. and Gorie,S. A new rhabdovirus isolated in Japan from cultured hirame (Japanese flounder) *Paralichthys olivaceus* and ayu *Plecoglossus altivelis*. Dis. aquat.Org., 1, 209-217, 1986.
8. Lannan,C.N., Winton,J.R. and Fryer,J.L. Fish cell lines : establishment and characterization of nine cell lines from salmonids. In Vitro, 20, 671-676,1984.
9. Pilcher, K.S. and Fryer,J.L. The viral disease of fish : a review through 1978. Part 1 : Disease of proven etiology. CRC Critical Reviews in Microbiology, 287-363, 1980.
10. Watanabe, T., Sano, M., Ishida, Y., Mizusawa, Y. and Michikawa, M. Establishment and properties of the continuous cell line derived from the embrionated eggs of chum salmon, *Oncorhynchus keta*. Bull.Jap. Soc. Sci. Fish., 46, 1203-1209, 1980.
11. Watanabe,T., Kobayashi, N., Sato,Y. and Ishizaki, Y. Continuous cell line derived from the kidney of Yamame, *Oncorhynchus masou*. Bull.Jap.Soc.Sci.Fish., 44, 415-418, 1978.
12. Winton,J.R.,Lannan,C.N.,Fryer,J.L.and Kimura,T.Isolation of a new reovirus from chum salmon in Japan. Fish Pathology, 15, 155-162, 1981.
13. Wolf, K. and Mann, J.A. Poikilotherm vertebrate cell lines and viruses : a current listing for fishes. In Vitro, 16, 168-179, 1980.

ENHANCED EXPRESSION OF ONCOGENES IN CULTURED MEDAKA TUMOR CELLS

C. Muraiso,[1] N. Nemoto,[2] K. Asami,[1] H. Etoh,[1] and H. Matsudaira[1]

1) National Institute of Radiological Sciences, Chiba 260, Japan
2) Cancer Institute, Tokyo 170, Japan

Cells from melanoma induced in HB32C strain medaka and from hepatoma in HO4C strain and normal fibroblastic cells derived from the fins of HB32C medaka were cultivated *in vitro* and established as cell lines. Presence and expression of some cellular oncogenes were examined by Southern and Northern blotting. After agarose gel electrophoresis or dot blotting, DNA and RNA samples were hybridized to ^{32}P-labeled human oncogene probes, such as v-Ki-*ras*, v-Ha-*ras*, c-*myc*, *fos* and *src*. Southern blotting analysis indicated that the above listed genes were present in normal and tumor cells and c-*myc* gene slightly amplified in melanoma cells. Restriction fragment length polymorphism (RFLP) was detected in the DNA of black and orange red medaka when c-Ha-*ras* and c-Ki-*ras* were used as probes. Northern blotting showed 3-fold enhancement in the expression of c-Ha-*ras*, c-*myc*, and *src* genes in tumor cells compared to normal cells.

Gene amplification and enhanced expression of oncogenes have been reported in various human and rodent tumors , and oncogenes were shown to be present also in the fish. Sequence homology of fish *ras* and *myc* genes to mammalian genes was found in goldfish (1) and rainbow trout (2). An enhanced expression of *ras* gene was observed in erythrophoroma cells of goldfish, although it was uncertain which *ras* gene was responsible for generation of the observed RNA (3). Enhanced expression was reported in various tumors in *Xiphophorus* fish (4). In our laboratry, the mechanism of chemical carcinogenesis has been studied in an aquarium fish, medaka. Four cell lines derived from normal and tumor tissues were established (5). Using these cell lines, amplification and expression of oncogenes were examined in this study.

Materials and Methods

Agents. Restriction endonucleases were obtained from Takara Shuzo Co., Ltd. (Kyoto, Japan.) Proteinase K was purchased from E.Merck, AG, Darmstadt, FRG. Kits for translation, α-^{32}PdCTP(3000Ci/mmol) and probe DNAs were from Amersham Japan Ltd.(Tokyo).

Cell lines. OLHE-131 is a cell line established from hepatoma induced in HO4C (inbred orange) strain medaka after treatment with a carcinogen, MAM acetate. OLME-104 is a cell line established from melanoma induced in HB32C (black) strain after treatment with MNNG. OLF-124 is a fibroblast line established from the fins of outbred orange medaka. OLF-136 is also a fibroblast line established from the fins of HB32C (5). JTC12P3, a monkey kidney epithelial cell line, was used as a reference.

DNA extraction. Cell pellets (0.5 g) were homogenized in 3.5 ml of buffer (20mM

Tris-HCl, pH 7.4, 0.1M NaCl, and 1mM EDTA. One ml of proteinase K (3mg/ml) and 1% SDS (sodium dodecyl sulfate) were added to the homogenates and the mixture was incubated at 37°C overnight. Then, DNA was extracted with buffer-saturated phenol and precipitated by ice-cold ethanol.

RNA extraction. Cell pellets (0.5 g) were homogenized in 5 volumes of solution containing 4M guanidinium thiocyanate, 0.5% Sarkosyl, 5mM sodium citrate, pH 7.0 and 0.1 M 2-mercaptoethanol. To the homogenates were added 1 g of cesium chloride and the mixture was overlayed on 1.2 ml of 5.7 M cesium chloride in 0.1 M EDTA, pH 7.5, in a polyallomer tube. Samples were centrifuged at 100,000g for 12h. The precipitates were dissolved in 10mM Tris-HCl, pH 7.4 containing 1% SDS and 5 mM EDTA. RNA was extracted with chloroform and precipitated by ethanol.

DNA and RNA analysis. Samples of each 20 μg DNA were digested with appropriate restriction endonuclease (Eco RI, Bam HI, or Hind III) and run on 1% agarose gel at 40V for 9 h. DNA was then transferred onto a nylon filter according to Southern(6). Dot blotting was performed as follows: DNA was denatured at 100°C for 8 min and more than 10 volumes of 20xSSC was added and the mixture spotted on a nylon filter which had been soaked in 20xSSC.

To samples of 30 μg RNA in MOPS buffer pH 7.0 were added 5.4 μl of HCHO and 15 μl of neutralized formamide. RNA was denatured at 60°C for 15 min and then run on 0.8% agarose gel containing 2.2 M formaldehyde. RNA was then transferred onto a nylon filter as above. The filters were dried at room temperature and baked at 80°C for 2h. Prehybridization was carried out at 65°C for 4h in 10xDenhardt's mixture containing 4xSSC. RNA samples on filter were then transferred into the newly prepared mixture to which were added ^{32}P-labeled DNA probes (v-Ha-ras,

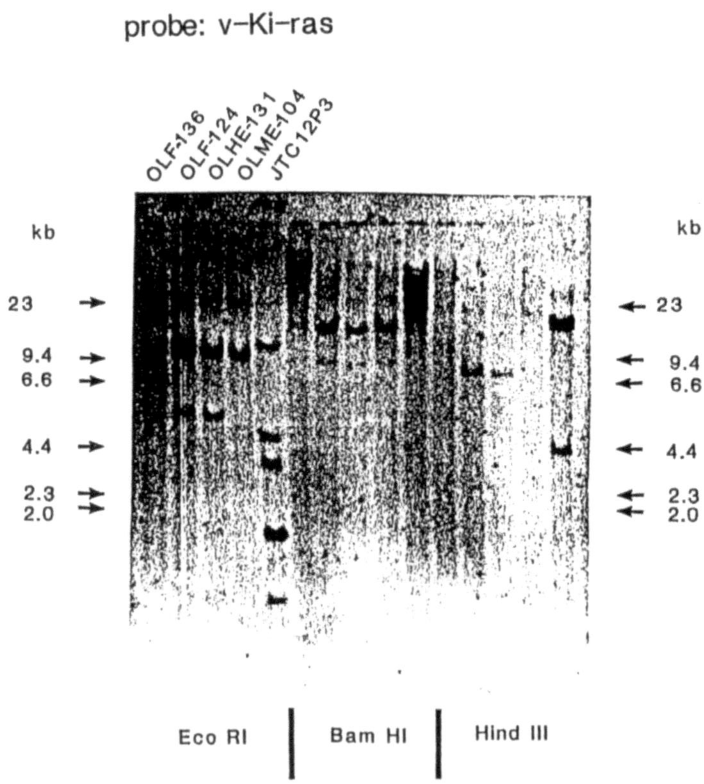

Fig. 1. Southern-blot hybridization of medaka DNA with viral Ha-ras DNA. The order of cells in lanes 6-10 and 11-15 is the same as lane 1-5.

v-Ki-ras, c-myc), E. coli DNA, and poly A. The filters were then kept at 60°C and then washed 3 times with 1xSSC containing 0.2% SDS each for 15 min at room temperature and once for 15 min with the same solution at 60°C.

Results

Presence of ras-related DNA. Fig. 1 illustrates the results of Southern hybridization analysis of the DNA from normal fibroblasts, hepatoma and melanoma cells. Each DNA was digested with either Eco RI, Bam HI, or Hind III. In the case of DNA probe v-Ki-ras, two or three bands appeared when the DNA samples were digested with Eco RI. Their molecular size was 23, 10, and 5.1 kilobase (kb). Digestion with Bam HI resulted in two bands of 12 and 8 kb. Hind III digestion gave only one band of 7.9 kb. After Bam HI digestion, a band of 12 kb was observed. It was slightly amplified in melanoma but no evidence for rearrangement was detected. Differences were observed in the migration pattern between hepatoma and melanoma DNA after Eco RI digestion. This may be due to the differences of the fish strains from which the cells originated, because the same pattern was observed in melanoma and fibroblast (OLF-136), both derived from the strain HB32C. These data suggest polymorphism in the fish DNA. In the case of DNA probe v-Ha-ras, Eco RI digestion gave three bands which were 23, 10.5 and 5.7 kb in molecular size. Migration patterns were similar to those obtained with v-Ki-ras probe. DNA polymorphism was also indicated in orange and black medaka. The results of Bam HI digestion showed an amplification of 14.4 and 10.5 kb bands in melanoma cells.

Gene amplification of c-myc family. Eco RI digestion gave one band of 1.5 kb and Hind III digestion gave two bands of 4.9 and 3.0 kb. To make clear gene

Table 1. Gene amplification and enhanced expression in medaka cells*

	v-Ki-ras		v-Ha-ras		c-myc		src		fos	
Cell	DNA	RNA	DNA	RNA	DNA	RNA	DNA	RNA	DNA	RNA
Hepatoma	-	-	-	+	-	+	-	+	-	+
Melanoma	+	-	+	+	+	+	-	+	-	?

*sis, mos and abl genes were not detected.

Table 2. Restriction fragment length polymorphism (Orange red medaka vs. black medaka)

Probe	Eco RI	Bam HI	Hind III
v-Ki-ras	+	+	+
v-Ha-ras	+	-	+
c-myc	-	-	-
src	-	-	-
fos	-	-	-

amplification in the tumor cells, dot blotting was performed. The results showed a two-fold amplification of c-myc gene in melanoma cells compared to normal fibroblast from the black HB32C strain.

Expression of oncogenes in the cells. To examine the changes in gene expression, Northern blotting was performed. In the case of c-myc, a three-fold amplification was detected in the RNA of hepatoma and melanoma cells. The molecular size of RNA was determined to be about 3.7 kb. When src was used as probe, a two-fold amplification was observed in the 3.7 kb RNA of OLHE-131 and OLME-104 cells. When v-Ha-ras was used as probe, a two-fold amplification was also detected in the 3.9 kb RNA.

Conclusion

The changes in oncogenes and their expression were studied in normal and tumor cells from different strains of medaka. The genomic DNA and total RNA were isolated and analyzed by Southern and Northern blotting. Dot blotting was also performed. An amplification but no rearrangement was detected in the DNA of melanoma cells, when examined by DNA-DNA hybridization with v-Ha-ras, v-Ki-ras, and c-myc as probes. C-sis, c-mos, and c-abl genes were not detected in the genomic DNA of the cells examined (Table 1). An indication of the restriction fragment length polymorphism was observed between outbred orange-red variety and inbred black strain medaka when Eco RI digests were analysed with v-Ki-ras and v-Ha-ras DNA probes. Differences were also found in Hind III digestion fragments (Table 2). A marked increase was observed in the expression of oncogenes, c-myc and c-Ha-ras, when examined by DNA-RNA hybridization, indicating alterations in the transcription in the tumor cells (Table 1). The results of oncogene analysis are generally in accord with those reported by Anders and his associates in Xiphophorus fish (4). Relevance of the amplification and enhanced expression to specific changes in a complex carcinogenesis process remains, however,to be studied.

Acknowledgements

We thank Dr. Furuno-Fukushi for her kind help in preparation of DNA from cultured cells. This work was supported in part by a special grant for cancer research from Science and Technology Agency, Japan.

References

1. Nemoto, N., Kodama, K., Tazawa, A., Prince Masahito and Ishikawa, T. Differentiation,32: 17-23, 1986.
2. Beneden,R.J.V., Watson, D.K., Chen, T.T., Lautenberger, J.A. and Papas, T. S. Proc, Natl. Acad. Sci. USA, 83: 3698-3702, 1986.
3. Nemoto, N. Kodama, K., Tazawa, A., Matsumoto, J., Prince Masahito and Ishikawa, T. J. Cancer Res. Clin. Oncol., 113: 56-60, 1987.
4. Anders, F., Anders, A., Schartl, M., Gronau, T., Lüke, W., Schmidt, C.R., and Barnekow, A. In NATO International Advanced Study Institute on New Experimental Modalities in Control of Neoplasia (Chandra,P., ed.). Plenum Press,New York and London (in press).
5. Etoh, H., Suyama, I., Hyodo-Taguchi, Y., and Matsudaira, H. (in this Proceedings)
6. Southern, E. M. J. Mol. Biol.,98:503-517, 1975
7. Lehrach, H., Diamond, D., Wozney, J.M., and Boedtker, H. Biochemistry, 16: 4743-4751, 1977.

SOMATIC CELL GENETIC STUDY OF DNA REPAIR IN CULTURED FISH CELLS

A.Shima, K.Isa, J.Komura, K.Hayasaka, and H.Mitani

Laboratory of Radiation Biology, Zoological Institute,
Faculty of Science, University of Tokyo,
Bunkyo-ku, Tokyo 113, Japan

We have been interested in evolution of DNA repair. As a practical approach to this problem, we decided to try to understand DNA repair from a phyletic point of view. For the purpose, comparison of fish cells with human cells has been made concerning growth and DNA repair characteristics. A brief summary of our work on the establishment of fish cell lines capable of colony formation and the characterization of growth and DNA repair will be given. With regard to DNA repair studies, discussion will be focused on ultraviolet radiation (UV) in order to avoid overlapping with another paper (Egami, Shima and Mitani, these Proceedings).

Establishment of Fish Cell Lines Capable of Colony Formation

In research on radiation biology, we believe that the first step is to quantitate the killing effect of radiation. For this purpose, individual survival is the most fundamental as well as error-free endpoint. However, in order to determine the events leading to the death of irradiated individuals, analyses at the cellular and molecular levels are obviously essential.

About 10 years ago, we decided to analyze radiation effects on fish not only at the individual level, which has been very well studied mainly by Egami and his colleagues particularly using Medaka and goldfish (1), but also at cellular and molecular levels using cultured cells of fish.

We therefore made a literature survey of fish cell line(s) capable of colony formation in order to quantitate the lethal effect of radiation using reproductive death as an endpoint. We found that the review by Wolf and Quimby (2) was very informative and the ATCC catalogue was useful in letting us know the presence of commercially available fish cell lines like FHM, RTG-2, BB and BF-2. However, to the best of our knowledge, the clonogenicity of these cells was very low and hence we reached a conclusion that no fish cell line was available for survival assay by colony formation. Therefore, we decided to establish fish cell lines capable of colony formation.

To begin with, we made an attempt to cultivate cells of the caudal fin of a goldfish, *Carassius auratus* (3):The procedures we used were completely the same as those routinely used for mammalian cell culture including the temperature of 37°C. We chose 37°C purposely, keeping in mind a future possibility of fusion study with mammalian and non-mammalian cells. The details are given in our original paper (3). Very briefly, fibroblast-like cells cultivated at 37°C (RBCF-1 line) attained exponential growth from the very beginning of culture, reached 200 population doubling number (PDN) at the 310th day and are still actively growing 10 years after the primary explant culture. The cloning efficiency of RBCF-1 cells was approx. 20 %. Subcutaneous injection of RBCF-1 cells into nude mice or X-irradiated goldfish failed to develop any observable tumor up to 5 months after injection. No significant change in the distribution of chromosome number (modal number=100) could be found. The population doubling time at

37°C was 12 hr and became longer with lowering culture temperature, e.g. 58 hr at 20°C, indicating that RBCF-1 cells remain "poikilothermal". The saturation cell density of RBCF-1 cells was almost the same at 20, 30 and 37°C, but higher at 27°C.

Not only from goldfish, but also from Amazon molly (*Poecilia formosa*) (4) as well as Medaka (*Oryzias latipes*) (5), we succeeded in establishing cell lines with cloning efficiencies high enough for use in survival assays.

Characterization of DNA Repair in UV-Irradiated Cultured Fish Cells

Making use of the high clonogenicity of RBCF-1 cells, we started to characterize DNA repair in 254 nm UV-irradiated cells (6). That is, cell survival and changes in the content of pyrimidine dimers after photoreactivation treatment with white fluorescent lamps were examined by colony forming ability and by a direct dimer assay. The D_o value, which is the UV dose necessary to reduce the cell survival to 37 % of the control, was 2 J/m^2, indicating that this fish cell line is considerably more sensitive to UV than other vertebrate cell lines. When UV-irradiated cells were illuminated by photoreactivating light, cell survival was enhanced up to a factor of 9 and the amounts of thymine-containing dimers were reduced in a photoreactivating fluence-dependent manner. On the other hand, keeping UV-irradiated cells in the dark did not result in any change in either parameter, indicating that there were no detectable levels of a dark-repair ability in the fish cells under our conditions. Thus, we concluded that fish cells are highly proficient in photoreactivation repair, but almost totally deficient in excision repair which occurs in the dark. Further studies on the effects of chronic sunlamp-UV together with daylight radiation revealed that the mechanism for cell killing, mainly formation of pyrimidine dimers by sunlamp-UV, was the same as that by 254 UV and suggested that photoreactivable organisms exposed to sunlight should display inactivation and photoreactivation repair simultaneously (7).

Establishment of the Conditions for Microcell-Mediated Gene Transfer from Fish Cells to Human Cells

Based upon the above-mentioned characterization of DNA repair in cultured fish cells, we started a somatic cell genetic approach to further analyze DNA repair mechanisms of fish cells from a phyletic point of view. As a clue, we first tried to establish a method for microcell-mediated gene transfer from fish cells (RBCF-1 line) to human cells (XP20S). The rationale of this gene transfer experiment is that RBCF-1 cell is highly proficient in photoreactivation repair but largely deficient in excision repair, while the XP20S cell is deficient in both repair abilities. Therefore, if photoreactivation-competent hybrid cells are obtained by fusion of microcell(s) from the fish cells with the human cells, then we can further look for the photoreactivation gene of the fish cells using such hybrids. We succeeded in establishing the conditions for micronucleation and enucleation of the RBCF-1 fish cells, fractionation of the microcells, quantitation of the DNA contents of microcells fractionated, and fusion of the microcells with the human cells (8). Despite our extensive attempts, using conditions established here, we have not yet been able to obtain hybrids which stably manifest the photoreactivability against 254 UV-irradiation.

Transfection of Cultured Fish Cells with Dominant Selection Markers

We, therefore, decided to give fish cells adequate selection markers which can be used for selection of hybrids between fish cells with one marker and human cells with another marker. As a dominant selection marker, a plasmid pSV2-*gpt* was chosen which can give the fish cells resistance to mycophenolic acid (MPA). By a slight modification of the method originally developed by Kawai and Nishizawa using Polybrene and dimethyl sulfoxide (9), we succeeded in obtaining clones of RBCF-1 cells with markedly elevated resistance against MPA and stable expression of the enzyme, xanthine guanine phosphoribosyltransferase, activity (10). Further studies are now under way using these transfectants.

Since we obtained evidence that the promoter in the early region of the SV-40 virus can function in the cultured fish cells, we have been making further attempts to transfect cultured fish cells with selection markers other than pSV2-gpt. As method for transfection, we are currently using electroporation more often than the previously used method (9). Very recently, transfectants were obtained using RBCF-1 cells and pSV2-neo, which afford the progenitor increased resistance against G418 (11).

The presence of different dominant selection markers in two different cell lines makes the production of hybrid cells much easier. Making use of these transfectants obtained in our laboratory together with those already obtained by others, studies are now in progress in our laboratory to analyze DNA repair mechanisms from a phyletic point of view.

Acknowledgements

The research summarized above was supported by a grant-in-aid from The Ministry of Education, Science, and Culture, Japan, and also by grants from The Institute of Physical and Chemical Research, Japan.

References

1. Hyodo-Taguchi, Y. and Egami, N. Comparative Cellular and Species Radiosensitivity (Bond, V.P. and Sugahara, T., eds.), pp.244-254, Igaku Shoin. 1969.
2. Wolf, K. and Quimby, M.C. Fish Physiology (Hoar, W.S. and Randall, D.J., eds.), Vol.3, pp.253-305, Academic Press. 1969.
3. Shima, A., Nikaido, O., Shinohara, S. and Egami, N. Exp. Gerontol., 15:305-314, 1980.
4. Shima, A. and Setlow, R.B. Zool. Sci., 2:477-483, 1985.
5. Komura, J., Mitani, H. and Shima, A. In Vitro (in press).
6. Shima, A., Ikenaga, M., Nikaido, O., Takebe, H. and Egami, N. Photochem. Photobiol., 33:313-316, 1981.
7. Shima, A. and Setlow, R.B. Photochem. Photobiol., 39:49-56, 1984.
8. Isa, K. and Shima, A. J. Fac. Sci., Univ. Tokyo, Sec.IV, 16:75-85, 1986.
9. Kawai, S. and Nishizawa, M. Mol. Cell. Biol., 4:1172-1174, 1984.
10. Isa, K. and Shima, A. J. Cell Sci., 88:219-224, 1987.
11. Hayasaka, K., Mitani, H. and Shima, A. (in preparation).

ESTABLISHMENT, CHARACTERIZATION AND APPLICATION OF 14 CELL LINES FROM WARM-WATER FISH

S.N. Chen and G.H. Kou

Department of Zoology, National Taiwan University,
Taipei, Taiwan

Fourteen permanent cell lines have been established from normal tissues of seven species of warm-water fish cultured in Taiwan. All the cell lines were routinely grown in Leibovitz's L-15 medium supplemented with 10-20% foetal calf serum at incubation temperatures ranged 28-31°C. Amongst these lines, EO-2 and BGK are fibroblastic and TK-1, TO-2, TM, BR, FI, GI, CF, LF, SK, BGF are epithelioid in morphology. However, both epithelioid and fibroblastic cells were observed in GON and EK-1 cell lines. The cell lines were demonstrated to be free of mycoplasmal, fungal and bacterial contaminations. With the exception of LF cells, thirteen cell lines were susceptible to IPNV of various serotypes, CV, LV-1, EVE,EVA and EVEX. Application of cell lines to the study of fish virology, cell physiology, biochemistry and toxicology are discussed.

In the past decade, cell lines have been obtained from a large number of fish species (1, 2, 3, 4, 5, 6, 7, 8). These cell lines were demonstrated to be useful for fish virology and environmental pollution studies (1, 6, 9). However majority of these studies were performed using cell lines derived from cold-water fish. In comparison with cold-water fish, cell lines from warm-water fish are rather scarce(1). The economic importance of several warm-water fish such as eel, tilapia, and carp in tropical or subtropical countries resulted in a significant increase of culture density. Concomitant with the high culture density, fish diseases occur more frequently, especially the viral infection which can probably initiate a high mortolity for the population. The limited number of reports on the viruses from warm-water fish compared with those from cold-water fish is probably due to the shortage of fish cell lines derived from warm-water fish.

Previous results showed that cell cultures from warm-water fish are easy to establish (4, 5). These cell lines not only reveal a wide range of viral susceptibility but also possess suitable characteristics for cell physiology (10, 11, 12, 13) and pollution studies (14).

This paper attempts to describe the development and characterization of 14 cell lines from fin, kidney ovary or gill of fresh- and sea-water fish cultured in Taiwan. The application of these cell lines to the studies of cell physiology and toxicology is also discussed.

Materials and Methods

Experimental Fish and Cell Line Establishment

Tissues from 7 warm-water fish species including eel (*Anguilla japonica*), tilapia hybrid (*Serotherodon mossambicus x Serotherodon niloticus*), top-minnow (*Gambusia patruelis*), color cap (*Cyprinus carpio*), common carp(*Cyprinus carpio*), snakehead(*Channa maculata*), loach(*Misgunus anguillicaudatus*) and banded grouper (*Epinephelus awoara*) were used for cell line establishment. Prior to the experiment all the fish were sterilized by immersion in freshly prepared 0.25% sodium hypochlorite solution for a few minutes. The surface of fish was then wiped with cotton saturated with 70% ethyl alcohol. Tissue as indicated in Table 1 was removed and washed 3-4 times in Hank's balanced salt solution (HBSS). HBSS supplemented with 0.15mM/ml NaCl was used for washing the tissues of banded grouper. Subsequently, the washed tissues were minced into small pieces using sterile scissors and then washed in HBSS or HBSS plus 0.15mM/ml NaCl, respectively. The tissue fragments were then inoculated into 25cm^2 Falcon plastic flasks containing culture media. The culture medium for tissues from fresh-water fish was GM L-15 which consists of Leibovitz's L15 (L-15) supplemented with 10% foetal calf serum (FCS), 250 µg/ml streptomycin, 250 IU/ml of penicillin and 2.5µg/ml Fungizone. For the cultivation of banded grouper L-15 supplemented with 20% FCS, 0.15 mM/ml NaCl and antiseptic compounds as described above was used.

The primary cell cultures were subcultured as described previously(4). The BGK cell culture at passages after 25 could be subcultured in GM-L15. All the cells were cultured at an incubation temperature of 28-31°C. L-15, FCS and antiseptic compounds were purchased from GIBCO Co, Grand Island, USA. Cell lines were tested for mycoplasma using a mycoplasma stain kit(Flow Laboratories, USA) and direct culture techniques as described by Russell(15). Thioglycollate broth and tryptic soy agar were used for the detection of bacterial contaminations. To examine the presence of fungus,potato dextrose agar was used.

Plating Efficiency, Karyology and Soft Agar Culture

Planting efficiency, chromosomal analysis and the ability to grow in soft agar were performed for each cell line.

Plating efficiency of the tested cell line at various subcultures as indicated in Table 2 was determined by making cell density less than 1x10^4 in a 25 cm^2 Falcon plastic flask as described previously (2). Cell line karyology was examined by the procedures as described by Chen and Kou (2) using a pre-treatment of 0.008% Colcemid (GIBCO, USA) in GM L-15 for 12 hours at 31°C.

The procedures described by Adams (16) were adopted to test the ability of cell lines to grow in soft agar.

Effect of Foetal Calf Serum (FCS) and Temperature on Cell Growth

To determine the effect of FCS and temperature on the cell growth, 24-well microplates were used. Into each well approximately 1x10^4-1x10^{5}cells were seeded in L-15 plus 2%, 5%, 10% and 20% of FCS, respectively. Following several hours' incubation, the cell growth was observed by removing the cells with trypsin in EDTA/PBS as described by Chen *et al.* (3). Cell lines were also tested over incubation temperatures of 4°C, 15°C, 18°C, 24°C, 28°C, 31°C and 37°C. Each well of microplates was seeded with 1x10^4-1x10^5 cells in GML-15 and the miorcplates were incubated at each temperature, respectively. After several hours post-incubation the cells were trypsinized and counted in a haemocytometer as described previously (2).

TO-2, LF, CF, BGK cells were tested for their ability to resist a prolonged incubation at 4°C. In this experiment, each cultured cell line was placed in 4°C. Three, 6, 9, or 12 months post-freezing, the cells were subcultured using GML-15 or GML-15+0.15 mM/ml NaCl and incubated at 31°C. The cell growth was then observed daily.

Viral Suceptibility

The lines were tested for their ability to replicate the following viruses: eel virus America (EVA), eel virus Europe (EVE), eel virus European X (EVEX), loach virus (LV-1) (17), clam virus (CV) (18), various serotypes(SP, AB and VR-299) of infectious pancreatic necrosis virus (IPNV),infectious hematopoietic necrosis virus (IHNV) and viral hemorrhagic septicemia virus (VHSV). The tested viruses were propagated in CHSE-214, RTG-2 and TO-2 cells respectively in GM L-15. Supernatants containing 100-5,000 $TCID_{50}$ of infective virus were used as inocula for viral susceptibility experiments.

Cell Storage

Each cell line was tested for its viability after liquid nitrogen storage. The confluent monolayers were trysinized, washed, pelleted and resuspended in GM L-15 to give a final concentration of 10^7-10^8cell per ml. Subsequently, approximately 10% glycerin or dimethylsulfoxide (DMSO) was added into the cell suspension, respectively. The cells were then stored in liquid nitrogen as described previously (2). Six to 12 months after storage, the cells were tested for their viability by re-culturing in GM L-15 at an incubation temperature of 28-31°C.

Effect of Cu^{++}, Cd^{++} and Hg^{++} on the Cell Line

The toxicity of $CuSO_4.5H_2O$, $3CdSO_4,8H_2O$ and $HgCl_2$ on cell survival was determined according to the procedures described by Fernandez-Pol et al (19) and Wang and Yew(14) with slight modification using 24-well tissue culture microplates (NLNC, Denmark). To each well 5×10^5 cells were seeded. Subsequently, the microplates were incubated for 24 hrs at 28°C to let them attach firmly to the plate. Compounds were then added into each well at various concentrations ranging from 0.1 to 100 ppm. Following 24 hours incubation at 31°C the cells were washed in GM L-15 twice and resuspended in GML-15. For the following 4, 8 hours after treatment the microplates were shaken gently to detach dead cells. After 12 hours post-incubation the attached cells were washed with PBS, trypsinized and cell numbers were counted with a haemocytometer. A fifty percent inhibitory dose of cells (ID_{50}) was then determined and used as an index of the toxcity of these compounds. For a control experiment DMSO at 0.4% (v/v) was added into the culture medium. All the heavy metal compounds were purchased from Merck, West Germany.

Results

Cell Line Establishment

Twelve cell lines (Figs. 1-12) from gonad, kidney, gill, fin and membrane covering brain of eel (A. japonica), tilapia (S. mossambica x S. nilotica), top-minnow (G. patruelis), common or color carp (C. carpio), loach (M. anguillicaudatus) and snakehead (C. maculata) were established by using L-15 plus 10-20% FCS (Table I). Using L-15 plus 20% FCS and 0.15 mM/ml sodium chloride, two cell lines from kidney (BGK) and fin (BGF) of banded grouper (E. awoara) (Table I; Figs. 13-14)were established. BGK could be subcultured in L-15 plus 20% FCS at a passage level after 25 under an incubation temperature of 28-31°C. Amongst these cell lines, only EO-2, and BGK showed a fibroblastic morphology. TK-1, TO-2, TM, BR, FI, GI, CF, LF, SK and BGF revealed an epithelioid morphology. Both epithelioid and fibroblastic cells were observed in EK-1 and GON cell lines.

Sterility test performed for each cell line showed that no bacteria, fungus and mycoplasma contamination was detected. Following frozen preservation in GM L-15+10% glycerin or DMSO for at least 6 months, all the tested cell lines remained viable upon rapid thawing and seeding in suitable medium, and cells grew readily into a confluent cell sheet.

The studies on viral sensitivity showed that most of our cell lines are susceptible to IPNV and viruses closely related to IPNV such as CV and LV-1. Rhabdoviruses, EVA and EVEX, isolated from eel (A. japonica) were also able to infect these cell lines. In contrast, no cell lines were found to be infected by IHNV and VHSV that are important for diseases of cold-water fish. These cell lines no doubt provide a valuable tool for the isolation and investigation of viruses for warm-water fish. Quite a few papers related to fish virology have been published using our cell lines (18, 20, 21, 22, 23, 24).

The results of toxicology studies showed that BGK cells are very sensitive to mercury chloride and resistant to copper sulfate. They also revealed that the

Table. 1 Origin, subculture, morphology and application of 14 cell lines derived from tissues of warm-water fish

Cell Line designatior	Species cf origin	Tissue of origin	Date initiated	Number of subcultures through Oune, 1987	Cell morphology	Application
EK-1	Eel *Anguilla japonica*	Kidney	October, 1979	225	E+F	Virus isolation
EO-2	Eel *Anguilla japonica*	Ovary	October, 1979	210	F	Virus isolation
TK-1	Tilapia hybrid *Serotherodon mossambica* X *S. nilotica*	Kidney	October, 1981	225	E	Virus isolation DNA repair after UV irradiation
TO-2	Tilapia hybrid *S. mossambica* X *S. nilotica*	Ovary	October, 1981	365	E	Virus Isolation Toxicology DNA repair after UV irradiation heat shock protein Recombinant DNA
TM	Top-minnow *Gambusia Patruelis*	Minced Tissue	September, 1983	121	E	Virus isolation Toxicology
BR	Color carp *Cyprinus Carpio*	Membrane covering brain	December, 1985	65	E	Virus isolation
FI	Color carp *C. carpio*	Fin	December, 1985	68	E	Virus isolation
GI	Color carp *C. carpio*	Gill	December, 1985	72	E	Virus isolation
GON	Color carp *C. carpio*	Gonad	December, 1985	115	E+F	Virus isolation
CF	Common carp *Cyprinus carpio*	Fin	October, 1981	175	E	Virus isolation
LF	Loach *Misgunus anguillicaudatus*	Fin	Cctober, 1981	295	E	Virus isolation Toxicology DNA repair after UV irradiation
SK	Snakehead *Channa maculata*	Kidney	July, 1986	54	E	Virus isolation
BGF	Banded grouper *Epinephelus awoara*	Fin	October, 1985	45	E	Virus isolation
BGK	Banded grouper *Epinephelus awoara*	Kidney	October, 1985	72	F	Virus isolation Toxicology heat shock protein

[a] The cells had been stored in liquid nitrogen for few years.
[b] Fibroblastic + Epithelioid
[c] Fibroblastic
[d] Epithelioid

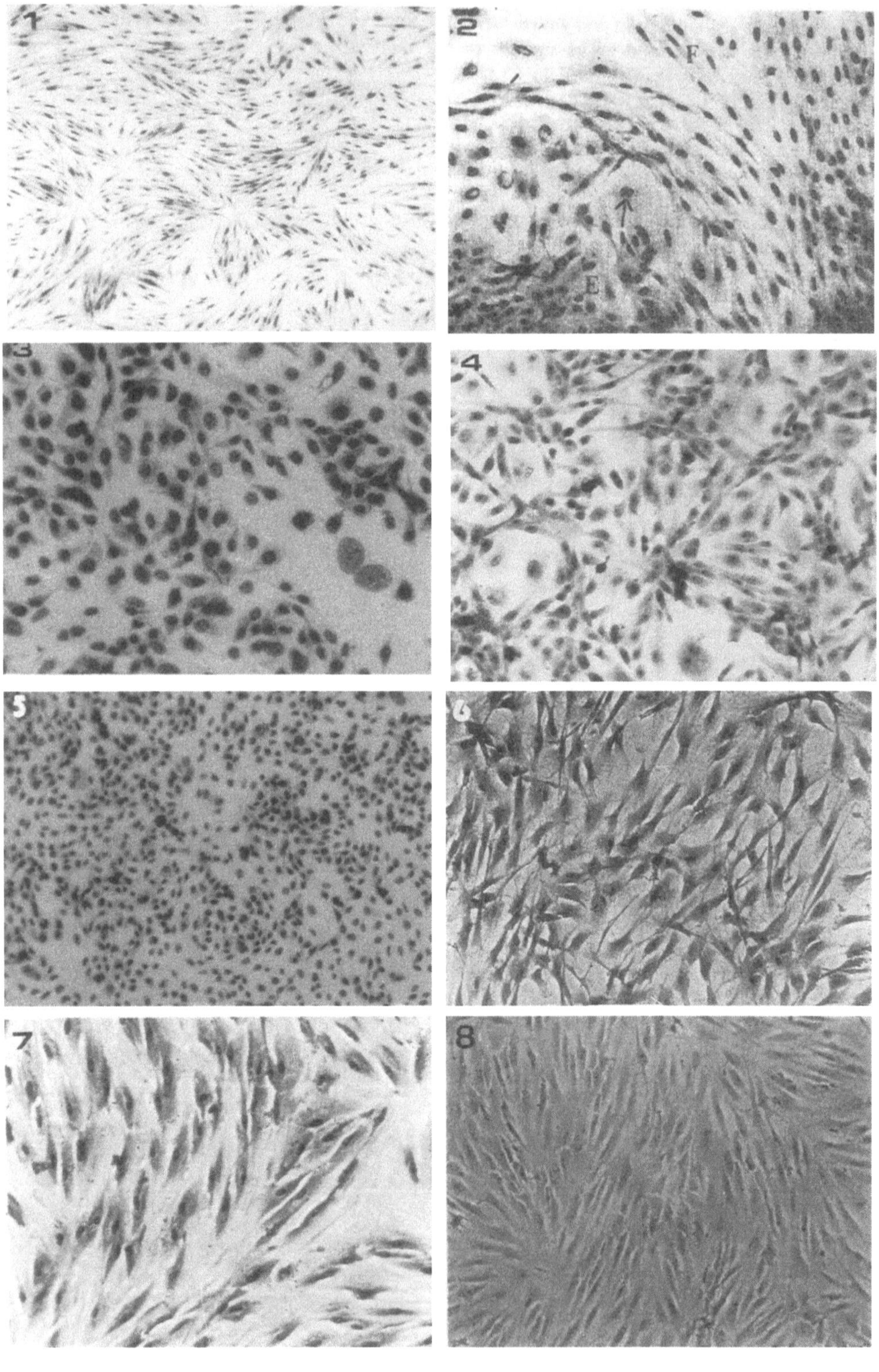
1
2
F
E
3
4
5
6
7
8

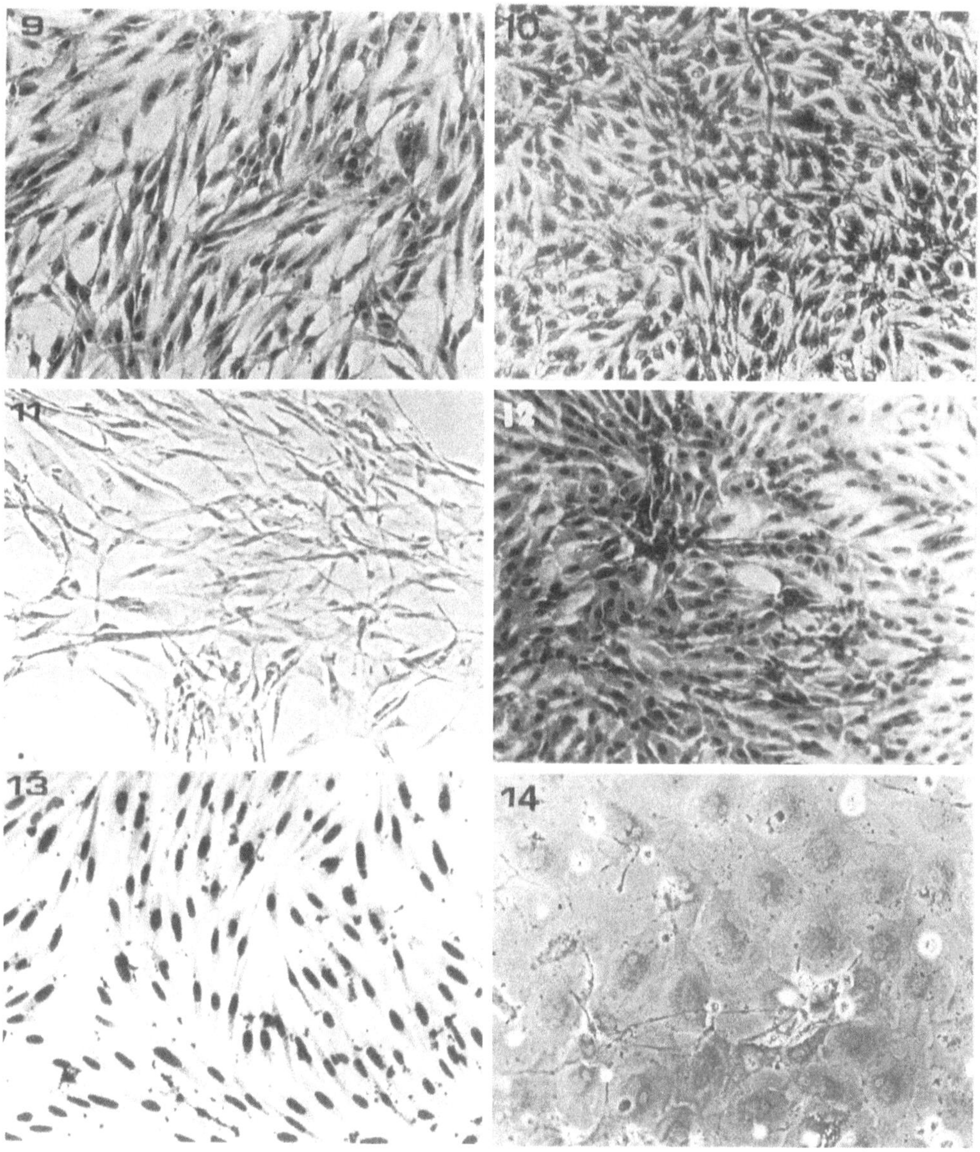

Figs. 1-14. (1-13,200X;14,400X) Cell Line derived from warm-water fish 1. TO-2 2. TK1 (E:Epithelioid cells; F: Fibroblastic cells) 3. TK-1 4. TO-2 5. CF 6. LF 7.SK 8. TH 9. BR 10. GON 11. FI 12. GI 13. BGk 14. BGF

Table 2. Model chromosome numbers, plating efficiency and ability to grow in soft agar of 14 cell lines

Cell lines	Model chromosome number (Passage No.)	Plating efficiency[a] (Passage No.)	Ability to grow in soft agar (Passage No.)
EK-1	38 (100)	17 (120)	ND
EO-2	38 (100)	25 (160)	ND
TK-1	38 (100)	20 (135)	ND
TO-2	40 (80)	13 (80)	ND
TM	38 (60)	12 (69)	ND
BR	ND[c]	7 (44)	-[b] (44)
FI	ND	12 (56)	- (50)
GI	ND	17 (67)	- (43)
GON	NC	25 (59)	- (65)
SK	ND	13 (46)	- (44)
LF	ND	27 (155)	- (125)
CF	ND	16 (90)	- (120)
BGF	ND	17 (41)	- (44)
BGK	ND	28 (60)	- (57)

[a] plating efficiency is (colony/cells seeded) X100.
[b] - No cell growth was observed on soft agar.
[c] ND: Not done.

Table 3. Susceptibility of 14 cell lines to viruses

Cell lines	Viruses									
	IPNV-AB	IPNV-SP	IPNV-VR299	LV-1	CV	EVE	EVA	EVEX	IHNV	VHSV
EK-1	+	+	+	+	+	+	+	+	-	-
EO-2	+	+	+	+	+	+	+	+	-	-
TK-1	+	+	+	+	+	+	+	+	-	-
TO-2	+	+	+	+	+	+	+	+	-	-
TM	+	+	+	+	ND	+	+	+	-	-
BR	+	+	+	-	ND	+	ND	ND	-	-
FI	+	+	+	+	ND	+	ND	ND	-	-
GI	+	+	+	+	ND	+	ND	ND	-	-
GON	+	+	+	+	+	+	ND	ND	-	-
SK	+	+	+	+	+	+	+	+	-	-
LF	-	-	-	-	-	-	-	-	-	-
CF	+	+	+	+	+	+	+	+	-	-
BGF	+	+	+	ND	ND	ND	ND	ND	-	-
BGK	+	+	+	+	+	+	+	+	-	-

Susceptibility of each cell line to viruses was based on the occurrence of cytopathic effect (CPE) in the infected cells.
+: CPE was observed.
-: no CPE was observed.
ND: Not done.

sensitivity of BGK cell lines towards heavy metal compounds varied extensively. Using TC-2, TM and LF cells, differential responses of these cell line toward butachlor, paraquat, saturn and azodrin were also obtained (14). The use of cell lines for toxicity tests is known to be a quick and sensitive method. Easy cultivation and differential responses toward various compounds may be valuable for the application of fish cell lines to water pollution studies.

Table 4. Inhibitory doses of C^{++}, Cd^{++} and Hg^{++} on growth of BGK and LF cells

Compounds	ID^{a}_{50} (ppm)	
	BGK	LF
Hg^{++}	0.32	6.81
Cd^{++}	6.28	7.15
Cu^{++}	10	>50

ppm: Parts per million

a: Concentration of compounds in growth medium that caused a 50% reduction in cell number after 24 hrs. of incubation at 31°C.

Cu^{++} = $CuSO_4.5H_2O$

Cd^{++} = $3CdSO_4.8H_2O$

Hg^{++} = $HgCl_2$

In addition to virology and toxicology studies, fish cell lines have also been used for studies on heat-shock protein, DNA repair and recombinant DNA as described in Table 1. Using TO-2 cells, heat-shock protein was induced after being incubated at 37°C for 10 minutes (13). TK-1 and TO-2 cells were also demonstrated to possess excision and post-replication repair after UV-irradiation (10, 13). Work is in progress on the studies of toxicology, cell physiology and recombinant DNA using our cell lines and the results will be published elsewhere.

Cell Line Characterization

In low seeding density of $1X10^2$-$1X10^4$ per 25cm Flacon flask, plating efficiency of less than 30% was obtained for the cell lines(Table 2). Chromosome number from randomly sampled 100 metaphase of EK-1, EO-2, TK-1 or TM showed a modal number of 38. The modal number for TO-2 was found to be 40 (Table 2).

Table 2 also showed that no growth was observed when soft agar was used to culture BR, FI, GI, GON, SK, LF, CF, BGF and BGK cells, respectively.

The effect of various concentrations of FCS was determined for each cell line. The results showed that FCS at a concentration level of 2-20% could enhance the cell growth. 20% and 10% FCS promoted better growth of cells than 5% and 2% FCS (Fig. 15).

Optimal growth temperature for all cell lines ranged from 24-32°C. At 37°C the cells grew slowly and eventually died. However, at a temperature between 15-24°C the cells remained viable and grew slowly (Fig. 16).

When TO-2, CF cell line was kept in 4°C for 3 month, subculture for the cells was a failure. However, when LF cells were maintained at 4°C for 6-12 months, the cells remained viable and cell growth continued after subculture. In the experiment performed on BGK cells, majority of the cells were dead after 4°C freezing for approximately 3 months. A few BGK cells remained viable and a new monolayer cell sheet was formed approximately 20 days after subculture.

Viral Susceptibility

Table 3 shows that except LF all the cell lines were susceptible to IPNV of various serotypes, LV-1, CV, EVE EVA and EVEX but refractory to IHNV and VHSV. No viral susceptibility was observed in LF cell line.

Effect of Heavy Metal Compounds on Cell Growth

Studies on LF and BGK cells exposed to heavy metal compounds were performed and the results are shown in Table 4. Copper sulfate ($CuSO_4.5H_2O$) did not result in any inhibition of growth in LF cells. However, copper sulfate at 10 ppm inhibits BGK cell number in 50%. When BGK and LF were exposed to cadmium

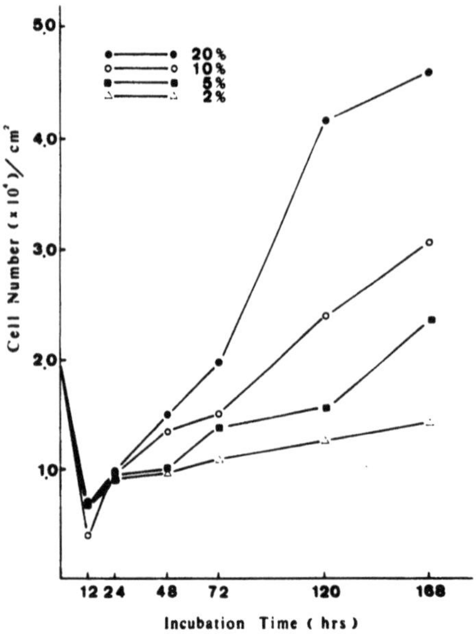

Fig.15

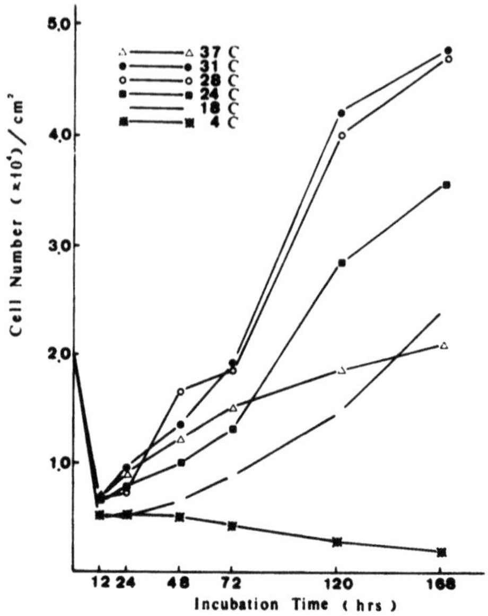

Fig.16

Fig.15. Growth Curve of LF Cells at Different Concentrations of Foetal Calf Serum. The cells were cultured in L-15 medium and inoculated in 24 well microplates at 28 C. Cell counts were determined using a haemocytometer. Each point was obtained from 4 wells.

Fig.16. Growth Curve of LF cells at Different Temperatures. The cells were cultured in L-15 (20% FCS) and inoculated in 24 well microplates. Cell counts were determined using haemocytometer. Each point was obtained from 4 wells.

sulfate($3CdSO_4.8H_2O$),ID_{50} values of 6.28 or 7.15 ppm were obtained, respectively. Mercury chloride ($HgCl_2$) showed highest toxicity to BGK cells with ID_{50} value of 0.32 ppm. However, this compound gave ID_{50} value of 6.81 ppm for LF cells.

Discussion

In the present study we have obtained the conditions for the establisment of cell lines from warm-water fish. Although minimal essential medium (MEM) is the most commonly used medium for poikilotherm tissue culture, our results suggest that Leibovitz's L-15 also possesses the potentiality for fish cell lines, especially for warm-water fish.

The optimal growth temperature for the established cell line ranged from 24-32 $^{\circ}$C which is similar to the preferred temperature of warm-water fish. Amongst 4 cell lines tested only LF tolerated 4°C. In contrast, most of the cell lines from cold-water fish remained viable after storage at 4°C for a minimum of 3 months without change of medium (6). In the work on BGK cells, a new population arose from surviving cells after incubation at 4°C for 3 months. Studies are in progress on the possible mechanism for the alteration of BGK cells through 4 $^{\circ}$C storage.

The results obtained from plating efficiency, karyology and soft agar culture suggest that none of the 14 cell lines have been transformed.

References

1. Wolf, K and Mann, J.A. In Vitro, 16: 168-179, 1980.
2. Chen, S.N. and Kou, G.H. Fish Pathol., 16(3): 129-137, 1981.
3. Chen, S.N. and Ueno, Y. and G.H. Kou. Proc. Natl. Sci. Counc. B. ROC., 6(1): 93-100, 1982.
4. Chen, S.N., Chi, S.C., Ueno, Y. and G.H. Kou. Fish Pathol., 18(1): 13-18, 1983.
5. Chen, S.N., Wen, S. C., Ueno, Y. and Kou, G.H. Bull. Euro. Fish Pathol., 3: 1-4, 1983.
6. Lannan, C.N., Winton, J.R. and Fryer, J.L. In Vitro, 20(9): 671-676, 1984.
7. Li, M.F., Marrayatt, V., Annada, C. and Odense, P. Can. J. Zoo., 63: 2867-2874, 1984.
8. Wang, S.Y., Chen, S.N. and Kou, G.H. Proc. Natl. Sci. Counc. B. ROC., 11 , 1987 (In press).
9. Li, M.F. and Clyburne, S. Environmental Protection Service Halifax, Tech. Rep. No. EPS-5-Ar-77-1, PP. 133-140, 1976.
10. Yew, F.H. and Chang, L.M. J. Cell Sci., 72: 213-226, 1984.
11. Lin, J.K. Hypersensitivity of a Loach Fin Cell Line. Master Thesis, Department of Zoology, National Taiwan University, 1984.
12. Yeh, P.Y., Chang, L.M. and Yew, F.H. J. Chinese Biochem. Soc., 15(1):17:23, 1986.
13. Chen, J.D. and F.H. Yew. J. Cell Sci, 75, 1987. (In press).
14. Wang H.C. and Yew, F.H. Bull. Inst. Zool., Academia Sinica, 26(1): 95-106, 1987.
15. Russell,W.C. Nature, 212: 1537-1540, 1966.
16. Adams, R. L. P. Laboratory Techniques in Biochemistry and Molecular Biology (Work, T.S. and Burdon, R.H. General eds.), PP. 102-103, Elsevier/North-Holl and Press, 1980.
17. Chen, S.N., Chi, S.C., Guu, J.J., Chen, J.C. and Kou, G.H. Report on Fish Disease Research, 6: 6-11, 1984.
18. Hong, Y. W. The Characteristic of the Virus Isolated from Gill of Clam (Meretrix lusoria). Master Thesis, Fu-Jen Catholic University, Taiwan, 1984.
19. Fernandez-Pol, J. A., Hamilton, P. D. and Klos, D.J. Cancer Res., 42:609-617, 1982.
20. Sorimachi, M. Fish Pathol., 17(2): 119-123, 1982.
21. Ueno, Y., Chen, S.N., Kou, G.H., Hedrick, R.P. and Fryer, J. Bull. Inst. Zool., Academia Sinica, 23(1): 47-55, 1984.
22. Chen J. J. S1 Nuclease Digestion of Partially Denatured IPNV Genome Double Strained RNA. Master Thesis, Fu-Jen Catholic University, Taiwan, 1987.
23. Lin, M.S. Viral Interference and Defective Interfering Particle of IPNV-HBI Isolated from Clam. Master Thesis, Fu-Jen Catholic University, Taiwan,1987.
24. Huang C. Y. Serological and Biochemical Properity of Birnavirus Isolated from Clam. Master Thesis, Fu-Jen Catholic University, Taiwan, 1987.

LONGEVITY OF RAINBOW TROUT ANTIBODY-PRODUCING CELLS AFTER IN VITRO IMMUNIZATION CULTURE

O. W. Dixon[1], D. P. Anderson[1], R. Kimenai[2], and E. F. Lizzio[3]

1) U.S. Fish and Wildlife Service, National Fish Health Research Laboratory, Kearneysville, West Virginia 25430, USA

2) Department of Experimental Animal Morphology and Cell Biology, Agricultural University, Wageningen, The Netherlands

3) Laboratory of Cell Biology, Division of Blood and Blood Products, Office of Biologics Research and Review, Center for Drugs and Biologics, Bethesda, MD 20892, USA

Spleens from large rainbow trout were excised and sectioned. Separate pieces were placed *in vitro* in EMEM, injected with dinitrophenyl (DNP)-Ficoll, and held at 15°C. The resultant immune response was measured by the passive hemolytic plaque assay to determine numbers of antibody-producing cells (PFC). The immunized spleen sections were held for as long as 42 days and periodically the PFC were monitored. The media was changed every other day by taking off half the total amount (10mℓ) and adding 5 mℓ of new. DNP-Ficoll stimulated PFC were still present at day 42 after immunization. The techniques of long term culture and immunization may be useful for the isolation of specific PFC.

In vitro immunization and culture is of great interest for many biological disciplines. Refinement of the techniques was begun by Globerson and Auerbach (4) who cultivated immunized organ sections from mice *in vitro* and showed the production of antibody to the antigens of sheep red blood cells. More widely used at present are the Mishell-Dutton assays (8) in which the individual cells are separated and the antibody-producing cells (PFC) occur in a suspension culture. These techniques have been important for development of hybridomas for the production of homologous monoclonal antibody used for immunodiagnostics and therapy.

Tissue culture and development of cell lines for fishes is of priority interest for pathogen detection, toxicity studies and research in fish physiology. Basic protocols for the development of cell lines (1, 9) and for primary culture of leukocytes (10) exist. In the following experiments we describe *in vitro* immunization and culture of spleen sections of rainbow trout immunized with dinitrophenyl-Ficoll (DNP-Ficoll) and the detection of the PFC by the passive hemolytic plaque assay.

Materials and Methods

Experimental fish. Rainbow trout (*Salmo gairdneri*) weighing 1.0-1.5 kg from the Fish Culture Research Laboratory, Kearneysville, West Virginia USA were held in 12°C spring water at the National Fish Health Research Laboratory. All trout were fed a pelletized dry feed obtained from Zeigler Brothers, Inc., Gardners, PA.

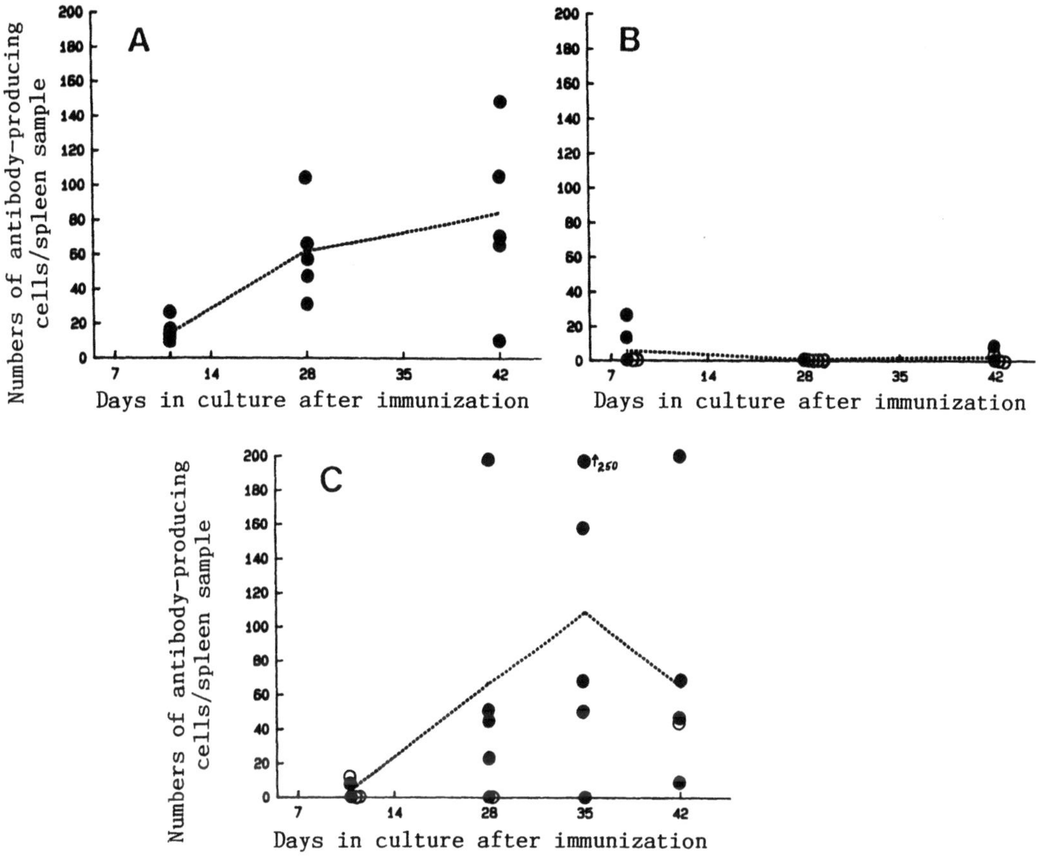

Figure 1. Numbers of antibody-producing cells from <u>in vitro</u> spleen section culture from 3 separate rainbow trout. A and C show data from 2 high responder fish; B shows a fish that did not respond well to the DNP antigen. The dotted line represents the mean of the 5 spleen samples taken for each time period.

\- - - - - - - - - - - -

<u>Antigens and Sheep Red Blood Cells</u>. Dinitrophenyl-Ficoll (DNP-Ficoll) and labeling of the sheep red blood cells (SRBC) were prepared according to the technique of Inman et al. (5).

<u>In vitro immunization and culture</u>. Trout were anesthetized in MS-222 and set under the laboratory hood where the peritoneal cavity was opened and spleen excised. The whole spleen was quickly rinsed in a sterile 100 mℓ beaker containing 50 mℓ of Eagle's minimum essential medium supplemented with 2% normal calf serum (EMEM-2). Within a few minutes, the spleen was aseptically cut into individual 5 mm diameter sections that were placed into individual 60 mm petri plates containing 10 mℓ EMEM-2. Sections were injected with 100 ug DNP-Ficoll in 0.1 mℓ phosphate buffered saline (pH 7.4). and incubated at 17°C. EMEM-2 (5 mℓ) was changed every other day.

<u>Sampling for immune response</u>. The passive hemolytic plaque assay (PHPA) or Jerne assay was used for demonstrating the numbers PFC of forming clear

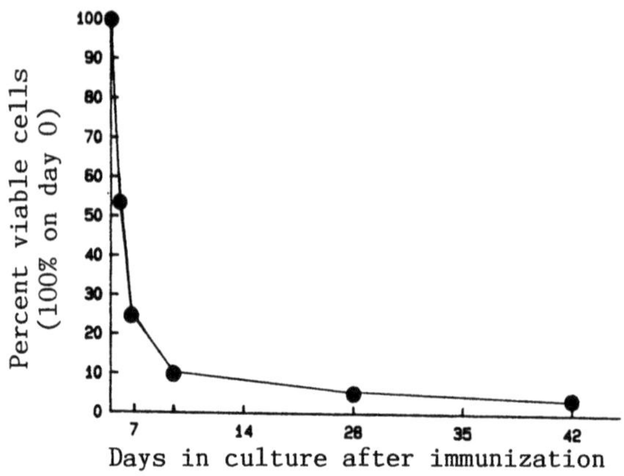

Figure 2. Numbers of viable white blood cells as determined by trypan blue vital staining from long term spleen section cultures. Day "0" is used for 100% viability to compare with samples on days 1, 5, 10, 28, and 42.

plaques in an agar matrix containing DNP-labeled SRBC, as described by Anderson et al. (3). Briefly, on days selected for sampling (given in Figures), organ sections for each fish sampled were placed in petri dishes with new media and scraped over a wire screen, separating the PFC into suspension. The isolated cells were then poured with the indicator SRBC onto a microscope slide and incubated for 4 hours after which homologous complement was overlaid for the induction of hemolytic lysis. Viable cells were counted with a hemacytometer by trypan blue staining on separate samples from the cell suspensions.

Results and Discussion

Standard protocols for in vitro culture and immunization in our laboratory give day 10 after injection as the time for analysis of the immune response in the rainbow trout organ sections. In these experiments we set up extra petri plates for holding organ sections for following the presence of the PFC throughout this period. In results shown here, two fish (Figure 1. A, and C) proved to be good immunological responders to the DNP antigen, whereas, a third fish did not give significant numbers of PFC (Figure 1. B). While the numbers of PFC were highly variable among sections, especially in the latter samples, they remained high throughout the culture period in the first two fish. In contrast, the numbers of total viable cells in the organ section were greatly reduced. Figure 2 shows the numbers of viable cells as determined by trypan blue staining after 0, 1, 5, 10, 21, and 42 days in section culture. Cell death occurred rapidly during the first 5 days in vitro; the rate declined thereafter.

Microsystems using separated cell suspensions have been developed for the primary and secondary immunization of channel catfish leukocytes, much as described by the organ section culture here with rainbow trout (7). Our preliminary experiments with cell suspension techniques have not yielded as high numbers of PFC as the organs sections, and the cells have not remained viable for the extended periods of time.

The medium used for the culturing of the spleen sections is the same used routinely for fish cell cultures. In each petri plate containing 10 mℓ, half of the medium was changed every other day, and growth promoters such as

Interleukin-II or conditioned media were not used. It has been reported (6) that fetal bovine serum used in the media can act as a mitogen, stimulating the numbers of nonspecific and specific PFC. Background PFC using the DNP antigen systems have always been problematical in vitro as well as in vivo; in contrast, the Yersinia ruckeri O-antigen preparations models give very little background (2). Indeed, if the DNP-SRBC assay slides were held beyond 4 hours after complement was applied, background numbers of PFC increased. There was an increase of specific PFC during these culture periods, but we don't know if the increase was due to derived clones from original PFC or new PFC were being stimulated constantly from residual DNP-Ficoll in the culture media. Nevertheless, immunization and long term culture of organ sections may have important implications as the ratio of PFC/viable cells is increased. Thus, the specific antibody-producing cells would be easier to isolate for making hybridomas for monoclonal antibody production, for use in toxicological assays, and for studying the cellular mechanisms and intercommunications of the fishes' immune response.

References

1. Ahne, W. and Bachman, P. A. Zbl. Bakt. Hyg., 228:447-464, 1974.
2. Anderson, D. P., Dixon, O. W. and Lizzio, E. F. Vet. Immunol. Immunopath., 12:203-211, 1979.
3. Anderson, D. P., Dixon, O. W., and Roberson, B. S. Dev. Comp. Immunol., 3:739-744, 1986.
4. Globerson, A. and Auerbach, R. J. Exp. Med., 124:1001-1016, 1966.
5. Inman, J. K., Merchant, B., Claflin, L. and Tacey, S. E. Immunochem., 10:165-174, 1973.
6. Kaattari, S. L. and Yui, M. A. Dev. Comp. Immunol., 11:155-165, 1987.
7. Miller, N. W. and Clem, L. W. J. Immunol. Methods, 72:367-379, 1984.
8. Mishell, R. I. and Dutton, R. W. J. Exp. Med., 126:423-443, 1967.
9. Wolf, K. and Quimby, M. C. Tissue Cult. Assoc. Manual, 2:441-443, 1976.
10. Wolf, K. and Quimby, M. C. Tissue Cult. Assoc. Manual, 2:471-474, 1976.

INVESTIGATIONS OF ANTIBODY-PRODUCING CELLS FROM SPLEENS AND ANTERIOR KIDNEYS OF RAINBOW TROUT, STRIPED BASS AND GOLDFISH

D.P. Anderson[1], T. Kitao , T. Yoshida[2], and O.W.Dixon[1]

1) U.S. Fish and Wildlife Service, National Fish Health Research Laboratory, Kearneysville, West Virginia 25430, USA
2) Department of Fisheries, Faculty of Agriculture, Miyazaki University, Kumano, Miyazaki 889-21, Japan

Rainbow trout, striped bass and goldfish were injected with Yersinia ruckeri bacterins. After 14 days, spleens and anterior kidneys were excised for enumerating antibody-producing cells (PFC). Numbers were higher in anterior kidneys of goldfish and striped bass when compared to the spleens, whereas, trout numbers were nearly the same for both organs. Goldfish and striped bass gave higher total numbers of PFC than trout. These assays are being expanded to investigate the locations of the PFC in tissues and organs during an immune response in fishes at different phylogenetic levels.

The passive hemolytic plaque assay (PHPA) or Jerne assay enumerates specific antibody-producing cells from a suspension by demonstrating a clear plaque in an agar matrix containing homologous, antigen-labeled sheep red blood cells (SRBC); these thus are called plaque forming cells (PFC). The assay has been used to determine the kinetics of the plasma response in rainbow trout against Yersinia ruckeri bacterins (2) or against Streptococcus sp. (6). Other fish species investigated include common carp (9), catfish (8), and Tilapia mossambica (5) immunized with SRBC and coho salmon with Vibrio anguillarum bacterins (7). These authors described variations of the PHPA method and discussed the timing and organs where the PFC are found after immunization. In addition, it is remarkable that while the complement used for final step of the assay can be used interchangeably among mammalian genera, fish complement is quite specific for a given genus. In the following studies we used the PHPA to further study the major organs of resident PFC in rainbow trout, striped bass and goldfish.

Materials and Methods

Experimental fish. Rainbow trout (Salmo gairdneri), weighing 200-250 g each were from the Fish Culture Research Laboratory, Kearneysville, West Virginia USA. Goldfish (Carassius auratus) weighing 150-170 g each were given courtesy of Dr. K. Hirano, Miyazaki University, Japan, and striped bass (Morone saxatilis) weighing 200-250 g each were from the Edenton National Fish Hatchery, North Carolina USA. All fish were held in the laboratory spring water (12°C) and maintained on pelletized feed obtained from Zeigler Brothers, Inc., Gardners, PA.

Antigens. Fish were injected with Yersinia ruckeri O-antigen preparation routinely used in our laboratory (2) with 1 mg (acetone-dried weight) suspended in phosphate buffered saline (pH 7.4). The bacterial antigen is a particulate, skeletal, bacterial complex prepared by a method described by Edwards and Ewing (3). The SRBC were labeled by incubation with 1 mg/mℓ O-antigen (10% SRBC) for 1 hr at 37°C.

Sampling procedure. Fish were sampled on day 14 after immunization with a brief immersion in anesthetic (MS-222), severing the tail and draining the blood into a test tube for later use as a complement source. The peritoneal cavity was opened and the spleen and anterior portion of the kidney excised and placed in individual 60 mm sterile plastic petri dishes containing 7 mℓ of Eagle's minimum essential media with 2% fetal bovine sera (EMEM-2). The organs were macerated on a chromium-coated wire screen and the cell suspension filtered through glass wool before a centrifuge run at 800 g. The cell pellet was then resuspended in 1 ml of EMEM-2 and briefly placed on the vortex. Test tubes containing 0.2 ml liquid agarose and 0.2 ml EMEM-2 (2x concentration) were held in an incubator at 47°C. For the individual samples, the tubes were taken out of the incubator bath and 0.2 ml of O-antigen labeled SRBC was quickly added. Finally, 0.1 ml of the cell

Table 1. Numbers of antibody-producing cells (PFC) in the spleens and anterior kidneys of fish immunized with *Yersinia ruckeri* O-antigen as detected by lysis of *Y. ruckeri*-labeled sheep red blood cells (SRBC) or negative control, unlabeled SRBC (PFC/10^6 white blood cells).

	Spleen		Anterior kidney	
	Labeled SRBC	Control SRBC	Labeled SRBC	Control SRBC
Rainbow Trout:				
Fish #1	10	1	9	0
#2	400	2	600	0
#3	350	0	159	0
#4	300	0	350	0
#5	210	0	250	0
#6	179	2	402	4
*#7	2	1	1	1
*#8	1	1	1	1
Goldfish:				
Fish #9	480	0	800	3
#10	61	1	216	3
#11	86	0	129	0
#12	10	0	9	0
*#13	1	0	1	0
Striped Bass:				
#14	11	0	120	5
#15	14	4	150	2
#16	47	1	104	0
#17	1	0	3	0
#18	71	2	40	0
#19	164	0	70	0
*#20	1	0	2	0
*#21	4	0	1	0

*Control fish; not immunized

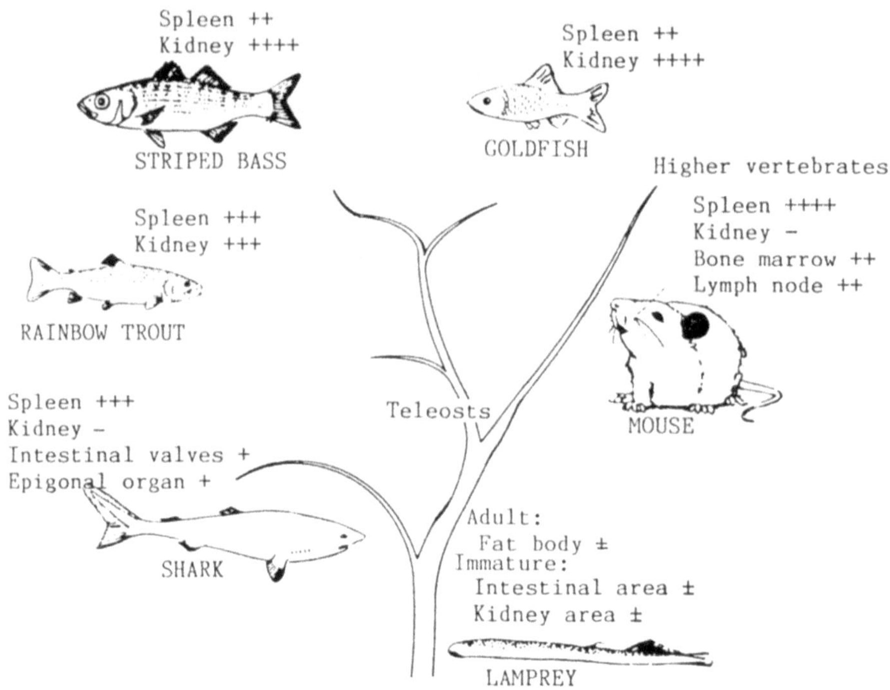

Figure 1. Relative numbers of antibody-producing cells (PFC) located in tissues and organs of selected species.

suspension was added and the mixture held briefly on the vortex before pouring onto a microscope slide. Slides were incubated for 4 hrs or overnight in trays over crushed ice before 0.9 ml of a 1:20 dilution of complement from the respective species was added. PFC were counted at 2, 4, and 24 hrs later for confirmation.

Results and Discussion

Fish generate PFC upon antigenic stimulation similar to the mammalian models, except that the appearance is dependent upon environmentally influenced body temperatures. In addition, for most fish species investigated, only one class of humoral antibody has been described, in contrast to a heavy (IgM) to light (IgG) molecular switch found in higher vertebrates. A third difference is the location and concentrations of PFC cells in various organs. The mouse models described show the spleen as a preferred source for obtaining mature plasma cells for the PHPA assays, the bone marrow and lymph nodes being generative hemopoietic areas and having fewer PFCs. In mammals, few PFCs are located in the kidney. In contrast, in each of the fish species studied, a major organ for PFC residency was the anterior kidney, the spleen being the second organ of importance for the striped bass and goldfish. In rainbow trout and other salmonids, the numbers of PFC seem about equally divided between the anterior kidney and the spleen.

In earlier experiments using PHPA assays, we immunized mature sea lampreys (Petromyzoan marinus) with the Yersinia ruckeri bacterin and found PFC in the residual intestinal hemopoietic areas. Other studies using immunocytoadherence have shown that in immature lamprey ammocoetes, the tyhlosole (midgut wall) and the nephric area contain antigen-binding cells, which are lost in the mature fish as the areas involute and the gut fat body remains the vestigial hemopoietic area (4). The spleen is the main location of immunoglobulin-forming cells in the bullhead shark (12) and skate (13); other lymphoid areas include the central area near the intestinal valves and the epigonal organ near the esophagus. The cells of these elasmobranchii were identified by using anti immunoglobulin sera rather than demonstrating specific antibody production; no immunopoiesis occurs in the kidney. Fishes have no bone marrow and the higher the species position is on the phylogenetic scheme the more reliance there seems to be on the anterior kidney hemopoietic areas for generation and storage of PFC (Figure 1). The complete picture of how and where each fish species generates PFC, and the migrations and kinetics of these cells will reveal an interesting phylogenetic story.

In each of these species, homologous complement had to be used, demonstrating the genetic divergence of fishes when compared to mammals. Indeed, osmotic levels, temperatures, and other physiological parameters may have to be altered for each fish species to optimize in vitro conditions for sampling the immune response.

References

1. Anderson, D. P., Dixon, O. W., and Lizzio, E. F., Vet. Immunol. Immunopathol., 12:203-211, 1987.
2. Anderson, D. P., Roberson, B. S., and Dixon, O. W., J. Fish. Res. Board Can., 36:636-639.
3. Edwards, P. R. and Ewing, W. H. Identification of Enteroacteriacease. Burgess Publishing Co., Minneapolis, Minnesota. 362 pp.
4. Fujii, T., Nakagawa, H., and Murakawa, S. Dev. Comp. Immunol., 3:609-620, 1979.
5. Jayaraman, S. R., Monan, R., and Muthukkaruppan, V. R. Dev. Comp. Immunol., 3:67-75, 1979.
6. Kitao, T., Aoki, T., and Kanada, M. Dev. Biol. Standard., 49:355-368, 1981.
7. Maule, A. G., Schreck, C. B., and Kaattari, S. L. J. Fish Aquat. Sci., 44:161-166, 1987.
8. Miller, N. W. and Clem, L. W. J. Immunological Meth., 72:367-379, 1984.
9. Rijkers, G. T., Fredrix-Wolters, E. M. H., and van Muiswinkel, W. B. J. Immunological Meth., 33:79-86, 1980.
10. Secombes, C. J., van Gronigen, J. J. M., van Muiswinkel, W. B., and Egberts, E. Dev. Comp. Immunol., 7:455-464, 1983.
11. Smith, A. M., Potter, M., and Merchant, E. B. J. Immunol., 99:876-882, 1967.
12. Tomonaga, S., Kobayashi, K., Hagiwara, K., Sasaki, K., and Sezaki, K. Dev. Comp. Immunol., 9:617-626, 1985.
13. Tomonaga, S., Kobayashji, K., Kajii, T., and Awaya, K. Dev. Comp. Immunol., 8:803-812, 1984.

Part 7

NEW CELL LINES AND THEIR CHARACTERIZATION

DEVELOPMENT OF CELL LINES FROM VARIOUS TISSUES OF LEPIDOPTERA

D. E. Lynn,[1] E. M. Dougherty,[1] J. T. McClintock,[1] and M. Loeb[2]

1) USDA/Agricultural Research Service, Insect Pathology Laboratory, Beltsville, MD 20705, U.S.A.
2) Insect Reproduction Laboratory, Beltsville Agricultural Research Center, Beltsville, MD 20705, U.S.A.

Continuous cell lines were developed from primary cultures of embryos and larval fat bodies of *Lymantria dispar* and larval testes of *Heliothis virescens*. Four morphologically distinct lines were obtained from the embryonic cultures through various selection techniques, a highly vacuolated line was obtained from the fat body tissues, and epithelial cells were obtained from the testes. These lines have been designated IPLB-LdEp, IPLB-LdEG, IPLB-LdEI, IPLB-LdEIt, IPLB-LdFB and IPLB-HvT1, respectively, and have been characterized by isozyme and chromosome analyses, growth kinetics, and susceptibility to various insect viruses.

Establishment of insect cell cultures has become relatively routine in recent years for many insects, particularly undifferentiated tissues from Lepidoptera and Diptera. Many of these lines have been useful for studies on insect viruses, including cells from gypsy moth (*Lymantria dispar*) ovaries developed by Goodwin et al.(1). However, compared to certain virus/cell systems [i.e. *Autographa californica* nuclear polyhedrosis virus, AcNPV, in the TN-368 cell line (2)], the ovarian gypsy moth cell lines were relatively poor producers of *L. dispar* nuclear polyhedrosis virus (LdNPV). We believed better lines for virus production could be obtained from embryonic and larval tissues since the primary effect of the virus is on the immature insect stages. Other studies at Beltsville have involved primary cultures of tobacco budworm (*Heliothis virescens*) testes to examine developmental processes. During some studies, cell outgrowths were noticed and attempts to culture these cells were undertaken. These tissues produced ecdysteroids *in vivo* and *in vitro* (3), so we believed these cultures could be a means of obtaining hormone synthesizing cells. Additionally, the *Heliothis zea* nuclear polyhedrosis virus (HzNPV) does not replicate well in currently available cell lines. Since HzNPV is infectious to *H. virescens*, new lines from this species may be more effective at replicating this virus. We were successful in obtaining four continuous cell lines from gypsy moth embryos, one from larval fat body and one from testicular sheath of the tobacco budworm.

Materials and Methods

Insects and medium. Laboratory reared insects were the source of tissues for primary cultures. Gypsy moth cultures were initiated in a 3:1 mixture of IPL-52B and IPL-76 (4) supplemented with 1% Nuserum (Collaborative Research, Lexington, MA), 9% heat treated fetal bovine serum (FBS)(K. C. Biological), and 50 µg/ml gentamicin (HEM Research, Rockville, MD). *H. virescens* testes cultures were initiated in modified Grace's medium supplemented with 14% FBS. The procedures for disinfection and dissections are as previously described (3,5).
Embryo cultures. *L. dispar* eggs at several different developmental stages were used during the primary culture attempts. Prior to disinfection, the eggs were "cleaned" of the extensive hairs covering the egg mass by rinsing 5 min in 0.025% NaOCl.
Larval cultures. Fourth and fifth instar larvae were dissected in culture medium. Primary cultures were initiated from imaginal discs and fat bodies in 0.1 ml standing drop cultures, with tissues cut into small fragments with a microscalpel. Cultures of larval hemocytes were initiated by bleeding surface-disinfected larvae into ice-cold medium or Hanks' phosphate buffered saline. Since gypsy moth hemo-

lymph rapidly melanizes in medium, the hemocytes were separated from the primary receiving media and resuspended in fresh medium. All cultures were sealed with parafilm and incubated at 28°C. Fourteen primary cultures of embryos, nine of hemocytes, two of imaginal discs and three of fat bodies were initiated from January through April, 1985. All subsequent handling of primaries and subcultures was as described previously (5). During media replacements, some of the media removed from several cultures was pooled. After 24 wk in culture, the cells in the pooled embryonic cultures could be regularly subcultured weekly at a 1:2 split ratio; the resulting cell line was designated IPLB-LdEp. The cells were suspended by flushing with medium from a pipet and 2 ml aliquots were transferred to 25 cm^2 flasks containing 2 ml culture medium. Gentamicin was deleted from the medium after the twelfth subculture. Subsequently, subcultivable lines were also developed from two of the primary embryo cultures (designated IPLB-LdEG and IPLB-LdEI) and one fat body culture (IPLB-LdFB). During efforts to grow these lines in other cell culture media, a subline of the IPLB-LdEI was developed which grows on TNM-FH (6) with different morphological and growth characterisics and was designated IPLB-LdEIt.

Chromosomes. Each cell line was prepared for chromosome analysis by the method of Earley (7). Chromosomes were stained with quinacrine mustard (8) and 10 spreads were counted with a Nikon Epiphot.

Biochemical Characterization. Homogenates of each cell line were run on agarose gels and stained for malic enzyme (ME), phosphoglucose isomerase (PGI), phosphoglucomutase (PGM) and isocitrate dehydrogenase (ICD) enzymes (5). Homogenates of each embryonic line were also tested for glucose-6-phosphate dehydrogenase (G6PD), malate dehydrogenase (MD), peptidase B (PepB) and asparagine transferase (AST).

Virus infectivity. Each cell line was inoculated with several different virus isolates, including plaque purified clones of AcNPV (6R)(9), LdNPV (5-7D) (10), and HzNPV (hvc2)(D. Lynn, unpublished data) and the "wildtypes isolates" of the Abington strain of LdNPV (11), Orgyia pseudotsugata nuclear polyhedrosis virus (OpNPV) (12), and Amsacta moorei entomopox virus (AmEPV)(13). Except for the inoculum of the Abington strain of LdNPV which was hemolymph from infected larvae, all inocula were supernatants from infected cell cultures. Infectivity was ascertained by the presence of polyhedral inclusion bodies (PIBs) as observed with phase contrast or Nomarski microscopy.

Results and Discussion

We have obtained five new cell lines from the gypsy moth (four from embryos and one of fat body origin) and a line from H. virescens testicular sheath. Growth curves on the gypsy moth lines revealed doubling times between 4 and 7 days for the embryonic lines and about 3.5 days for the fat body line after about five months of regular subcultures. After a longer time in culture (about 1.5 years), all lines displayed faster growth rates (with split ratios of 1:3 to 1:10 weekly).

Chromosome analysis on the gypsy moth lines revealed typical lepidopteran chromosomes ranging in number from 40 to 100 for IPLB-LdEIt to 55 to 143 for IPLB-LdFB. Means for each line are listed in Table 1. Isozyme analyses showed distinct differences between the gypsy moth and tobacco budworm cell lines (Table 1). While differences were not detected between the embryonic lines for any of the enzymes tested, the banding pattern for IPLB-LdFB showed an additional allozyme for isocitrate dehydrogenase. This same allozyme was present in extracts of fat body obtained directly from L. dispar larvae. Staining with malic enzyme revealed an additional band in the fat body extracts not observed in any cell line. While these results are supportive evidence that the IPLB-LdFB is actually fat body, we have not tested other larval tissues. However, the most likely "contaminant" of fat body from insects are hemocytes, and our failure to obtain such cell lines, even after many more attempts (nine tries compared to two with fat body primaries) suggests that these cells are not hemocytes. We have also shown by selective staining (Oil Red O, a lipophilic stain) that the cytoplasmic vacuoles in IPLB-LdFB cells are largely composed of lipids (Figure 1).

Susceptibility of these cell lines to various viruses reveals some promising results (Table 2). All five L. dispar cell lines supported LdNPV replication. In the previous tests, a cloned isolate of this virus (5-7D) had shown the best infectivity for an ovarian gypsy moth line used in our laboratory. However, in our preliminary tests, this cloned virus did not produce large numbers of PIBs or non-occluded virus in any of the new lines. Recently, however, we have tested the Abington strain of LdNPV (11) which produces many more PIBs in all of these lines, especially IPLB-LdFB (Figure 2). Of great interest is the ability of the four embryonic lines to replicate both LdNPV and AcNPV. While Goodwin et al. (14) reported a L. dispar ovarian line capable of replicating AcNPV, it was unable to plicate LdNPV. The new lines should provide a tool to determine factors responsible for infectivity of L. dispar by these viruses. The gypsy moth lines were

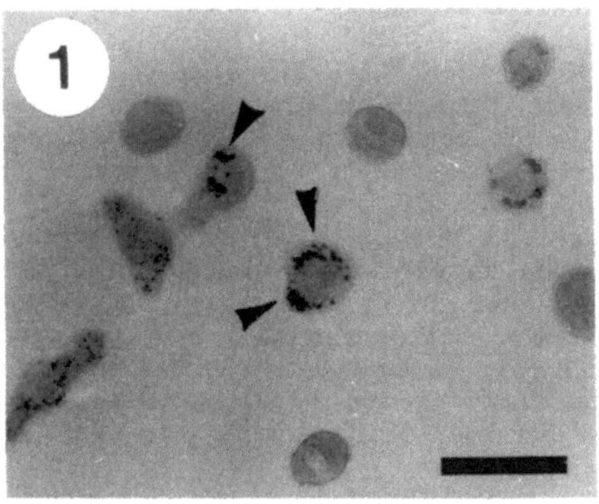

Figure 1. Oil red O stained IPLB-LdFB cells showing intensely stained cytoplasmic vacuoles (arrows). Marker bar is 50 µm.

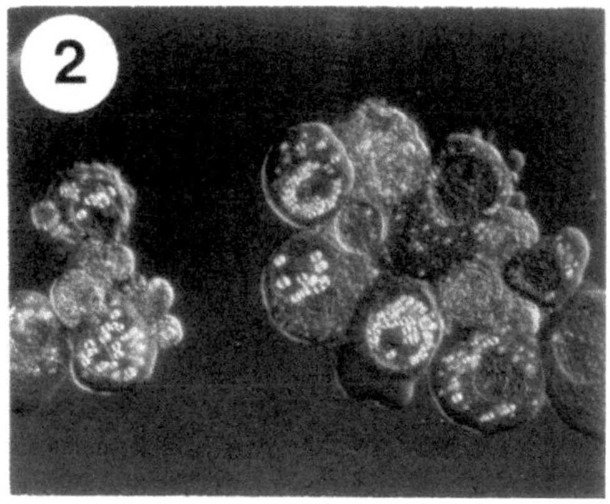

Figure 2. IPLB-LdFB cells four days post-infection with Abington strain of LdNPV photographed with Nomarski optics.

Table 1. Chromosome and isozyme analysis

Cell line IPLB-	Mean Chrom.#	Isozyme[1] (Migration from origin in mm) PGM	ICD	PGI	ME	PepB	G6PD	AST	MD
LdEp	nd	15.5	8.0	16.0	-6.5	21.5	4.7	-6.0	-7.0
		19.7			9.0			-1.5	9.7
					18.0			16.8	
LdEG	80.9	15.5	8.0	16.0	-6.5	21.5	4.7	-6.0	-7.0
		19.7			9.0			-1.5	9.7
					18.0			16.8	
LdEI	82.7	15.5	8.0	16.0	-6.5	21.5	4.7	-6.0	-7.0
		19.7			9.0			-1.5	9.7
					18.0			16.8	
LdEIt	61.0	15.5	8.0	16.0	-6.5	21.5	4.7	-6.0	-7.0
		19.7			9.0			-1.5	9.7
					18.0			16.8	
LdFB	89	15.5	6.0	16.0	-6.5	nd	nd	nd	nd
			8.0		9.0				
					18.0				
larval fat body	nd	15.5	6.0	16.0	-6.5	nd	nd	nd	nd
		19.7	8.0		9.0				
					11.8				
					18.0				
HvT1	nd	9.5	3.6	7.0	10.2	nd	nd	nd	nd
		18.6	9.8		12.5				

[1]Abbreviations of isozymes are as listed in the text. nd: not determined.

Table 2. Infectivity of cell lines to insect viruses

Cell Line	LdNPV	AcNPV	OpNPV	AmEPV	HzNPV
LdEp	+	+	-?	+	-
LdEG	+	+	+	+	-
LdEI	+	+	+?	+	-
LdEIt	+	+	-?	+	-
LdFB	+	-	+?	+	-
HvT1	-	+	nd	+	+

also susceptible to AmEPV and possibly OpNPV, although further studies are necessary to confirm the response to this latter virus. The IPLB-HvT1 line supported HzNPV, AcNPV and AmEPV replication. Further studies are underway to quantitate the response of these cell lines to virus infection.

Preliminary tests by RIA on the capacity of the IPLB-HvT1 cells to produce ecdysteroids have indicated no apparent synthesis. Possibly a stimulatory hormone is required. Alternatively, the IPLB-HvT1 cells may not be the ecdysteroid synthesizing cells of the testes. We are also testing these cells for the presence of other biologically active materials.

Ultimately, the new lines may be important in a variety of research areas. Our preliminary results indicate their usefulness in virology studies. The IPLB-LdFB line should also prove useful in many studies on insect physiology, since its morphology is maintained in a differentiated form in vitro. Since fat body performs a number of functions in the insect, such as storage of nutrient reserves and synthesis of specific products, (hemolymph storage proteins and vitellogenins), and conversion of important developmental hormones, it will be interesting to learn if these functions occur in the cell line.

Acknowledgements

We acknowledge the technical assistance of Mrs. V. F. Bray in preparation and quality control of cell culture media and Ms. S. M. Kragh for cell and virus studies. The mention of a commercial product does not constitute an endorsement by the U. S. Department of Agriculture.

References

1. Goodwin, R. H., Tompkins, G. J. and McCawley, P. In Vitro 14:485-494, 1978.
2. Vail, P. V., Jay, D. L. and Hink, W. F. J. Invertebr. Pathol. 22:231-237, 1973.
3. Loeb, M. J., Woods, C. W., Brandt, E. P. and Bôrkovec, A. B. Science 218:896-898, 1982.
4. Goodwin, R. H. and Adams, J. R. Invertebrate Systems In Vitro (Kurstak, E., Maramorosch, K., and Dübendorfer, A.,eds.), pp. 493-509, Elsevier, 1980.
5. Lynn, D. E. and Stoppleworth, A. In Vitro 20:365-368, 1984.
6. Hink, W. F. and Strauss, E. Invertebrate Tissue Culture, Applications in Medicine, Biology, Agriculture (Kurstak, E. and Maramorosch, K., eds.), pp. 297-300, Academic Press, New York, 1976.
7. Earley, E. M. TCA Manual 1:31-35, 1975.
8. Lin, C. C., Biederman, B. and Jamro, H. TCA Manual 4:937-940, 1978.
9. McClintock, J. T., Dougherty, E. M. and Weiner, R. M. Virus Res. 5:307-322, 1986.
10. McClintock, J. T., Dougherty, E. M. and Weiner, R. M. J. Virol. 57:197-204, 1986.
11. Shapiro, M., Robertson, J. L., Injac, M. G., Katagiri, K. and Bell, R. A. J. Econ. Entomol. 77:153-156, 1984.
12. Hughes, K. M. and Addison, R. B. J. Invertebr. Pathol. 16:196-204, 1970.
13. Granados, R. R. and Naughton, M. Intervirology 5:62-68, 1975.
14. Goodwin, R. H., Adams, J. R., Vaughn, J. L. and Louloudes, S. J. Proc. 1st Int. Coll. Invertebr. Pathol., pp. 94-102, Queen's Univ. Press, Kingston, Ont, 1976.

ESTABLISHMENT OF CELL LINES FROM BOMBYX MORI AMD B. MANDARINA, AND ATTEMPT OF CELL FUSION

O. Ninaki[1)], K. Nagayasu[1)], H. Kameoka[2)], H. Fujiwara[2)], N. Miyajima[2)], N. Takada[2)] and H. Maekawa[2)]

1) Sericultural Experiment Station, Tsukuba, Ibaraki 305, Japan
2) National Institute of Health, Shinagawa-ku, Tokyo 141, Japan

New cell lines were established from mulberry silkworm, Bombyx mori, and mulberry wild silkworm, Bombyx mandarina. In order to reintroduce the characteristics from wild silkworm to Bombyx mori, we attempted the cell fusion between them. The condition of the cell fusion was determined by use of the Bombyx and Antheraea cell lines. The fused hybrid cells between Bombyx and Antheraea surviving in the selection medium have been passaged about 4 months. Under the same condition we tried the cell fusion experiment between the two species of Bombyx genus. We practically observed the fused cells but could not clearly distinguish the hetero type cells derived from fusion.

B. mandarina, so-called mulberry wild silkworm, characteristically produces only a small amount of silk, while the domesticated silkworm, B. mori, has lost the characteristics of adaptation to its environment, e.g. resistance to lower temperature, etc. Therefore, it is significant to develop a method of reintroducing the genes or chromosomes of the useful characteristics of B. mandarina to B. mori. We first determined the condition of cell fusion, and tried to fuse B. mori and Antheraea eucalypti or B. mori and B. mandarina cell lines.

Materials and Methods

The SES-Bm-e 21 and SES-Bm-l 30 cell lines were derived from B. mori strains, e 21 and l 30, respectively. They carry several mutant genes(1) which are indicated as e 21(pe, re, ch, Sph) and l 30(pM, Ze, L, q), and were offered from Institute of Silkworm Genetics, Kyushu University. The Bma-01 cell line was derived from B. mandarina which was collected and reared in the Sericultural Experiment Station.

The embryos were aseptically removed from eggs one day before hatching. Each embryo was dissected into several parts and primarily cultured in IPL-41 medium(2; Goodwin, personal communication) containing 15% of fetal bovine serum(FBS) for two weeks at 25 °C in the dark. A half of the medium was exchanged with new IPL-41 including 10% of FBS every two weeks. Following the primary culture, the tissues were subcultured for one year or more and stable cell lines were obtained(3). The toxicity of polyethylene glycol(PEG) to several cell lines and the optimum density of PEG for fusion were examined before cell fusion experiments.

Results and Discussion

We have established new cell lines of B. mori (3) and B. mandarina. All of them are derived from matured embryos (stage 27-29) and designated as SES-Bm-e 21, SES-Bm-l 30 and SES-Bma-01 cell lines (Table 1). Each cell line has been separated into several sublines by shape (spindle, round shape and fibroblast) and so on. They have been passaged 3 to 4 years since the primary culture. These cell lines were characterized by the method of blot hybridization with a gene probe specific to the genus of Bombyx (Ninaki et al., in preparation).

Table 1. Newly established cell lines of Bombyx

Cell line	Shape	Date of initiation	Subculture number	Origin
SES-Bm-e 21A	mixture	Jan.12,1983	86	B. mori, e 21
Bm-e 21S	spindle		118	mature embryo
Bm-e 21R	round		88	
*SES-Bm-l 30A	mixture	Jan.27,1984	58	B. mori, l 30
Bm-l 30R	round		71	mature embryo
SES-Bma-01A	mixture	May 24,1984	44	B. mandarina
Bma-01R	round		52	mature embryo

Subculture interval: two weeks, Dilution at subculture: 1/11 - 1/61, Maximum cell density : $2x10^6$-$4x10^6$ cells/ml, Medium: IPL-41 containing 10% FBS.
*This cell line grows in an aggregated form.

We attempted experiments of the cell fusion by PEG. The treatment of the 50% PEG whose density is usually efficient for the mammalian cell fusion was harmful to the proliferation of the Bombyx cell lines. In order to examine the toxic effect of PEG, several kinds of PEG and its density were checked on the cell growth and survivability. The permissible density for the cell proliferation under the continuous exposure to PEG was 1% of Peg 1500 (M.A.Bioproducts, M.W.=500-600) which is less toxic than other polyethylene glycols, but toxic at 5%. The similar effect was also observed in Grace's Ae(4) cell line derived from the ovary of Antheraea eucalypti, while the Bm 36 cell lines(5) survived even in 5% of the PEG about two weeks. Since good results were expected by use of the cell lines with a similar response to PEG, newly established Bombyx cell lines and Ae cell line are good candidates for the cell fusion experiment. From the results of several experiments, we selected the condition of a 5 minute treatment of PEG for fusion. Furthermore, the increase of the cell density and the appearance of large cells were observed one to four days after the treatment with 5 to 25% PEG. As shown in Table 2, the ratio of the large cells to the total living cells came to over 40% only in the group treated with the 12.5% PEG.

Then we tried to fuse the SES-Bm-l 30R and Ae cell lines, under the condition of a 5 minute treatment by 12.5% of PEG. The reason why we selected these two cell lines is that these cells can be distinguished by their shapes, the proliferating

Table 2. Optimum density of the polyethylene glycol for cell fusion

Treatment	Cell number at treatment ($x10^5$cells/ml)	Observation	Days in culture		
			1	2	4
Control		Cell nos.	3.75	4.02	5.33
(mixed)	3.17	L.C. nos.	0.39	0.47	1.06
		L.C. %	10.4	11.7	19.9
		Cell nos.	2.19	3.28	3.20
25%	3.17	L.C. nos.	0.55	0.70	1.09
		L.C. %	25.1	21.3	34.1
		Cell nos.	2.73	2.81	3.20
12.5%	3.17	L.C. nos.	1.09	0.94	1.33
		L.C. %	39.9	33.5	41.6
		Cell nos.	3.28	2.81	3.83
5%	3.17	L.C. nos.	0.94	0.78	0.94
		L.C. %	28.7	27.8	24.5

Used PEG: MAB, Peg 1500(M.W.= 500-600), Treatment time: 5 min.,
Cell number = $x10^5$cells/ml, L.C.= Large cell

forms and the selectivity of the medium as shown in Table 3. Therefore, the spindle shaped cells surviving in the culture of the IPL-41 are probably identified as the fused cells if they reveal the expected characteristics (Table 3). Two weeks after PEG treatment, the majority (70-80%) of the large spindle shaped cells in the IPL-41 medium were heterokaryons(Fig.1 C). Four months later, practically fused cells which

Table 3. Characteristics of Bombyx and Antheraea cell lines on the cell fusion

Item	Bm-1 30R cell line	Ae cell line	Fused cell
Shape	round	spindle	spindle-like
Proliferated form	aggregated	not aggregated	not aggregated
Selectivity of medium			
IPL-41*	growth	not growth	growth
Silkworm hemolymph	not essential	essential	not essential

*IPL-41 medium contains 10% of FBS.

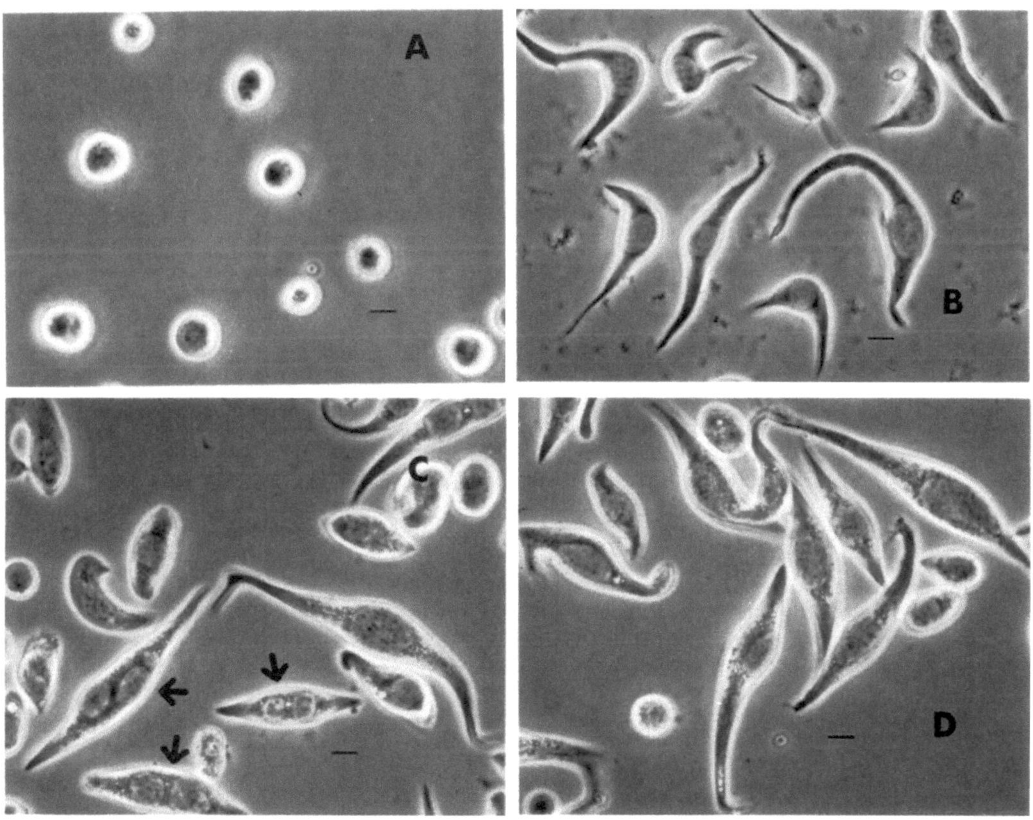

Fig.1 Cell fusion between SES-Bm-1 30 and Ae cell lines. A, *Bombyx mori* cell line, SES-Bm-1 30R; B, *Antheraea eucalypti* cell line, Ae; C, Fused cells two weeks after PEG treatment. Most fused cells are heterokaryons (arrows); D, Fused cells from Bm-1 30 and Ae cell 4 months after treatment. One nucleus is observed in most fused cells. Bars indicate 10 μm of length.

were not heterokaryotic cells including two nuclei had increased over 60%(Fig.1 D). These results indicate that characteristics, such as the spindle shape and non aggregated form of the Ae cells were introduced to the Bombyx cells or that, vice versa, the selection characteristics of the Bombyx were introduced to the Antheraea cells.

Next, we tried to fuse similar type cells, which are Bm-e 21S cell lines from B. mori and Bma-01R from B. mandarina. In order to confirm the cell fusion, we carried out flow cytometry by cell sorter, Becton & Dickinson FACS 440. The horizontal line shows the amount of DNA stained with the ethidium bromide and the vertical line shows the cell number in Fig. 2. When a 1:1 mixture of the two cell lines was applied to the cell sorter, one peak was observed although it was broad due to the heterogeneous shaped cells(Fig.2 A). On the flow cytometry 24 hours after PEG treatment, we could detect three major peaks (Fig.2 B). These peaks probably correspond to the diploid cells, the fused cells from two cells and more than three cells, respectively. The cell fusion was successful as revealed in the sorting pattern but we did not attempt to determine whether these fused cells were derived from the same or different type cells. This sorting machine is convenient to detect the fused cells.

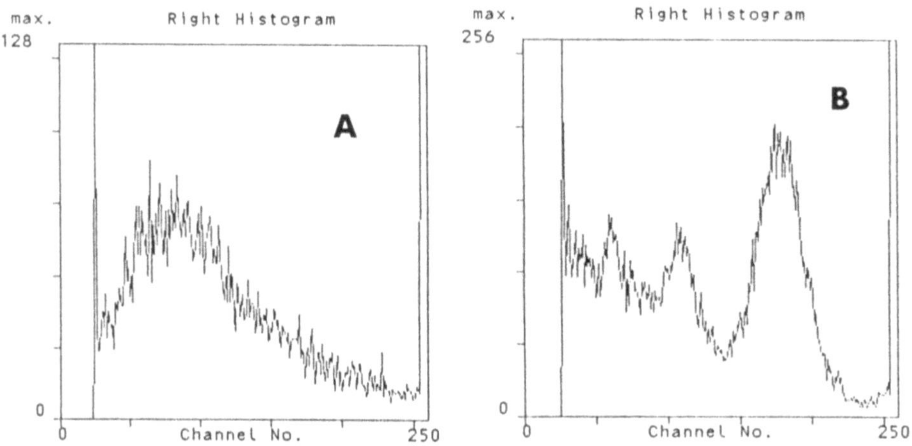

Fig. 2 Flow cytometry of cells stained with ethidium bromide.
A, 1:1 mixture of SES-Bm-e 21S and SES-Bma-01R cell lines;
B, 1:1 mixture cells 24 hours after 12.5% PEG treatment.

Then we tried to label the cytoplasm and nucleus, respectively, by the incorporation of ^{3}H-uridine and ^{3}H-thymidine. The cells labeled in the cytoplasms were not clear because of the lower incorporation of ^{3}H-uridine. Strictly speaking, we could not detect the cells fused between the different cells. However, as we observed cells containing two nuclei, one of which was labeled, the cell fusion between B. mori and B. mandarina also seemed successful in this preliminary work.

References

1. Tazima, Y., Doira, H. and Akai, H., Handbook of Genetics, Vol. 3 (King, R.C.,ed.), pp. 63-124, Plenum Publishing Corp. 1975.
2 Goodwin, R. H., In Vitro, 11:369-378, 1975.
3. Ninaki, O., Takada, N., Fujiwara, H., Ogura, T., Miyajima, N., Kiyota, A. and Maekawa, H. Invertebrate Cell Systems in Applications (Mitsuhashi, J., ed.) CRC Press (in press).
4. Grace, T. D. C., Nature, 195:788-789, 1962.
5. Quiot, J. M., Sericologia, 22:25-31, 1982.

MORPHOLOGICAL AND FUNCTIONAL STUDIES ON HEMOLYMPH CELLS OF LAND SLUG, INCILARIA BILINEATA, IN VIVO AND IN VITRO

K. Yamaguchi[1], E. Furuta[2], and A. Shimozawa[2]

[1] Laboratory of Medical Sciences, Dokkyo University School of Medicine, Mibu, Tochigi 321-02, Japan

[2] Department of Anatomy, Dokkyo University School of Medicine, Mibu, Tochigi 321-02, Japan

In the land slug, Incilaria bilineata, three morphologically distinct hemolymph cells were found; Type I, II and III. Spreading Type I cells could recognize and phagocytise foreign materials. However, whether from circulating hemolymph or from a hemopoietic organ, the cells could be maintained for only 4 days in cultivation. After 3 or 4 days of cultivation, the cells gradually divided into several pieces and seemed to become platelet-like structures. The platelet-like structures seem to be derived from Type I cells and seem to play a role in the fixation of foreign materials.

The circulating hemolymph cells of gastropods play an important role in the defense reactions against parasites and pathogenic or non-pathogenic microorganisms, which might enter the hemocoel through wounds in the skin.

In 1986, we found three morphologically distinct cell types and platelet-like structures in the circulating hemolymph of the land slug, Incilaria fruhstorferi: Type I cells were macrophage-like, Type II cells were lymphocyte-like, and Type III cells were fibroblast-like(1). The platelet-like structures had long fine fibers but lacked a nucleus(1). They seem to play a role in the fixation of foreign materials, the same as mammalian platelets(2).

In the present study, we investigated whether or not the same 3 types of cells and platelet-like structures existed in another species of land slug, Incilaria bilineata.

Materials and Methods

To culture the hemolymph cells, the hemolymph was collected from the heart and mantle cavity using a 1.0 ml syringe with a 26 gauge needle, and put into plastic dishes containing a slug saline-antibiotics(SA) solution. The heart-kidney region(the hemopoietic organ reported by Sminia and van der Knaap in 1986(3)), was also cut out and placed in a SA solution. After the heart-kidney region was maintained in the solution for 24 hrs, it was transferred into SH7 slug culture medium made by Furuta and Shimozawa(4). The cultures were kept at 25°C in a 5 % CO_2-incubator, so as to adjust the pH to 7.6. The medium was renewed every three days.

In preparation for scanning electron microscopic(SEM) observations, the cells which adhered to the glass were fixed with 0.08 % glutaraldehyde in a slug buffer solution of pH 7.4, for 12-16 hrs at 25°C. They were dehydrated in a graded ethanol series, placed in isoamylacetate, and dried to the critical-point using liquid CO_2. They were then sputtercoated with Au-Pd and examined with a Hitachi S-650 electron microscope at 10 KV.

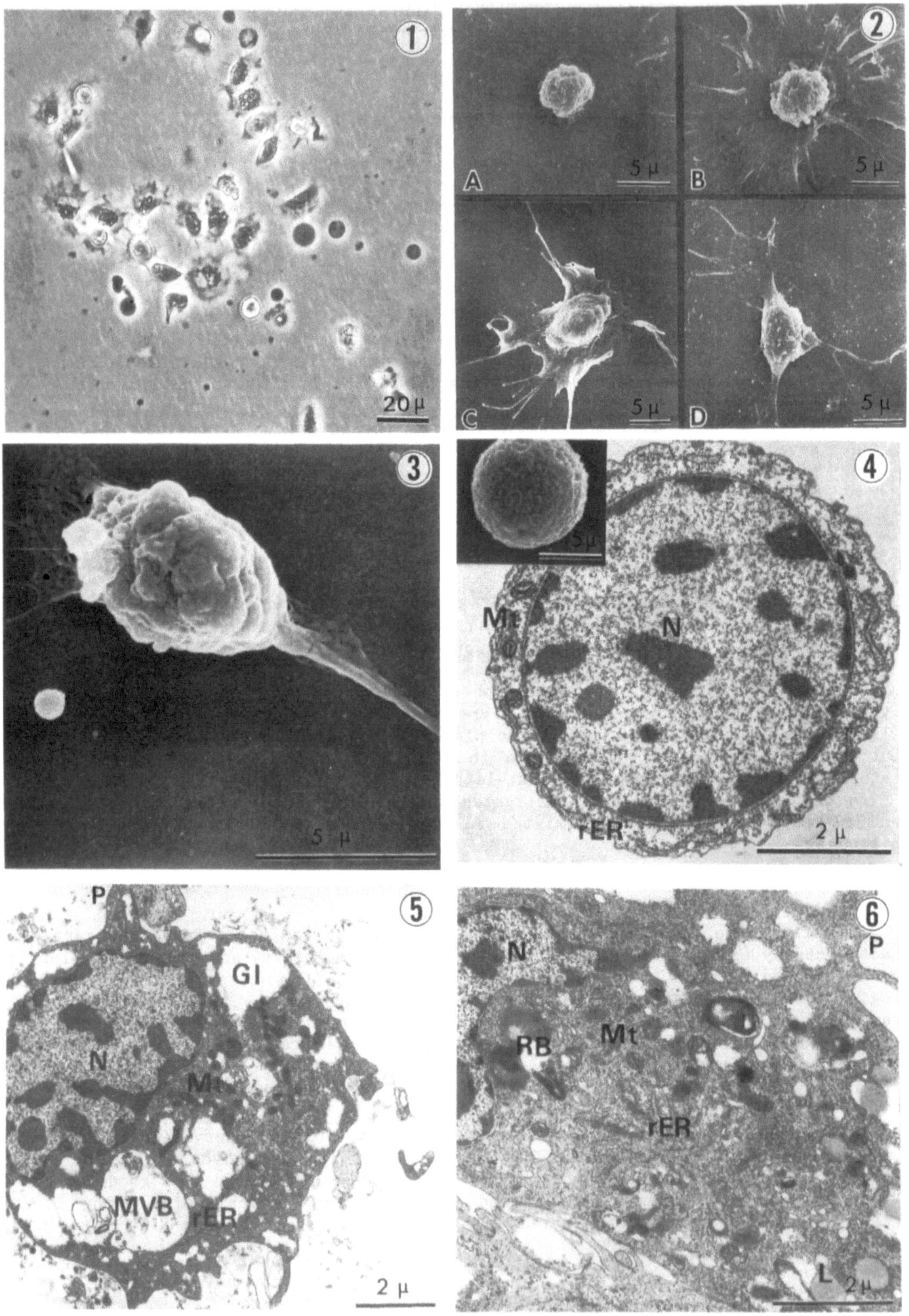

Figs. 1~6. Phase contrast micrograph of Type I cells (Fig. 1) and electron micrographs of Type III cells (Figs. 2 and 3), Type II cells (Fig. 4) and Type I cells (Figs. 5 and 6) of <u>Incilaria bilineata</u>.

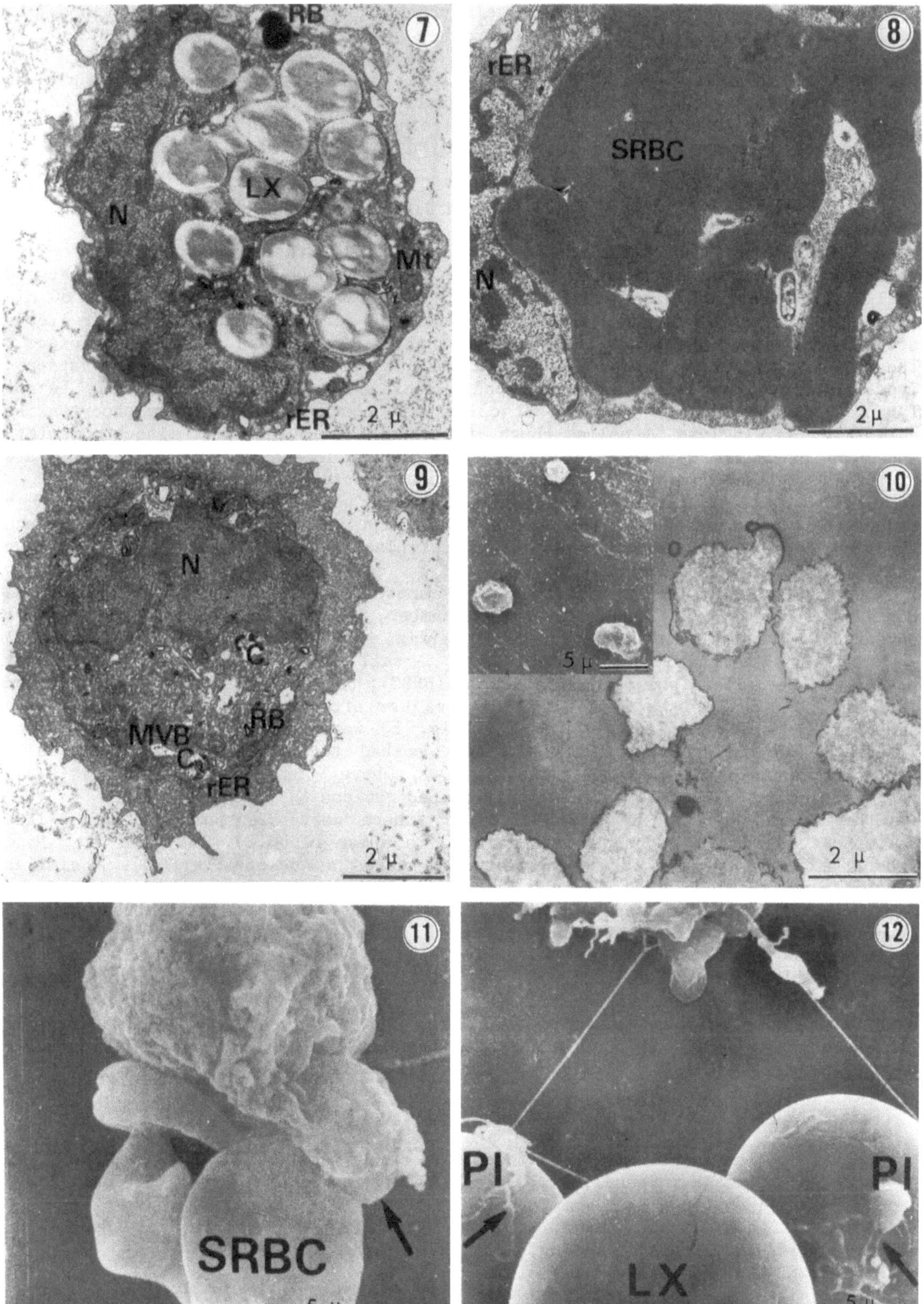

Figs. 7~12. Electron micrographs of phagocytosing Type I cells (Figs. 7~9) and Type II and III cells (Fig. 10) of *Incilaria bilineata* and Type I cells of *I. fruhstorferi* (Figs. 11 and 12).

In preparation for transmission electron microscopic(TEM) observations, the cells were fixed with 0.08 % glutaraldehyde for 24 hrs at 25°C and post-fixed with 1 % OsO_4 in PBS for 40 min. Epon 812 was poured directly into the dishes. Later the Epon blocks were detached from the plastic dishes by breaking the dishes with cutting pliers. The cells embedded in Epon were horizontally sectioned on a LKB-4800 ultramicrotome. Ultrathin sections were stained with uranyl acetate for 30 min and lead citrate for 5 min and examined with a JEOL-200CX electron microscope at 100 KV.

Results and Discussion

As shown in the phase contrast micrograph(Fig. 1), the hemolymph Type I cells, which migrated from the heart-kidney region after 24 hrs of cultivation, possessed veil-like pseudopodia with supporting ribs. The cell migration began about 60 min after the initiation of the culture and continued for more than 6 hrs. Three days after cultivation, the cells became round and finally divided into several pieces. The fate of the cells seemed to be platelet-like structures. In SEM observations, Type I cells, whose veil-like pseudopodia diminished with time, gradually became Type III cells(Figs. 2, 3). Type II cells were observed as round lymphocyte-like cells(Fig. 4). In Type II cells, rough endoplasmic reticulum(rER), mitochondria(Mt) and free ribosomes were scattered around a large nucleus(N). The origin or fate of the Type II cells may be clarified in future investigation.

TEM observations revealed that Type I cells possessed pseudopodia(P) and a kidney-shaped nucleus(N). Mitochondria(Mt), rough endoplasmic reticulum(rER) and glycogen deposits(Gl) were found in the cytoplasm. Characteristic multivesicular bodies(MVB) were also observed(Fig. 5).

When cultivated for 4 days, the cell contour became round and the residual bodies(RB) in the cytoplasm increased in number, while psuedopodia(P) of the cell appeared to fade away in the peripheral area(Fig. 6).

To study the phagocytosis of a Type I cell, 0.2 ml suspensions of latex beads(LX), formalized sheep red blood cells(SRBC), or carbon black(C) were injected into the hemocoel of the land slug. Twelve hrs after injection, specimens were prepared. Numerous LX(Fig. 7) and SRBC(Fig. 8) were phagocytised into the Type I cells. The nucleus(N) appeared to have pushed to the periphery. C was also incorporated into the Type I cells(Fig. 9), where multivesicular bodies(MVB), residual bodies(RB) and a typical kidney-shaped nucleus(N) were seen.

Other than Type I, II and III cells, there were also small platelet-like structures, similar to the platelets found in human blood(2). They had long fine fibers and were covered with membrane, but lacked a nucleus(Fig. 10). A similar finding was reported by Carter and Green in 1987(5). According to their finding, the insect, <u>Tipula paludosa</u> larvae possessed vast numbers of membrane-bounded granule-containing fragments which appeared to be derived from granular cells.

In another species, <u>Incilaria fruhstorferi</u>, Type I cells phagocytosed several sheep red blood cells(SRBC)(Fig. 11). We also observed an aggregation or clamping of platelet-like structures to foreign materials. The platelet-like structures(Pl) with long fine fibers attached to latex beads(LX) and had filopodia-like processes(arrows)(Fig. 12). This result suggests that platelet-like structures seem to play a role in the fixation of foreign materials.

References

1. Furuta, E., Yamaguchi, K. and Shimozawa, A. Anat. Anz., 162:215-224, 1986.
2. Haque, A., Cuna, W., Bonnel, B., Capron, A. and Joseph, M. Parasite. Immunol., 7:517-526, 1985.
3. Sminia, T. and van der Knaap, W. P. W. Immunity in Invertebrates (Berehelin, ed.) C9:112-124, Springer-Verlag,Berlin Heidelberg. 1986.
4. Furuta, E. and Shimozawa, A. Zool. Mag., 92:290-296, 1983.
5. Carter, J. B. and Green, E. I. J. Morphol., 191:289-294, 1987.

CONTINUOUS CELL LINES FROM IMAGINAL DISCS OF DROSOPHILA MELANOGASTER

K. Ui*, R. Ueda, and T. Miyake

Laboratory of Cell Biology, Mitsubishi-Kasei Institute of Life Sciences, 11 Minamiooya, Machida-shi, Tokyo 194, Japan
*Predoctoral Fellow from Graduate School of Science and Engineering, Waseda University, Tokyo, Japan

We established 20 cell lines, designated as ML-DmD1~20, from wing, haltere, antenna, eye-antenna discs and disc mixtures (1). The profiles of primary cultures and successful frequencies of cultures seemed to differ depending on disc types. The cell aggregates formed in the primary culture were examined for their differentiation capacities by injecting them into metamorphosing larvae. At the early stage of *in vitro* culture, they differentiated into adult cuticular structures containing bristles and/or trichomes which were indicative of disc origin.

A method for culturing imaginal discs indefinitely in adult female abdomen was developed by Hadorn in 1963 (2). This method revealed that imaginal disc cells can proliferate maintaining their determined state, and without undergoing adult differentiation. These studies allowed detailed examination of regulation of pattern formation and transdetermination, which made imaginal discs the most useful material to investigate these problems.

However, there are some limitations in *in vivo* culture of imaginal discs. It is practically impossible to prepare a large amount of disc cells cultured for a long period. And in the course of proliferation *in vivo*, it is difficult to operate culture conditions artificially. In addition, it has not been possible to culture a single cell *in vivo*. These problems may be excluded if the *in vitro* culture conditions for the disc cell proliferation are found. Davis and Shearn reported that the cells of wounded discs slightly proliferated *in vitro* (3), and Wyss succeeded in making the dissociated disc cells form colonies *in vitro* (4). Referring to their results, we checked the effect of various hormonal supplements and the supernatant of primary embryonic culture for conditioning the medium (5), and established cell lines from dissociated disc cells (1). Under the culture conditions we employed, the profiles of primary cultures were different depending on the disc types. Differentiation capacities of cell aggregates cultured *in vitro* were examined by injecting them into metamorphosing larvae.

Primary Cultures of Various Kinds of Imaginal Discs

Culture method of imaginal discs was described previously (1,5). When the dissociated wing disc cells were cultured, they were tightly assembled in

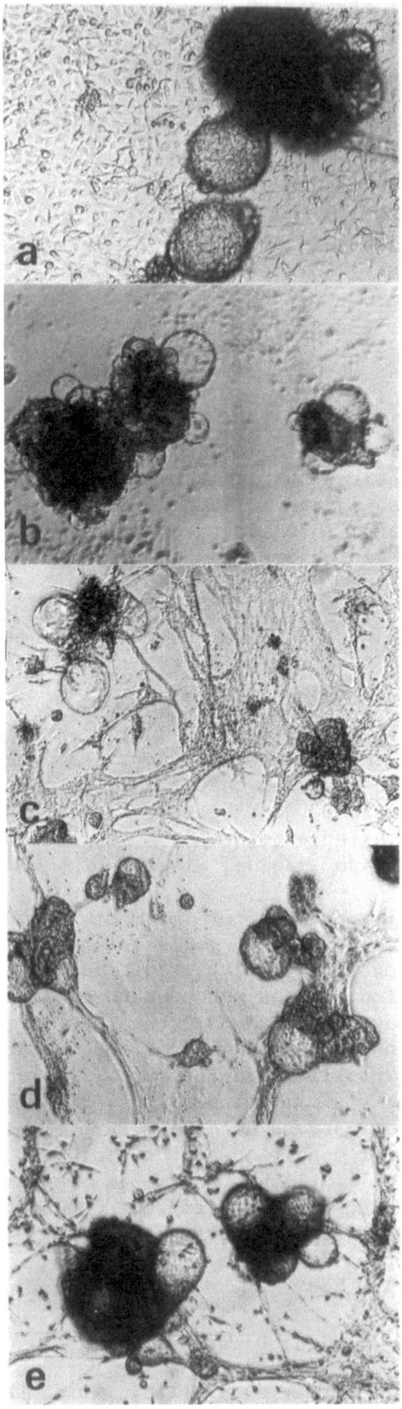

aggregates (dia. 100-300 μm) at about 1 week after initiation of primary culture. Cellular vesicles were formed on the surface of aggregates, and in most cases adhesive cells around the aggregates were observed, and gradually and fully spread over the substratum at about 2 weeks (Fig. 1a). In 27 out of 50 wells inoculated, small round cells around the aggregates rapidly proliferated and piled up to form large foci at about 3 to 4 weeks. These cells can be transferred into the other wells at 1/2 or 1/4 dilution and established as cell lines in 2 to 4 months. But in other cases (the remaining 23 wells) the cell proliferation did not occur. In these cases, the frequency of successful culture (f.s.c.) was represented as 54 % (27/50).

When the haltere disc cells were cultured, the profiles were very similar to those of the wing disc cell cultures. But the cells around the aggregates were more adhesive and polygonal compared to the cells from wing discs. The cells proliferated in 4 out of 10 wells (f.s.c.=40%). We established 4 cell lines from haltere discs.

When eye parts of eye-antenna discs were cultured, the cell aggregates with many cellular vesicles were formed on the substratum, like wing or haltere disc cell cultures (Fig. 1b). But no vivid cell was observed around the aggregates and cell debris was observed all over the substratum. This profile was very characteristic for eye disc cell culture. Then the aggregates were detached from the substratum and floated at about 2 weeks. No cell proliferation was observed in 10 wells tested (f.s.c.=0%).

When only antenna parts of eye-antenna discs were cultured, the dissociated cells formed aggregates with many cellular vesicles. Thin layers of very sticky and polygonal cells were spread around the aggregates forming the sheet-like structures until about 2 weeks (Fig. 1c). This seemed to be a characteristic of antenna disc cell culture. In 4 out of 11 wells tested, these cells began to proliferate (f.s.c.=36%) and we established 2 cell lines from antenna disc cells.

Figure 1. In vitro culture of cells from various kinds of imaginal discs at about 2 weeks after initiation. a: wing discs, b: eye discs, c: antenna discs, d: eye-antenna discs, e: leg discs. x93.

When eye-antenna disc cells were cultured, the profile of primary cultures was intermediate between eye and antenna discs at about 2 weeks (Fig. 1d). Cell aggregates with cellular vesicles were formed, and around them thin layers of very adhesive cells were spread, which were similar to the cells observed in antenna disc cell cultures. This seemed to show that antenna disc cells were more easily adapted to our culture conditions even when whole eye-antenna disc cells were cultured. And this probably results in the preferential proliferation of antenna disc cells. In 2 out of 16 wells tested, cells began to proliferate (f.s.c.=12%) and they are now being submitted to dilution-transfer.

Leg disc cells (mixture of 1st, 2nd and 3rd leg discs) were also cultured under our culture conditions. The cell aggregates were formed with many cellular vesicles, and around them spindle shaped cells were spread all over the substratum at about 2 weeks (Fig. 1e). But in almost all cases of leg disc cultures, these cells gathered to the aggregates gradually, detached from the substratum and degenerated at about 3 to 4 weeks. These profiles are similar to those of eye disc cell cultures. No cell proliferation has been observed for 2 months after initiation (f.s.c.=0%).

Differentiation Capacity

The cell aggregates from the primary cultures of disc cells were tested for the differentiation capacities by the method of Ursprung (6). The aggregates cultured *in vitro* for 1-3 weeks were transplanted into host larvae ready to pupate, and the implants were recovered from the abdomen of the metamorphosed adult host and examined under a microscope.

The cell aggregates of wing disc cells were transplanted into 68 host larvae, and 25 implants were recovered from 30 eclosed hosts. Among them, 19 implants differentiated into adult cuticular structures which formed vesicles with bristles and/or trichomes on their inner surfaces. The ratio of differentiated adult structures to total recovered implants (ratio A) was 76 % (19/25). The wing blade specific structures with many trichomes were often observed among these samples (Fig. 2a). The other 6 implants (24%) formed simple cuticular structures (usually plain cuticular vesicles with neither bristles nor trichomes). The differentiation capacities of cell aggregates from eye disc cell cultures were tested at about 1 to 3 weeks. From 9 eclosed hosts, 6 implants were recovered and 3 were differentiated into adult structures (ratio A=50%), one of which contained trichomes, bristles and also pigmented cells (Fig. 2b), the color of which was similar to that of adult ommatidial cells; but the complete ommatidial structure

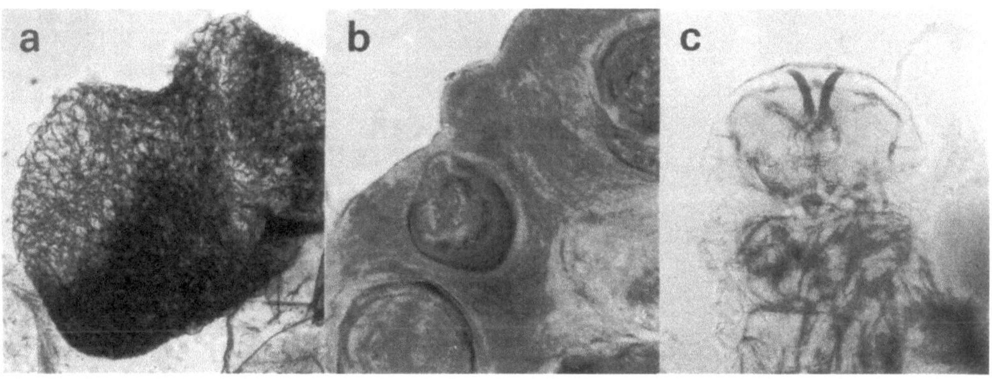

Figure 2. Differentiated adult cuticular structures derived from (a) wing disc cells (x200), (b) eye disc cells (x4267) and leg disc cells (x373) cultured for 2-3 weeks in *in vitro* cells.

has not been formed. And when cell aggregates of eye disc cells were tested at about 5 weeks, the bract structure was differentiated which was specific to wing or leg. So transdetermination seemed to occur in eye disc cells. We observed 4 more examples of transdetermination in the culture from other discs. The differentiation capacities of the aggregates of antenna discs were also tested. From 4 eclosed hosts, 3 implants were recovered. All have differentiated into adult cuticular structures which contained trichomes and/or bristles (ratio A=100 %). When cell aggregates of eye-antenna discs were tested, 7 implants were recovered from 19 eclosed hosts, and only 1 differentiated into adult structure (ratio A=14%)and the remaining were simple cuticle structures. When cell aggregates of leg discs were transplanted into host larvae, 9 implants were recovered from 16 eclosed hosts. Two implants were differentiated into adult structures (ratio A=22%), one of which was a claw-like structure (Fig. 2c). The results show that ratio A for cell aggregates from wing and antenna disc cells, which showed high f.s.c., was very high. In contrast, ratio A for eye-antenna, eye and leg disc cells, which showed low f.s.c., was very low.

Conclusion

We have developed a culture method of dissociated imaginal discs of Drosophila melanogaster, and established 20 continuous cell lines from various kinds of imaginal discs. The profiles of primary culture were different depending on the kinds of imaginal discs. The differentiation capacities of cell aggregates in the culture was checked. At the early stage of culture, the differentiated adult structures were indicative of disc origin. Wing and antenna disc cells, which proliferated easily, maintained differentiation capacities relatively well.

References

1. Ui, K., Ueda, R. and Miyake, T. In Vitro, in press.
2. Hadorn, E. Dev. Biol., 7:617-629, 1963.
3. Davis, K.T. and Shearn, A. Science, 196:438-440, 1977.
4. Wyss, C. Exp. Cell Res., 139:297-307, 1982.
5. Ui, K., Ueda, R. and Miyake, T. Invertebrate Cell Systems in Applications (Mitsuhashi J. ed.), CRC Press Inc. in press.
6. Ursprung, H. Methods in Developmental Biology (Wessells, N.K. ed.), pp. 485-492, Thomas Y. Crowell Co., 1967.

CHARACTERISTICS OF CELL LINES DERIVED FROM EMBRYOS IN THE SILKWORM, BOMBYX MORI

S. Imanishi[1], S. Sato[2], and T. Mori[3]

1) Sericultural Experiment Station, Tsukuba, Ibaraki 305, Japan
2) Nippon Medical School, Bunkyou-ku, Tokyo 113, Japan
3) Kinjyou Gakuin University, Moriyama-ku, Nagoya 463, Japan

Embryonic cells from the Japanese, the Chinese, the European and the tropical and mutant strains of the silkworm, Bombyx mori, were cultured in vitro. Migration and proliferation of the cells in the primary culture varied with the strains. Two cell lines, designated as SES-BoMo-C129 and SES-BoMo-J125, were established from embryos. Their population doubling times ranged from 3 to 3.5 days. Two cell lines consisted mainly of cells with 91- 110 chromosomes. Both cell lines showed best condition at the culture temperature of 28°C. The optimum cell growth was observed at 10 % fetal bovine serum. The cultured cells contained lipids and glycogen granules, and were comparatively undifferentiated.

In general in Japan, the silkworm, B. mori has been geographically classified into the Japanese strain, the Chinese strain, the European strain,the tropical strain and so on. The silkworm is a holometabolic insect and one life cycle takes about 50 days when silkworms are reared at about 25°C. Extensive studies using silkworms as an experimental material have been performed and contributed to the development of modern genetics and elucidation of physiological mechanisms.

Successful culture of silkworm cells was achieved by Trager (1). At present there are several cell lines of silkworm, such as the cell line established by Grace (2), S.P.C.Bm 36 (3), Bm-N (4), SES-BoMo-15A (5) and SES-BmN-e 21 & SES-BmN-1 30 (6).

In the present study, we examined cell migration from embryo tissue in primary cultures of 15 silkworm strains and then studied the characteristis of two cell lines newly established.

Materials and Methods

Cell culture : Eggs were surface-sterilized by submersion in 70 % ethyl alcohol for 5 minutes,washed in sterile distilled water twice and transferred into Carlson´s fluid (7). The embryos were dissected into several parts and placed in Falcon type 50 ml culture flasks containing MGM 448 culture medium (8) with 10 % fetal bovine serum (FBS). The cultures were maintained at 25°C by renewing half of the medium once a week during the primary culture.

Karyology : Karyological examination was performed by the procedure of Earley (9) using a pretreatment of 500 μg colchicin/ml. Fifty spreads were counted after staining with Giemsa.

Temperature and serum concentration in culture : After the subculture, flasks were kept at the temperature of 15, 18, 25, 28, 33 and 36°C and also in media with 0, 2, 5, 10 and 20 % FBS by adjusting the FBS volume of the MGM 448 culture medium. The cells were counted by hemocyto-counting chamber (Bürker-Türk) every day.

Observation : For the observation with a transmission electron microscope, the cells were fixed for 30 minutes in cold 2.5 % glutaraldehyde in 0.1 M phosphate buffer (pH 7.4). After fixation the cells were rinsed once with a buffer solution, and then they were postfixed in 1 % osmium tetroxide in the same buffer. The materials were dehydrated in graded concentration of ethyl alcohol and finally embedded in Epon 812. The sections were cut with a diamond knife on a LKB ultratome and were stained for 10 minutes in a saturated solution of uranyl acetate in distilled water, and then for 3 minutes in lead citrate (10). The micrographs were taken using a HITACHI electron microscope.

Results

Primary culture : Cell migration from the embryo tissues was observed within 24 hours after the culture was established. Epithelial-like cells of the first group migrated actively for about one month. The fibroblast-like cells formed the networks around explants quickly. The cells of the second group divided actively about two months after the culture was established. But the division activity did not continue in the " Cambodge ", " E18 " and mutant " Ia " strains. After the cells proliferated continuously for the Chinese strains " C25 ", " C108 ", " C140 " and " C146 ", the cells of the Japanese strain " J145 " and the cells of the mutant " pnd " and " U " strains died after 4 months. Only the cells of the Japanese " J125 " strain and of the Chinese " C129 " strain proliferated rapidly about 5 months after culture initiation. On the other hand, the cells of the " J137 " strain and the mutant " pere " strain died accidentally at near 5 months.

Karyotype : The cell lines of both SES-BoMo-J125 and SES-BoMo-C129 were subcultured 70 times. The chromosome number was counted. The number ranged between 31 and 230 in SES-BoMo-J125 and between 31 and 310 in SES-BoMo-C129. Cells of both lines contained mainly 91-110 chromosomes. (Fig.1)

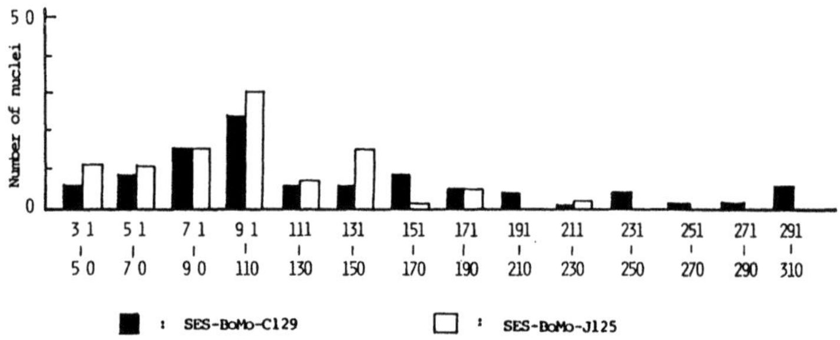

Fig. 1 Number of chromosomes in cell lines

Relationship between temperature and cell proliferation : In SES-BoMo-C129 the growth curve was identical when the cells were cultured at a temperature from 18 to 33°C, but in SES-BoMo-J125 the cells proliferated as the temperature rose from 15 to 28°C. The optimum temperature for both cell lines was 28°C. The population doubling time ranged from 3.0 to 3.5 days. The number of cells did not increase at 15°C and 36°C, and the temperature of 36°C damaged the cells. Some cells changed their forms to large type.

Relationship between FBS concentration and cell proliferation : The SES-BoMo-J125 showed similar growth rates in the media containing 5 %, 10 % and 20 % FBS, while the growth of SES-BoMo-C129 was optimal at 10 % FBS. In both cell lines, the cells were able to grow continuously at the minimum concentration of 2 % FBS. However, when FBS was omitted from the medium, no multiplication occurred. (Figs. 4,5)

Electron microscope observations : In the nucleus, several condensed chromatin bodies with a characteristic globular shape and a large nucleolus were located. The nucleus size and shape varied. Several lipid granules were detected in the cytoplasmic regions. Higher concentrations of glycogen granules were recognized in the cytoplasm. Each granule was composed of fine sub-units. However, structural changes of the nuclear envelope were clearly detected in the nuclear region. One part of the cytoplasm contained mitochondria of various sizes, granular endoplasmic reticulum, a small number of lysosomes and residual vacuoles. But in another part, there were large areas of cytoplasmic matrix that contained few organelles. (Figs.6,7)

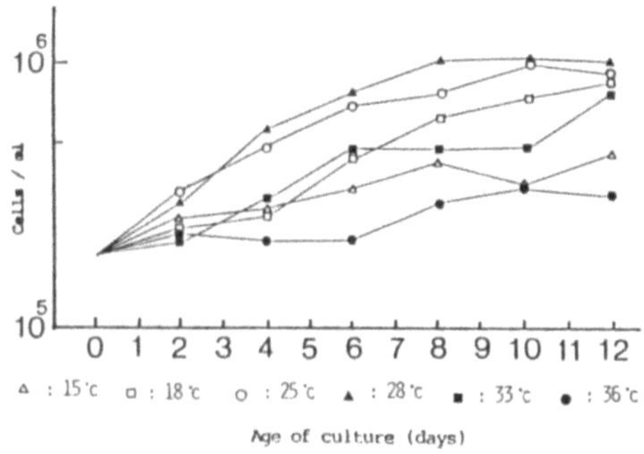

Fig. 2 SES-BoMo-J125

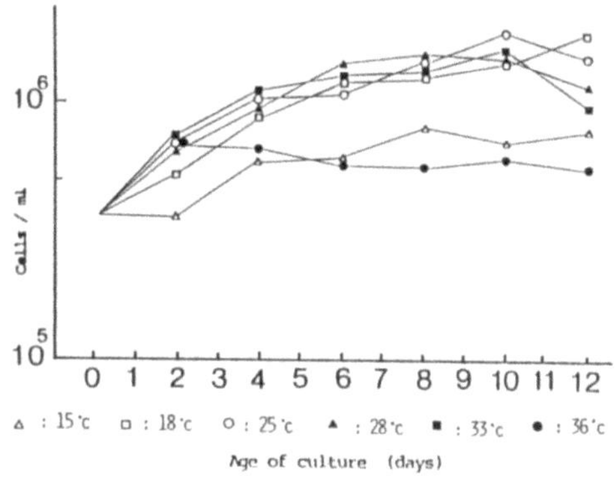

Fig. 3 SES-BoMo-C129

Effect of temperature on cell proliferation

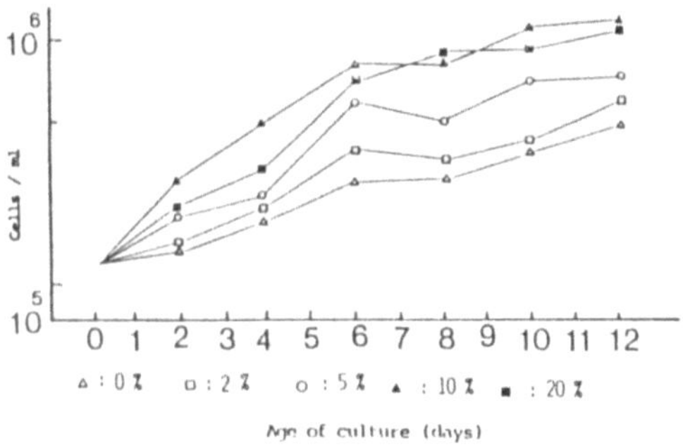

Fig. 4 SES-BoMo-J125

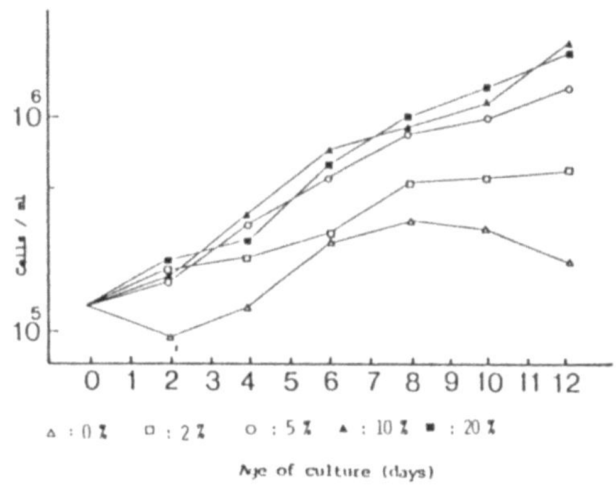

Fig. 5 SES-BoMo-C129

Effect of FBS concentration on cell proliferation

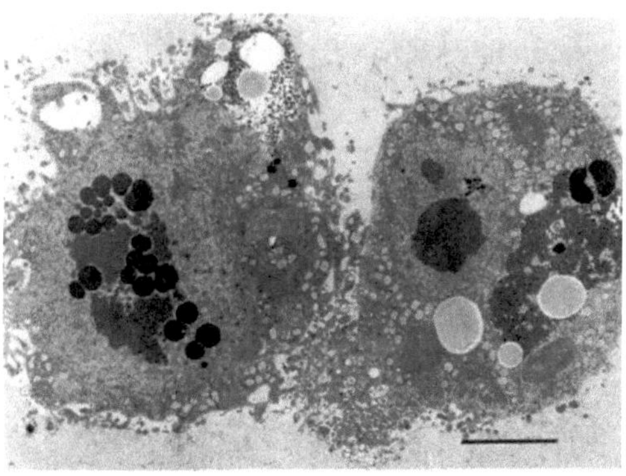

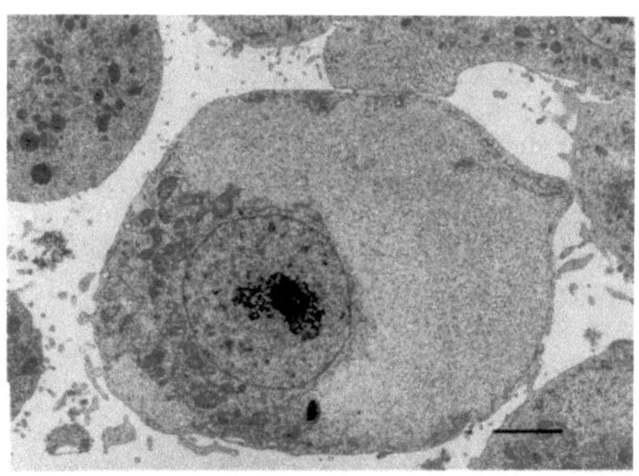

Fig. 6 Fig. 7

An electron micrograph of cultured cells

The bar represents 5 μm.

Discussion

The strains of B. mori are classified into the Japanese, the Chinese, the European, the tropical and mutant strains in Japan. Their geographical distribution was determined based on the locus of isozyme genes by Matsumura (11) and Yoshitake (12). In addition, it is suggested on the basis of the results obtained in this study that the strains can be classified by the pattern of migration and growth of the cells.

Lepidopteran cells contain microchromosomes with diffuse centromeres (13). SES-BoMo-C129 and SES-BoMo-J125 which were established from B. mori were polyploid. The chromosome number was characteristic in each cell line.

These cell lines derived from embryos, so each cell line is composed of a few types of cell. The rate of this composition seems to affect the cell growth.

Temperature ranging between 25°C and 28°C afforded the optimum conditions for rearing the silkworm larvae, but at temperatures above 36°C and as low as 15°C, they did not develop actively. This tendency was quite similar to the tendency of the larval growth. FBS is most often used at concentration ranging from 5 % to 20 % volume in insect cell culture. The optimum concentrations of FBS for cell growth varies from cell line to cell line. The cells from A. eucalypti and B. mori multiplied most rapidly at concentrations of 5 % and 20 %, but 30 % caused a reduction in cell growth(14). However, newly established cell lines multiplied actively at a concentration of 10 % FBS, and at 2 % and 20 % the cells multiplied continuously. The cells were observed in the electron microscope. They developed from embryos in early organogenesis and were subcultured for 70 passages. The SES-BoMo-J125 showed that some cells were in the phase of undifferentiation. It is difficult to discuss directly the relationship between the cultured cell derived from embryos in vitro and the state of cells, but the ultrastructural changes in cultured cells seems to be an interesting problem in cell culture.

References

1. Trager, W. J. J. Exp. Med. 61, 501-513, 1935.
2. Grace, T. D. C. Nature 216, 613, 1967.
3. Quiot, J. M. Sericologia 22, 25-31, 1982.
4. Volkman, L. E. and Goldsmith, P. A. Appl. Environ. Microbiol. 44, 227, 1982.
5. Inoue, H. and Mitsuhashi, J. J. Seric. Sci. Jpn. 53, 108-113, 1984.
6. Ninagi, O. et al., In Invertebrate Cell Systems in Applications, J. Mitsuhashi, ed. CRC Press. in press.
7. Carlson, J. G. Biol. Bull. 90, 109-121, 1946.
8. Mitsuhashi, J. Zool. Sci. 1, 415-419, 1984.
9. Earley, E. M. Chromosomal preparations from monolayer cell cultures. TCA Manual 1, 31-35, 1975.
10. Sato, S. J. Electron Micr. 17, 158-159, 1987.
11. Matsumura, T. Jpn. J. Genet. 25, 256, 1951. (Japanese)
12. Yoshitake, N. J. Seric. Sci. Jpn. 37, 83-87, 1968.
13. Ennis, T. and Sohi, S. S. Canad. J. Genet. Cytol. 18, 471-477.
14. Sohi, S. S. and Smith, C. Canad. J. Zool. 48, 427-432, 1970.

ESTABLISHMENT OF CELL LINE FROM EMBRYOS OF THE SILKWORM, BOMBYX MORI

Chen Q., Li L., Yu Z., and Peng J.

Insect Virus Research Laboratory, Department of Biology, Huazhong Normal University, Wuhan, China

A new cell line, designated Bm-21E-HNU5, has been established from embryos of the silkworm, Bombyx mori. Since the first subculture was made, the cells have been subcultured at 1:2 split ratio at an interval of 4-5 days. Preliminary studies showed that the new cell line is very promising and was susceptible to many baculoviruses, such as NPVs of Heliothis armigera, Autographa californica, Philosamia cynthia ricini and GV of Plutella xylostella. However, it seemed to be neither susceptible to the NPV nor to the GPV of the homologous host insect, Bombyx mori.

Since Grace (1) reported the first continuous cell line from ovaries of silkworm in 1967, several cell lines derived from this insect have been established (2,3,4). The cell lines have been used extensively to study in vitro replication of insect viruses and expression of recombinant DNA carrying eukaryotic genes (2,3,4,5). Here we report the establishment and characterization of a new cell line derived from the embryos of the silkworm and infective experiments with baculoviruses, including NPV and CPV.

Primary culture: A primary culture was initiated on June 12, 1985. The eggs of the silkworm (a strain of Huahe 671 Dongfei) were incubated at room temperature for 4 days and then sterilized in 75% alcohol for 15 min. After mincing with

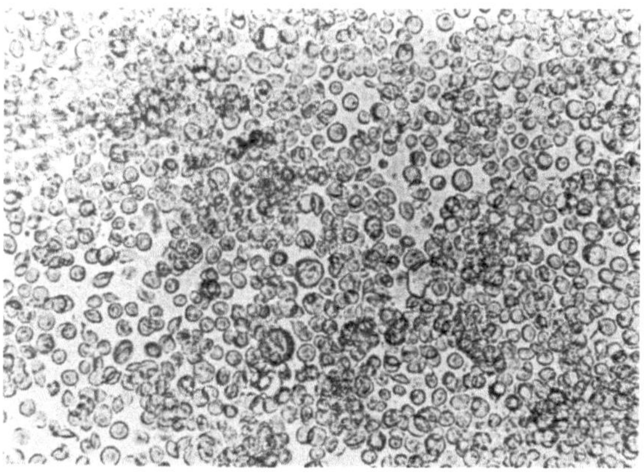

Fig. 1. The floating cells of Bm-21E HNU5 in passage 31

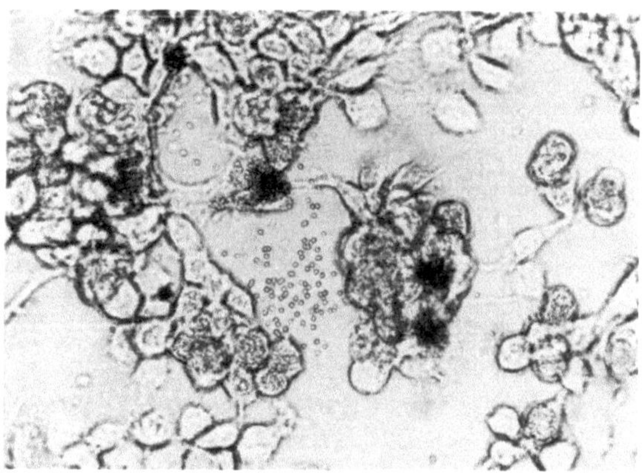

Fig. 2. The cells infected with H. armigera NPV 6 days post infection.

scissors and trypsinizing for 15 min, the minced embryos were cultured in TC-199-MK medium supplemented with 10% bovine serum and 0.75% lactalbumin hydrolysate as well as antibiotics in glass flasks at 28°C.

Subculture: First subculture was made 144 days later. Since then it has been subcultured at 1:2 split ratio at an interval of 4-5 days. The cell line has now passed 110 generations and is designated Bm-21E-HNU5 (Bm).

Morphology and growth: Most Bm cells are suspended or slightly adhere to the flask. They do not require any trypsin treatment for their removal. Therefore subculture can be easily performed by agitation. The cells are mostly round and spindle-shaped, and the number of spindle-shaped cells seems to excell the round cells at logarithmic stage. The cells that possess only a protoplasmic extension are few. The population doubling time during 72 hr to 96 hr after subculture is 48 hr. A maximum population is 8.3×10^5 cells/ml 6 days post subculture.

Karyotype analysis: For chromosome analysis, the cells were prepared conventionally. The chromosomes were short, condensed rods and spheres. Their numbers were numerous. The chromosomes were typical of Lepidoptera.

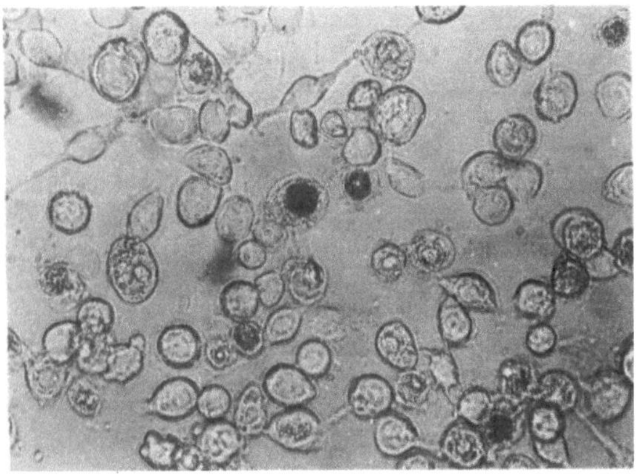

Fig. 3. The cells infected with P. cynthia ricini NPV.

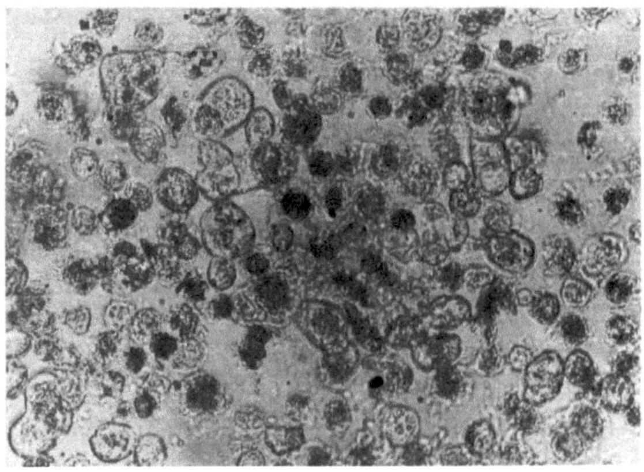

Fig. 4. The cells infected with A. californica NPV 5 days post infection.

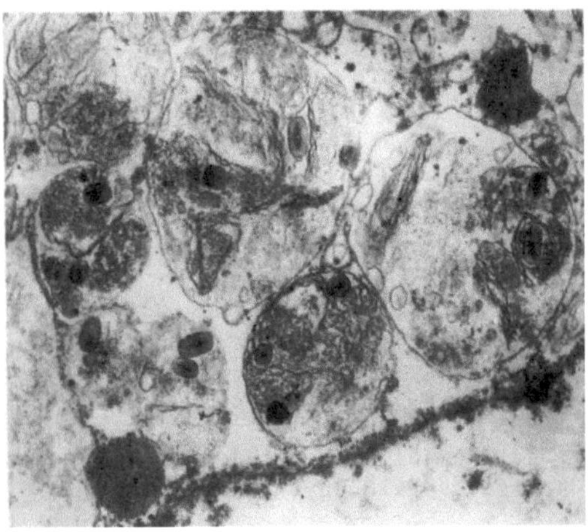

Fig. 5. The special vacuoles containing capsules in the cytoplasm of the cells infected with P. xylostella GV.

Biochemical marker: The analysis of isoenzyme pattern indicated that the cell line was different from the cell lines derived from other insect species, especially markedly different from the cell line of BCIRL-PX2-HNU3 of Plutella xylostella (6), which is vitally growing and frequently used in our laboratory.

Storage and revival of the cells: When the 8th and the 13th passage cells were stored in liquid nitrogen, 90% and 98% of the cells were respectively revived. The viabilities of the cells were respectively 80% and 84% when recovered 6 months later. The revived cells have now been subcultured more than 30 generations.

Virus susceptibility: Bm-21E-HNU5 cell line could replicate the NPVs of Autographa californica, Philosamia cynthia ricini and VHA-273, a strain of NPV of Heliothis armigera isolated in China. The hypertrophy of nuclei was observed 24 hrs after challenging the culture with AcNPV, VHA-273 or PcrNPV, respectively. By 48 hrs post infection, the inclusion bodies appeared in the nuclei of the cells. Interestingly, the cell line could support the replication of granulosis virus of Plutella xylostella (PrGV). All of the cells tightly attached to the bottom of the flask after infection with PrGV. With the increase in infecting time, marked pathological changes were visualized by light microscopy. By 72 hrs post infection virogenic stroma, virions and capsules were observed in a transmission electron microscope. However, the cell line seemed to be neither susceptible to its homologous NPV nor to CPV.

References

1. Grace, T. D. C. Nature (London) 216:613, 1967.
2. Quiot, J. M. Sericologia 22:25-31, 1982.
3. Inoue, H. and Mitsuhashi, J., J. Seric. Japan 53:108-113, 1984.
4. Maeda, S., J. Seric. Sci. Tokyo 53:547-548, 1984.
5. Maeda, S., et al., Nature 315:592-594, 1985.
6. Chen Quhou, et al., J. Central China Teacher's College (Natural Science Edition), No. 3:99-107, 1983.

FISH CELL LINES AS MATERIALS FOR STUDYING CELLULAR RADIATION BIOLOGY

N. Egami[1], A. Shima[2], and H. Mitani[2]

[1] National Institute for Environmental Studies, Tsukuba, Ibaraki 305, Japan

[2] Zoological Institute, Faculty of Science, University of Tokyo, Tokyo 113, Japan

To study the radiosensitivity of fish cells _in vitro_, CAF-MM1, RBCF-1 and some other cell lines derived from goldfish and medaka have been established. Responses of these fish cells to ionizing radiation and to ultraviolet rays were examined. Unique characteristics of fish cells as compared with mammalian cultured cells were found: (1) Fish cells were very resistant to ionizing radiation. (2) The amount of DNA strand breaks was almost the same both in mouse L cells and fish cells. (3) Initial DNA breaks were able to be rejoined in fish cells. (4) The colony forming ability of fish cells was sensitive to UV and these cells showed enzymatic photoreactivation. (5) Unique temperature dependency of these responses was found.

In a series of experiments we have demonstrated that teleost fish are in general more resistant to the lethal effects of ionizing radiation than are mammalian species. Main purpose to cultivate fish cells _in vitro_ was to compare radiosensitivity of fish cells between _in vivo_ and _in vitro_ conditions. In 1980 we succeeded in establishing a cell clone, CAF-MM1, from the caudal fin of the goldfish, _Carassius auratus_, which has a high colony-forming ability(1). Since then, several fish cell lines have been established. Here we review some data on the characteristics of the responses of fish cells to radiation.

Standard Culture Conditions

Cultured fish cells (CAF-MM1) were grown at 26-33°C attached to plastic bottles in a HEPES-buffered medium, TC199, supplemented with 10% fetal bovine serum and 100μg/ml streptomycin sulfate, 100I.U./ml penicillin(1). Colony forming ability _in vitro_ was used to measure survival.

Resistance to Ionizing Radiation

Fish cells _in vitro_ were much more resistant to ionizing radiation than mouse L cells. Figure 1 shows survival curves of CAF-MM1 and mouse L cells after gamma-irradiation(1). Radiosensitivity in terms of Do and Dq of various mammalian cells reported by other authors and goldfish cells is given in Figure 2 and Table 1.

Recovery of Sublethal Damage

Figure 3 shows the survival curve of the fish cells after high (●) or low (o) dose-rate irradiation at 33°C(a), 26°C(b), 20°C(c) and 6°C(d), respectively(2). From the data, it was concluded

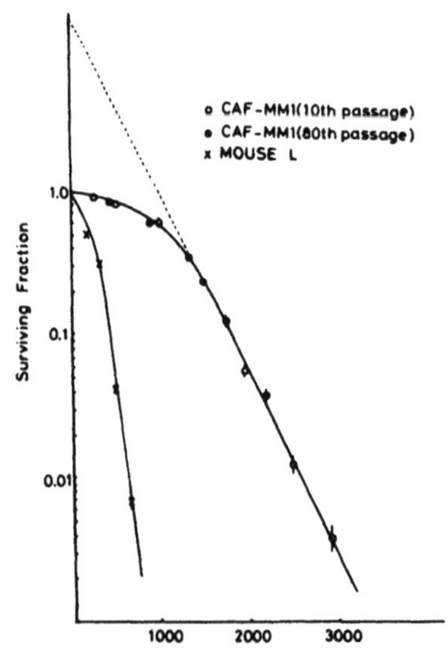

Fig. 1

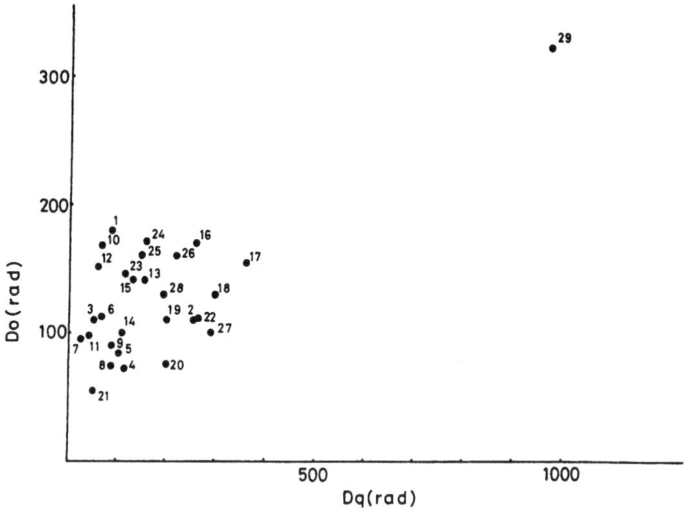

Fig. 2

Table 1

No*	Origin	Cell	Irradiation	Do(rad)	Dq(rad)
1	Human	HeLa S3	γ-ray	180	85
2		HeLa	X-ray	110	255
3		HeLa S3	X-ray	110	52
4		HRT18	X-ray	71	116
5		HCT8	X-ray	85	104
6		HT29	X-ray	112	66
7		SZC	X-ray	95	25
8		MEWO	X-ray	74	87
9		HF19	X-ray	91	87
10		HL21	X-ray	168	69
11		AG1877	X-ray	98	40
12		AG1518	X-ray	152	61
13		embryonic liver cell	X-ray	140	154
14		fibroblasts	X-ray	100	110
15		TCF145	X-ray	140	130
16		U-118MG	γ-ray	170	260
17	Mouse	L	X-ray	155	359
18		L-929	X-ray	130	301
19		L5	X-ray	110	201
20		L	γ-ray	75	200
21		L5178Y	X-ray	55	51
22	Chinese hamster	V79-379-A	X-ray	111	257
23		H-1	X-ray	145	115
24		H-2	X-ray	170	157
25		CHL	X-ray	160	150
26		CHO-K1	X-ray	160	220
27	Ratkanga-roo	PKK	X-ray	100	291
28	Xenopus	A8W243	X-ray	130	196
29	Goldfish	CAF-MM1	γ-ray	325	975

that the lethal effect of gamma-rays was decreased at low dose-rate at 20-33°C, but not at 6°C. Similarly, split-dose experiments showed that the recovery of sublethal damage occurred at the high temperature,but not at 6°C(2). These results are consistent with the *in vitro* data on the effect of temperature on the radiosensitivity and recovery of sublethal damage reported previously for live fish.

DNA Strand Breaks after Irradiation

We have measured amount of DNA strand breaks after irradiation in fish cells and mouse L cells. It was found that the amount immediately after irradiation was almost the same in both cells in spite of their large difference in radiosensitivity. The initial DNA strand breaks were able to be rejoined during the further incubation at 26°C,but this process did not take place at 4°C in the fish cells. The amount of the residual DNA strand breaks at 7 hours after irradiation was higher in the fish cells than in the mouse L cells,and the residual DNA strand breaks did not correlate with the radiosensitivity of the fish and mouse cells(3).

Radiosensitivity of Fish Cells with Different Ploidy

There are many varieties of goldfish, and radiosensitivity of animals as a whole was different among different varieties. For instance,common goldfish(Wakin)is more resistant to X-rays than black telescope eye goldfish(Demekin). Furthermore,although heteroploid cells are rather rare in vertebrates,diploid, triploid and tetraploid varieties of *Carassius auratus* are found in the wild in Japan. Since it has been reported that cellular

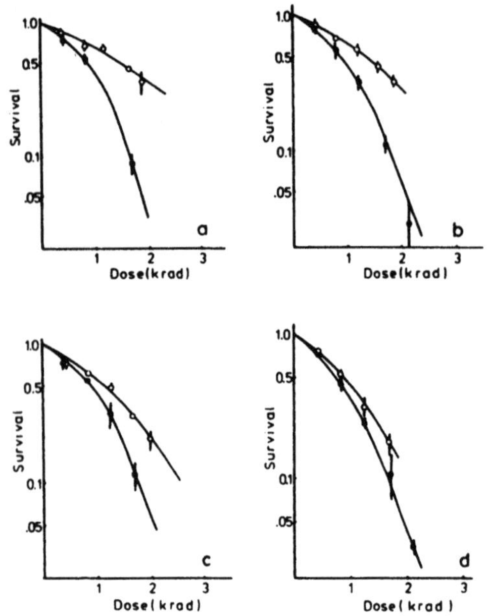

Fig. 3

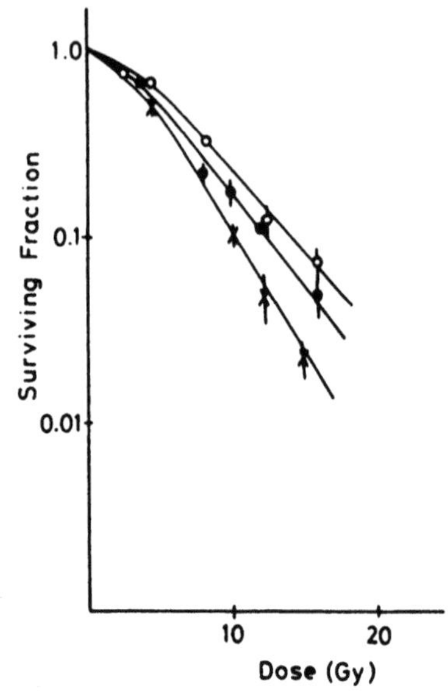

Fig. 4

radiosensitivity is correlated with the ploidy number in plants and in yeast, we have compared radiosensitivity of the primary cultured cells of these fish. No significant difference in radiosensitivity was found between Wakin and Demekin cells. As shown in Figure 4,which shows dose-survival curves of diploid(•),triploid(o), and tetraploid(x) cells,no correlation between radiosensitivity and ploidy number was recognized(4).

Lethal and Mutagenic Effects of Radiation and Chemicals

For examining these effects,GEM199 cells derived from an erythrophoroma of goldfish were used. N-methyl-N'-nitro-N-nitrosoguanidine(MNNG) was effective in inducing ouabain-resistant mutation,while 4-nitroquinoline 1-oxide(4NQO) and X-rays did not induce mutation(5). Responses of the cells were different from mammalian cells. The comparative study is very interesting for an analysis of mutation mechanism in cultured cells.

Sensitivity of Fish Cells to Ultraviolet Rays

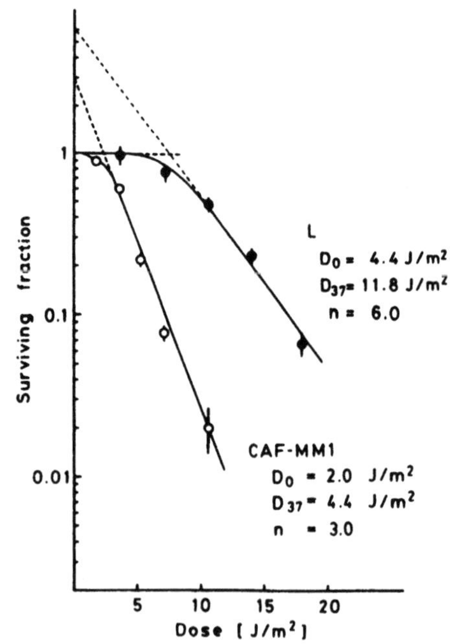

Fig. 5

The colony forming ability of these fish cells was sensitive to ultraviolet rays(5). Figure 5 shows the UV dose-survival curves of CAF-MM1 cells(o) and mouse L cells(•)(6). To study the high sensitivity of the fish cells, direct assay of thymine dimers was made. The ratio of thymine dimers to thymine immediately after UV-irradiation was proportional to UV dose(7,8).

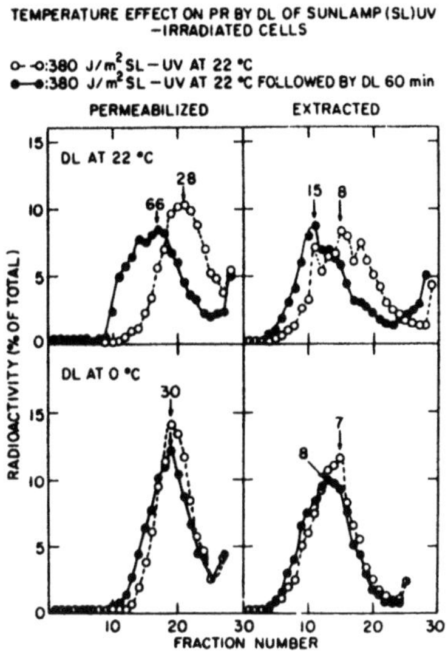

Fig. 6

Survival and Pyrimidine Dimers in UV Irradiated Fish Cells

Fish cells irradiated with filtered sunlamp ultraviolet rays were exposed to daylight radiation and photoreactivation took place. In RBCF-1 cells derived from fin of goldfish, the photoreactivable fraction reached 0.52 at 22-25°C,while almost no change was observed in pyrimidine dimers at 0°C,indicating that this PR process is enzymatic(9)(Fig.6).

Usefulness of Medaka Cells

There is no doubt that recent progress in modern biology is a result in some part of the developments in laboratory animal science, particularly the establishment of inbred strains of mice, since the genetic background of inbred animals is uniform. However,in general,inbred strains of animals are difficult to establish. Recently Hyodo-Taguchi succeeded to establish some inbred strains of Oryzias latipes(10). Furthermore, cell lines derived from these strains have been established(11,12). Radiobiological studies with these cell lines have been carried out by members of the Zoological Institute of University of Tokyo and the National Institute of Radiological Sciences,Japan,and important results have been obtained.

Conclusions

Many fish cell lines have been established and these cells are useful for comparative cellular radiation biology. Some of the unique responses of cultured fish cells are summarized in this review.

References

1. Mitani,H., Etoh,H. and Egami,N. Radiation Research, 89:334-347, 1982.
2. Mitani,H. and Egami,N. Radiation Research, 97:48-54, 1984.
3. Mitani,H. and Egami,N. Int. J. Radiat. Biol., 41:85-90, 1982.
4. Mitani,H., Egami,N. and Kobayashi,H. J. Radiat. Res., 27:284-290, 1986.
5. Mitani,H. Mutation Research, 107:279-288, 1983.
6. Mano,Y., Mitani,H. and Egami,N. Radiation Research, 84:514-522, 1980.
7. Mano,Y., Katoh,K. and Egami,N. Radiation Research, 90:501-508, 1982.
8. Mano,Y., Katoh,K. and Egami,N. Photochem. Photobiol., 35:49-56, 1984.
9. Shima,A. and Setlow,R.B. Photochem. Photobiol., 39:49-56, 1984.
10. Hyodo-Taguchi,Y. and Egami,N. Zoological Science, 2:305-315, 1985.
11. Komura,J., Mitani,H. and Shima,A. (submitted)
12. Etoh,H., Hyodo-Taguchi,Y. et al.(in preparation)

ESTABLISHMENT AND CHARACTERISTICS OF VARIOUS CELL LINES FROM MEDAKA (TELEOSTEI)

H. Etoh, I. Suyama, Y. Hyodo-Taguchi, and H. Matsudaira

National Institute of Radiological Sciences, Chiba 260, Japan

Four permanent cell lines, OLME-104 from melanoma of HB32C medaka, OLHE-131 from hepatoma of HO4C, OLF-136 from the normal fins of HB32C and OLF-124 from those of outbred medaka of orange-red variety, were established. From their growth curves, doubling times were estimated to be 43 h for OLF-136, 77 h for OLME-104 and 67 h for OLHE-131. Nearly diploid amounts of DNA for OLF-136 and OLME-104, but nearly tetraploid for OLHE-131 cells were observed on cytofluorometrical analysis and confirmed on metaphase preparations. The cultured cells were injected ip into respective host fish. No nodule was observed in HB32C medaka injected with OLF-136 cells. The fish injected with OLME-104 and with OLHE-131 cells died within a month due to tumor growth. The OLME-104 cells were transplantable to HB32C medaka but not to HO4C and outbred medaka. The OLHE-131 cells were transplantable only to HO4C. The tumor cells have maintained their malignant characters for many generations in _in vitro_ culture. It seems from UV-irradiation experiments that these cell lines from medaka, as well as a cell line from goldfish, are more sensitive to UV-radiation compared to the cultured mammalian cells.

Many cell lines from various normal tissues of fish have been established for various purposes in the fields of virology, radiation biology and others (1-4). However, cell lines derived from tumors in fish are limited (5, 6). Several inbred strains of medaka, _Oryzias latipes_, were established in our laboratory (7). Moreover melanoma was induced in HB32C strain after treatment with N-methyl-N'-nitro-N-nitrosoguanidine (MNNG) and hepatoma in HO4C, another inbred strain, with methylazoxymethanol (MAM) acetate (8, 9). These tumors were transferred into _in vitro_ culture, and cell lines were established. In the present paper, characteristics of cultured tumor cells will be reported.

Materials and Methods

Cultivation of the cell lines : Melanoma was induced in HB32C medaka with MNNG and hepatoma in HO4C with MAM acetate. Small pieces of tumors were transplanted first into the anterior eye chamber of the respective strains (8). After serial transplantations _in vivo_, portions of the tumors were transferred into _in vitro_ cultures. Cultures were performed by the method reported in the previous paper (2). Briefly, sterilized tissues were minced with scissors and plated into plastic culture bottles. HEPES-buffered TC-199 medium supplemented with 20 per cent fetal bovine serum (FBS) and antibiotics were used. When the cells made a confluent sheet, the cells were dispersed by a rubber scraper in early passages and by treatment with D-PBS containing trypsin (0.25 %) and EDTA (0.02 %) in later passages, and then subcultured with a split ratio of 1:2. Normal fibroblastic cell lines were also established from the fins of HB32C and outbred medaka of orange-red variety by the method mentioned above.

Cell cycle analysis : To obtain their growth curves, the cells were inoculated into plastic dishes and were harvested on 1 to 24 days. Cell numbers were counted using the blood cell counting chamber (Thoma).

When the cells were growing in the log phase, 5 ml of medium and 1 ml of the staining solution were added after washing with fresh medium. The staining solution consisted of 45 mg of propidium iodide, 600 mg of RNase, 3 ml of triton-X 100 and 97 ml of distilled water. Culture bottles were kept cold on ice for 2 hours after incubation at 37°C for 1 minute. Fluorescence profiles were obtained by a cytofluorometer (Ortho Diagnostic Co., USA) operated with a 250 mW argon laser at 480 nm. Each parameter was analyzed from the ratios of different phases of the cell cycle. Chromosome preparations were also made by means of routine air-drying techniques.

Transplantation of tumor cells : Cultured cells were collected and resuspended in D-PBS at appropriate concentrations. Fish were injected i.p. with 0.02 ml of cell suspension containing $0.4\text{-}1000\times10^3$ cells (Table 1). These fish were fixed with Bouin's solution at death or 2 months later and sagittal sections were made for observation of tumor nodules.

UV-irradiation : Methods for ultraviolet (UV) irradiation and photoreactivating light (PRL) illumination were described in detail in a previous paper(4). At various times after treatment, cells were collected with a rubber scraper and counted on the counting chamber, because no colonies were formed in these cell lines.

Table 1. Number of medaka with tumor after injection of cultured cells

Cell lines	Strains injected	Number of cells injected ($\times10^3$)	Number of fish tested	Number of fish with tumors
OLME-104	HB32C	1000-50	15	13
		10	5	2
		4	6	3
		1	5	2
		0.4	5	2
	HO5	500	3	3
		250	3	3
		50	6	3
	HO4C	500	5	1 *
	outbred	500-50	19	0
OLHE-131	HO4C	500	5	5
OLF-136	HB32C	1000	5	0

* Very small nodule was found microscopically.

Results

Establishment of cell lines : Two different cell lines, derived from melanoma induced with MNNG in HB32C medaka and hepatoma induced with MAM acetate in HO4C, were established and designated OLME-104 and OLHE-131, respectively.

Two other cell lines, designated OLF-136 and OLF-124, were established from the fins of HB32C and outbred medaka, respectively. These cell lines have been successfully subcultured for many generations. The OLME-104 and OLHE-131 cells grew in crisscross fashion and formed dense foci after confluency. These characteristics were never found in OLF-136 and OLF-124 cells having fibroblastic characters. From their growth curves, doubling times of OLME-104, OLHE-131 and OLF-136 cells were estimated to be 77, 67 and 43 h, respectively. Since OLF-124 cells were used mainly for oncogene analysis (Muraiso et al., in this volume), cell characters were not presented here.

Table 2. Cell cycle times in hours of cultured fish cells

Cell lines	G_0+G_1	S	G_2+M	T
OLF-136	23.5	14.4	5.1	43
OLHE-131	27.8	9.4	9.4	67
OLME-104	46.9	23.6	6.5	77

Cell cycle times : On the basis of cytofluorometrical analysis, and their growth curves, fractions of different phases of cell cycle of these cultured cells were estimated (Table 2). Mean channel numbers of G_0 + G_1 peak were 156 for OLF-136, 326 for OLHE-131 and 186 for OLME-104 cells. These data, confirmed with chromosome analysis by light microscopy, showed a near diploid in nature of OLF-136 and OLME-104 cells, and near tetraploid for OLHE-131 cells.

Transplantability of tumor cells : Table 1 shows the number of medaka with tumor after injection of cultured cells. The fish, injected with more than 1×10^5 OLME-104 cells and 5×10^5 OLHE-131 cells, died within a month due to tumor growth. When HB32C medaka were injected with $0.4\text{-}50 \times 10^3$ OLME-104 cells, half of the fish were found to bear tumors on serial histological sections of the body. When HO5 medaka were injected with $5\text{-}50 \times 10^4$ OLME-104 cells, tumors were found in almost all fish. The tumor was found in HO4C medaka injected with OLHE-131 cells. On the contrary, when OLME-104 cells were injected into HO4C and outbred medaka, no tumor was found 2 months later.

Effects of UV-radiation : The cell number was counted on 0 to 8 days after UV only or UV followed by PRL (Table 3). In all cell lines tested, cell growth occurred after UV-irradiation at 5 J/m^2, but no growth was observed after 10 J/m^2 or more. When PRL was administered after UV, cells survived and grew even after the cells received as much as 15 J/m^2.

Table 3. Responses to UV only and UV+PRL

Dose (J/m^2)	OLF-136		OLME-104		OLHE-131	
	UV	UV+PRL	UV	UV+PRL	UV	UV+PRL
5	+	+	+	+	+	+
10	-	+	-	+	-	+
15	-	+	-	+	-	+
20	-	-	-	-	-	+
30	-	-			-	-

Cell growth occurred (+) and did not occur (-) after the treatments.

Discussion

Although many permanent cell lines from various tissues of fish have been established, reports on cell lines derived from tumors of fish are restricted (5, 6). It has been reported that induced tumors are transplantable in vivo (8, 10). However, no report was made on the transplantation of cultured tumor cells. Cultivation of tumor cells is useful to investigate their characters. Since the inbred strains of medaka have been established in our laboratory, it is possible to perform the transplantation of cultured cells in the syngeneic medaka. The fact that HO5 strain, an inbred strain of orange-red variety, accepted the OLME-104 cells from HB32C, is peculiar and remains to be explained. The fish, when injected with cultured tumor cells, bore tumors and some of them died eventually due to tumor growth.

Measurements of DNA/cell showed nearly diploid amounts of DNA for OLF-136 and OLME-104 cells, but nearly tetraploid amounts for OLHE-131 cells. Polyploidization was observed in medaka liver tumor induced with chemical carcinogen in vivo and liver tumor cells cultured in vitro (5, 11), but not in normal aging fish liver (12). If DNA index is expressed as a ratio of the amount of DNA/cell of OLF-136 cells to that of tumor cells at the mean channel of G +G peak, the values are 1.2 for OLME-104 and 2.1 for OLHE-131 cells. These values seem to indicate hyper-aneuploidy in those tumor cells. The human tumors showing hyperaneuploidy were reported to be malignant (13). From these results, The OLHE-131 and OLME-104 cells maintain their malignant characters even after they had been cultivated in vitro for many generations.

The CAF cells, derived from the fins of goldfish, are sensitive to UV-radiation (4). Generally speaking, fish cells may be more sensitive to UV compared to the cultured mammalian cells, because all cultured cells from medaka are as sensitive to UV as CAF cells.

Acknowledgements

We thank Dr. I. Watanabe, NIRS, for his kind help in cytofluorometrical analysis. This work was supported in part by a special grant for cancer research from Science and Technology Agency, Japan.

References

1) Watanabe, T., Kobayashi. N., Sato, Y. and Ishizaki, Y. Bull. Jpn. Soc. Sci. Fish., 44: 415-418, 1978.
2) Suyama, I. and Etoh, H. Zool. Mag., 88: 321-324, 1979.
3) Shima, A. and Setlow, R.B. Photochem. Photobiol., 39: 49-56, 1984.
4) Mano, Y., Mitani, H., Etoh, H. and Egami, N. Radiat. Res., 84: 514-522, 1980.
5) Mitani, H. and Egami, N. J. Fac. Sci., Univ. Tokyo, Sec IV, 14: 391-398, 1980.
6) Wakamatsu, Y., Oikawa,A., Obika, M., Hirobe, T. and Ozato, K. Develop. Growth and Differ., 26: 503-513, 1984.
7) Hyodo-Taguchi, Y. Zool. Mag., 89: 283-301, (in Japanese) 1980.
8) Hyodo-Taguchi, Y. and Matsudaira, H. JNCI, 73: 1219-1227, 1984.
9) Unpublished data.
10) Schultz, M.E. and Schultz, R.J. Exp. Mol. Pathol., 42: 320-330, 1985.
11) Egami, N., Kyono-Hamaguchi, Y., Mitani, H. and Shima, A. In "Phyletic Approaches to Cancer" (Dawe, C.J. et al. eds.), pp. 217-226, Jpn. Sci. Soc. Press, Tokyo, 1981.
12) Shima, A. and Egami, N. Exp. Geront.,13: 51-55, 1978.
13) Heliö, H., Karaharju, E. and Nordling, S. Cytometry,6: 165-171, 1985.

ESTABLISHMENT OF A CELL LINE FROM UMBRA LIMI (UMBRIDAE; PISCES)

I. Suyama and H. Etoh

National Institute of Radiological Sciences, Chiba 260, Japan

A fibroblast-like cell line (designated ULF-23) was established from the fins of Umbra limi. The cell doubling time was estimated to be 72 hours from the growth curve. It did not form colonies at all. The modal chromosome number shifted to 23 after 50 passages, but it still kept almost all other karyological characters.

ULF-23 cells were irradiated with X-rays for chromosome aberration analysis or treated with MNNG for sister chromatid exchange (SCE) analysis. Both yields were lower than in cultured mammalian cells, but their usefulness in in vitro fish cytogenetic studies was confirmed.

Generally speaking, fish have a large number of small chromosomes and their mitotic indices are too low to get enough chromosome spreads for cytogenetic studies. The central mudminnow, Umbra limi, however, has a low number of large meta- or submetacentric chromosomes (2n=22), and was recommended as a model for the study of chromosome aberrations in fish by Kligerman et al(1). The present authors established a method of cultivating a small amount of Umbra blood for chromosome preparations and applied it to radiation-induced chromosome aberrations and sister chromatid exchange (SCE) analyses(2,3). In this report the establishment and characteristics of a cell line from the fins of Umbra limi were presented. Also presented were two cases of applications of the cells to clastogens, X-rays and MNNG.

Materials and Methods

The fins of Umbra limi which were originally imported from the United States and later bred in our Institute were used. The fins were sterilized in 1 % sodium hypochloride and 70 % alcohol for one minute each, and then washed once with sterilized distilled water and twice with TC-199. Thereafter, the fins were minced with a pair of scissors as finely as possible, and were planted in a 25-cm^2 plastic flask containing 5 ml HEPES-buffered culture medium (pH 7.2). The culture medium consisted of TC-199, 80 % supplemented with fetal bovine serum, 20 %, streptomycin sulfate, 100 mcg/ml, and penicillin, 100 IU/ml. Cells were incubated in a closed system at 25°C, and were divided into two new flasks at confluency using 0.25 % EDTA-added trypsin solution.

For the cell cycle analysis, the cells at log phase were stained with fluorescent dye solution. The stock solution consisted of propidium iodide, 45 mg, RNase, 600mg, triton-X 100, 3ml, and distilled water, 97 ml. The working solution was a mixture of 5ml of TC-199 and 1 ml of the stock solution. After washing with TC-199, 1 ml of the working solution was added to the culture bottle which was kept cold on ice for 2 hours after incubation at 37 °C for one minute. Then a fluorescence profile was obtained using a cytofluorometer (Ortho Diagnostic Co., USA) operated with a 250 mW argon laser at 480 nm. Percentage of each cell cycle time was determined from the ratios of different phases of the cell cycle.

For chromosome preparations, six hours before fixation colcemid was added to the culture bottles at the concentration of 0.03 mcg/ml. After harvesting, chromosome preparations were done by a routine hypotonic treatment-air-drying method. Slides were stained with 5 % Giemsa in Sorensen's phosphate buffer (pH 6.8). For sister chromatid exchange analysis, the slides were stained by modifying the FPG technique of Perry and Wolff (4).

Results

1) Characteristics of the ULF-23 cells

The cell line obtained was designated ULF-23. The cells were fibroblastlike. In spite of repeated attempts, the cells did not form colonies at all. The growth of the cells was very slow. From the straight portion of the growth curve the doubling time was estimated to be 72 hours. The fluorescence profile was shown in Fig. 1. From the analytical data of this profile and the cell doubling time of 72 hours the cell cycle times were estimated as follows; $G_0 + G_1 = 62$ hours, S = 7 hours, $G_2 + M = 3$ hours.

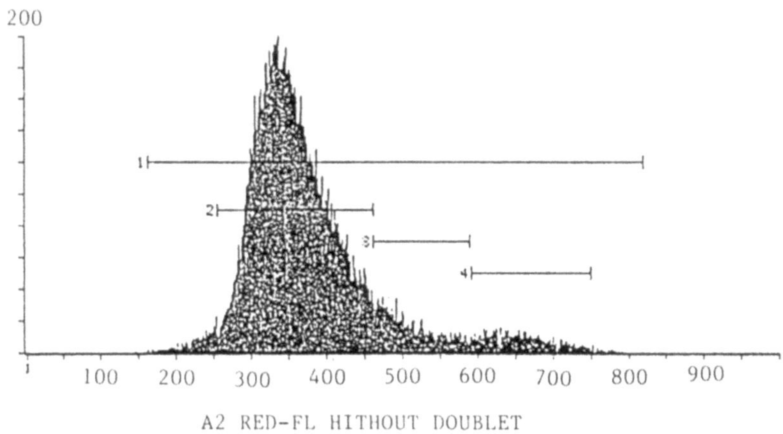

HIST	LC	UC	COUNT	%TOT	PEAK	MEAN	SD	CV	CV(HM)
A2 1	164	820	26116	99.9	335.0	381.8	95.4	25.0	11.5
A2 2	256	461	21776	83.3	335.0	353.6	45.4	12.0	11.5
A2 3	462	590	2230	8.5	467.0	510.9	36.1	7.1	20.2
A2 4	591	749	1435	5.5	611.0	656.4	41.1	6.3	.5

Fig. 1. A fluorescence profile of ULF-23 cells stained with propidium iodide solution

2) The chromosomes of ULF-23 cells

The diploid chromosome number of Umbra limi is 22. After about 50 passages the modal number shifted to 23 with its maximum at 25 (Table 1). The karyological characters of the chromosomes, however, remained almost unchanged, being meta- or submetacentrics. Some of the cells have one or two satellite chromosomes, whereas none of the cells obtained from Umbra lymphocyte cultures do.

Table 1. Chromosome frequency distribution (T.N.=200)

Chromosome number	20	21	22	23	24	25
Frequency	5	10	28	142	13	2

Responses of ULF-23 cells to clastogens

1) X-ray irradiation

Just before confluency the cells were irradiated with various doses of X-rays at the dose rate of 100 R/min. The doses employed were 0, 50, 100, 150, 200 and 300 R. Forty-eight hours after irradiation the cells were harvested and fixed with methanol-acetic acid (3:1) fixative. At 48 hours after irradiation about 80 % of the chromosome spreads observed were known to be in their first postirradiation division. Chromosome preparations were done by the method mentioned in the section Materials and Methods. All types of chromosome aberrations were observed, but ring chromosomes were excluded from the scoring because it was impossible to tell whether they were centric or not and because their frequency was very low. The frequencies of chromosome aberrations are shown in Table 2. The dose-response relationship was calculated to be $Y = 0.42 + 5.36 \times 10^{-4} X + 3.73 \times 10^{-6} X^2$ using a least-squares method.

Table 2. Frequencies of chromosome aberrations in ULF-23 cells irradiated with X-rays

X-Ray Dose (rad)+	Cells Examined	Aberrant Cells(%)	Chromosome Aberrations* CG	CB	ICG	ICB	CE	D	Aberrations /cell+S.E.
0	600	22(3.7)	3	11	0	14	0	0	0.047+0.009
47.5	401	22(5.5)	2	6	0	9	0	5	0.067+0.013
95	600	55(9.2)	5	22	0	21	0	13	0.123+0.014
142.5	600	82(13.7)	0	21	1	30	0	37	0.210+0.019
190	600	107(17.8)	2	11	0	45	0	52	0.270+0.021
285	600	166(27.7)	4	16	1	56	0	111	0.498+0.029

\+ Absorbed doses calculated by using a conversion factor of 0.95.

* Abbreviations;

CG:Chromatid gap, CB:Chromatid break, ICG:Isochromatid gap, ICB:Isochromatid break, CE:Chromatid exchange, D:Dicentric chromosome

2) Treatment with MNNG

The cells were treated with various concentrations of MNNG (1-methyl-3-nitro-1-nitrosoguanidine) in TC-199 for 2 hours in log phase. After treatment the

Table 3. Frequencies of SCEs induced in ULF-23 with MNNG

Concentration of MNNG (mol)	No. of Cells	No. of Chromosomes	No. of SCEs	SCEs/Cell	SCEs/ Chromosome
0	50	1150	599	11.98	0.52
10^{-7}	50	1140	705	14.10	0.62
10^{-6}	50	1143	1180	23.60	1.03
2×10^{-6}	50	1142	1440	28.80	1.26
10^{-5}	--*				
2×10^{-5}	--*				

* No M_2 cells were observed.

cells were washed with TC-199 twice and then continued to be cultured in normal culture medium containing BrdUrd (5-bromodeoxyuridine) at the concentration of 1 mcg/ml for 96 hours in the dark.

Colcemid treatment and chromosome preparations were done in the same way as in the X-irradiation experiment. The slides were stained in the above-mentioned method for sister chromatid differentiation (SCD).

A typical chromosome spread stained for SCD was shown in Fig. 2 and the results in Table 3. Frequencies of SCE were increased with increasing concentrations of MNNG up to 2 x 10^{-6} M. At 10^{-5} M or higher no M_2 cells were encountered.

The spectrum of chromosome aberrations observed in irradiated ULF-23 cells was similar to that found in cultured mammalian cells. The yields, however, were rather lower in this system than in mammalian ones. As for SCE induction, the control value for SCE in this system was 11.98/cell, and almost the same as those of CHO cells which were reported to range from 9 to 15.2 SCEs/cell (4,5,6). However, the increase of SCE in MNNG-treated ULF-23 cells was not marked compared to that of cultured mammalian cells treated with the same agent (6)

The aquatic environment has been polluted with various kinds of wastes. Under such circumstances, the present system as well as the microculture technique of Umbra lymphocytes will become useful for detecting effects of waterborne pollutants.

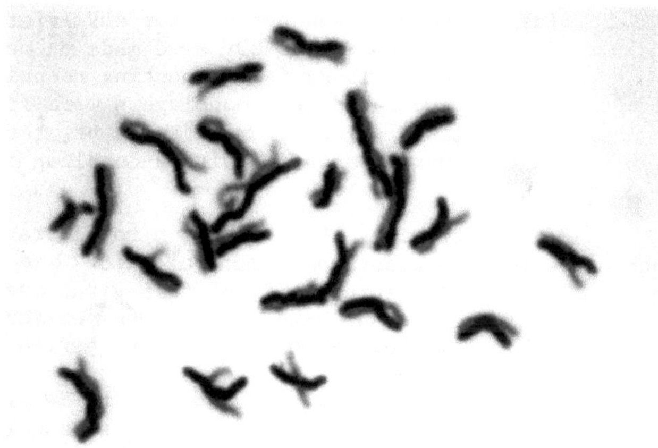

Fig.2. A typical chromosome spread of ULF-23 cells stained for SCD

References

1) Kligerman,A.D.,Bloom, S.E. and Howell, W.M. Mutat. Res., 31:225-233, 1975
2) Suyama, I. and Etoh, H. Mutat. Res., 107:111-118, 1983
3) Suyama, I. and Etoh, H. Proceedings of the Second Workshop on Tritium Radiobiology and Health Physics (NIRS-M-52), 146-156, 1985
4) Perry P. and Wolff, S. Nature, 251:156-158, 1974
5) Wolff S. and Perry, P. Chromosoma (Berl.), 48,341-353, 1974
6) Perry,P. and Evans, H.J. Nature, 258,121-125, 1975

ESTABLISHMENT OF CELL LINES OF VECTOR INSECTS, AND INFECTION OF THESE CELL MONOLAYERS WITH RICE DWARF VIRUS (RDV)

I. Kimura

National Institute of Agrobiological Resources,
Department of Cell Biology,
Yatabe, Tsukuba, Ibaraki 305, Japan

Six NC, four NN and five RD cell lines, derived from embryos of Nephotettix cincticeps, N. nigropictus and Recilia dorsalis, respectively (leafhopper vectors of RDV), have been established employing Liu and Black's medium modified slightly (1). All cell lines were epithelial cell monolayers attached to the bottom surface of plastic flasks, and very susceptible for RDV infection.

Infectivity assays of RDV were made on vector cell monolayers by the fluorescent antibody focus counting technique. Infective progeny virions in inoculated cell monolayers were first detected 12 hr after the initial inoculation. From 12 to 20 hr, the growth rate of the virus in the cell was exponential with a doubling time of 96 min.

Rice dwarf virus (RDV) is known to be transmitted by three species of leafhopper vectors to plant hosts, none of which has been successfully inoculated mechanically (2). The infectivity of this virus has been assayed only by artificial injection of small quantities of virus suspension into abdomens of the leafhoppers (3). However, this method required a long period, about 50 to 60 days, before final results were obtained.

The fact that the virus was shown to be propagative in its insect vectors as well as in host plants following Fukushi's pioneer discovery in 1940 (4), led to the conclusion that its pathogen possesses the properties of both plant and insect viruses. From this point of view, one can expect the establishment of monolayer cell cultures of the insect vectors to be an effective experimental system. Chiu and Black (1967) first succeeded in culturing cell monolayers of Agallia constricta, a vector of clover wound tumor virus, and demonstrated infection of the insect cell monolayers with virus (5). This paper describes monolayer cell cultures derived from Nephotettix cincticeps (NC), N. nigropictus (NN) and Recilia dorsalis (RD), leafhopper vectors of RDV, and the inoculation method of the virus to the cultured cell monolayers.

Monolayer Cell Cultures of Leafhopper Vectors

To obtain tissue culture explants, embryonic fragments were dissected from NC, NN and RD leafhopper eggs which had been oviposited 5 to 11 days earlier in leaf sheaths of rice plants. The eggs were removed from the leaf sheaths by needles under a binocular microscope, and were surface-sterilized by immersion in 70% ethanol for 3 min, followed by rinsing 3 times in sterile Tyrode's solution. Each young embryo removed from the eggs, was cut into a few fragments in Tyrode's solution on a hollow slide. The embryonic fragments were treated with 0.25% trypsin (Difco Laboratories, Detroit, Mich., U.S.A.) in Tyrode's solution (pH 7.5) for 10 to 20 min at 30°C, and then added with growth medium (Table 1) to stop the enzyme activity. About 30 fragments were placed together in a small amount of medium (about 0.3 to 0.5 ml) in the

Table 1. Medium composition for Nephotettix leafhopper cells

1. Schneider's Drosophila medium (revised)	500 ml
2. Medium 199 (X 10 concentrate) with Hanks' salt and glutamine, without sodium bicarbonate	50 ml
3. Medium CMRL 1066 with glutamine	25 ml
4. Fetal bovine serum*	150 ml
5. 0.05 M histidine solution	500 ml
6. Penicillin G potassium	120,000 unit
7. Streptomycin sulfate	120 mg
8. Neomycin sulfate (10,000 ug/ml)	6 ml
9. Fungizone (250 ug/ml)	4 ml

Total volume of the medium was about 1,240 ml, and the pH value of the medium was adjusted to 6.50 to 6.60 with 2N HCl.
* : Heat treatment at 56°C for 30 min.

center of a small plastic dish (55 mm in diameter) enclosed in a sealable glass dish (80 mm in diameter). Additional medium was placed around the bottom inside rim of the plastic dish to maintain a suitable vapor pressure and thus avoid drying of the separate medium containing the fragments. The cultures were kept at 25°C in an incubator. The medium was changed at intervals of 7 to 10 days.

In the primary cultures, most explants of embryonic fragments attached to the bottom surface of the cultured vessel after a few hours, and cell growth began to occur around the explanted tissues within 48 hr after the cultures were set up. The cells increased in number to form cell sheets and expanded gradually to engulf the original tissue explants. Such cultures could be maintained for several months by changing the medium once a week. When the cells grew almost confluently in the dish, they were transferred to new dishes. These subcultured cells were then passed to plastic culture flasks (25 cm^2 in bottom area) for further subculturing. The monolayered cells were usually composed of 2 types of epithelial cells. The majority were spindle shaped cells, 7 to 10 μm wide and 15 to 60 μm long in NC cells. A minority were round shaped cells, 6 to 20 μm in diameter in NC cells. Eventually, 6NC, 4NN and 5RD cell lines were established and these have been passed through 50 to 165 subcultures at intervals of 4 to 14 days (Table 2).

Table 2. Establishment of some cell lines of leafhopper vectors (NC, NN & RD)

Cell line	Date primary culture set up	No. of eggs employed	Ages of eggs	Max. No. of passages*
NC - 15	April 27, '78	30 eggs	8 - 10 days	98 passages
NC - 19	Jan. 24, '79	40 "	8 - 11 "	85 "
NC - 20	Feb. 14, '79	24 "	9 - 11 "	98 "
NC - 21	Feb. 15, '79	40 "	7 - 10 "	102 "
NC - 24	Feb. 28, '79	40 "	7 - 9 "	147 "
NC - 25	March 2, '79	40 "	6 - 9 "	165 "
NN - 1	April 29, '80	36 eggs	8 - 10 days	102 passages
NN - 2	April 30, '80	25 "	5 - 9 "	122 "
NN - 5	May 9, '80	36 "	7 - 9 "	109 "
NN - 7	May 14, '80	27 "	6 - 8 "	108 "
RD - 5	May 14, '84	15 eggs	7 - 10 days	59 passages
RD - 6	May 17, '84	29 "	7 - 10 "	62 "
RD - 7	May 21, '84	18 "	8 - 11 "	50 "
RD - 8	May 23, '84	15 "	7 - 10 "	65 "
RD - 9	May 26, '84	13 "	7 - 10 "	66 "

* : Maximum number of passages in each cell line was checked on Dec. 25, 1986.

Inoculation of Virus to Vector Cell Monolayers

Preparation of inoculum : All the established cell lines of NC, NN and RD leafhoppers were subjected to inoculation with virus extracts from infected leaves of rice plants. The infected leaves were surface-sterilized with 70% ethanol for 3 min and then rinsed with sterile distilled water. The sterilized leaves were ground with 4 times the leaf weight of sterilized phosphate buffer solution (0.1 M, pH 7.3). The slurry was then clarified in a Hitachi 20PR centrifuge at 1,470 x g for 10 min. The supernatant was used as a source of inoculum. Further dilution of the inoculum was made in a solution containing 0.1 M histidine and 0.01 M $MgCl_2$ (His-$MgCl_2$, pH 6.0)

Preparation of monolayers : Cell monolayers for infectivity assays were grown on coverslips of 15 mm diameter, previously washed in ethanol in an ultrasonicator. About 2 to 3 days after subculturing, cells were harvested from cultured flasks, and sedimented by low speed centrifugation (about 250 x g, 3 min). The cell pellets were resuspended in growth medium to give a dilution of about 3.5-5.5 x 10^6 cells/ml and thoroughly dispered. Cell suspensions of 0.12 ml were spread over the entire area of two or three coverslips in a plastic dish (55 mm in diameter) enclosed in a sealable glass dish sealed with parafilm M (American Can Company, Greenwich, CT, U.S.A.). The freshly seeded coverslips were allowed to remain undisturbed on a lab. bench for a minimum period of 2 hr to allow cell attachment before removal to a 28°C incubator. The cell monolayers on coverslips were used for virus inoculation less than 48 hr after seeding.

Virus inoculation and incubation of inoculated cell monolayers : For infectivity assays, confluent cell monolayers on coverslips were washed twice with His-$MgCl_2$ solution and then covered with 0.05 ml of inoculum. Inoculated monolayers were incubated for 1-3 hr at 28°C to accomplish inoculation. At the end of the inoculation period, the inoculum was removed, and the monolayers were washed twice with growth medium and then each coverslip was covered with 0.12-0.2 ml of the medium. Inoculated monolayers were then incubated at 28°C for 40 to 48 hr.

Detection of virus infection : The direct fluorescent antibody technique (6) was used for detection of virus infections. The cell monolayers were washed by dipping in a beaker of phosphate-buffer saline (PBS : 0.01 M phosphate, 0.15 M NaCl, pH 7.3). After fixation with cold acetone for 5 min, they were stained with fluorescent antibody (7) at 37°C for 40 to 50 min. Excess stain was removed by dipping in two successive washes of PBS followed by a 30 to 40 min soak in PBS. The stained specimens, mounted in PBS containing 50% glycerol (pH 7.3), were examined under a Nikon fluorescence microscope at x 200 magnification.

Fluorescent antibody focus counts : The method of counting infection units was 'the focus count method' (8,9). An infected cell and any adjoining cells were counted as one infection unit. In order to use the method, the factor for relating diametral zone counts to counts over the whole monolayer on a coverslip was determined. Stained cell foci on 10 coverslip monolayers were counted. The diametral factor for coverslip was determined by dividing the number for the whole coverslip by the average number for horizontal and vertical diametral zones. The data were treated statistically, and the average diametral factor with 95% confidence limits was 10.45 ± 0.45. Number of foci per diametral zone counts were considered satisfactory if between 4 and 150, but when the number was less than 4, foci over the coverslip were counted and used instead (10).

Infection and Multiplication of Virus in Vector Cell Monolayers

Growth of RDV in vector cell monolayers : The growth of RDV in cell monolayers was studied by preparing virus extracts from the inoculated monolayers at various intervals and assaying these on further monolayers.

Monolayers in flasks were inoculated with 10^{-3} inoculum for 60 min at 28°C. The monolayers were then washed three times with medium, covered with 4 ml medium, and then incubated at 28°C. At various intervals, the inoculated monolayers were harvested. Usually, the harvest from one flask was taken as a sample for each time point, and the volume of harvested cells was measured by sedimentation in a 5 ml centrifuge tube with a capillary tip graduated to 0.01 ml and accurate to $\pm$ 0.002 ml. The cells were frozen and stored at - 80°C with His-$MgCl_2$ solution of a volume 20 times that of the cell volume. Later, all cell samples in an experiment were thawed at 25°C in a water-bath at the same time, the cell debris was removed after centrifugation at 1,200 rpm for 5 min, and the infectivities of the supernatant fractions were compared by assays on coverslip monolayers.

No infectivity was detected in supernatant fractions prepared during the first 12 hr, indicating that residual virus inoculum had been effectively removed by washing three times with medium. Progeny virus was first detected at 12 hr but not at 8 hr after inoculation, indicating an eclipse period. From 12 to 20 hr, the growth rate of RDV in cells was exponential with a doubling time of about 96 min. From 20 to 28 hr, there was little or no further increase of infective virus.

Comparison of the sensitivity of focus count and vector injection methods for virus assay : Infectivity assays of the same serially diluted inocula were carried out by focus counting on cell monolayers and by vector injection methods. The number of foci were counted as mentioned above. Under the experimental conditions, the relationship between RDV concentration and number of foci of infected cells was linear in the dilution range from 10^{-6} to 10^{-4}. The dilution end point by focus counting was between $10^{-6.5}$ and $10^{-6.0}$.

Using the insect vector injection method, the relationship between virus concentration and infectivity (number of transmitters/number of inoculated insects) was almost linear in the range of virus concentrations 10^{-4} to $10^{-2.5}$, and the dilution end point of the comparable inoculum by this assay was between $10^{-4.5}$ and $10^{-4.0}$.

In this comparison of the two methods of infectivity assay, the focus count assay detected about 100 times fewer virus than the injection assay.

References

1. Liu, H. Y. and Black, L. M. Proc. Amer.Phytopathol. Soc., 3 : 234 (Abstr.), 1976.
2. Fukushi, T. J. Facult. Agr. Hokkaido Imp. Univ., 37 : 41-164, 1934.
3. Kimura, I. and Fukushi, T. Ann. Phytopathol. Soc. Japan, 25 : 131-135, 1960.
4. Fukushi, T. J. Facult. Agr. Hokkaido Imp. Univ., 45 : 83-154, 1940.
5. Chiu, R. J. and Black, L. M. Nature (London), 215 : 1076-1078, 1967.
6. Reddy, D. V. R. In Methods in Virology (Maramorosch, K. and Koprowski, H., eds.), 6 : 393-434, 1977.
7. Chiu, R. J. and Black, L. M. Virology, 37 : 667-677, 1969.
8. Spendlove, R. S. In Methods in Virology (Maramorosch, K. and Koprowski, H., eds.), 3 : 475-520, 1967.
9. Kimura, I. and Black, L. M. Virology, 46 : 266-276, 1971.
10. Kimura, I. Japan Agr. Res. Quarterly, 19 : 109-114, 1985.

CHARACTERIZATION OF *BOMBYX* AND RELATED INSECT CELL LINES BY SDS-PAGE AND TWO-DIMENSIONAL GEL ELECTROPHORESIS

K. Asaoka

Matsumoto Branch, Sericultural Experiment Station, Matsumoto, Nagano 390, Japan

Characterization of the cell lines derived from *Bombyx* and related insects was undertaken using SDS-PAGE and O'Farrell's two-dimensional gel electrophoresis as a means of discriminating between cell lines derived from different strains and species of insects. The four *B. mori* cell lines from embryos gave nearly identical protein patterns. In the *B. mandarina* cell line from embryos, the pattern of most proteins was similar to that of the proteins of these cell lines, but there were a few species-specific spots. The two *M. brassicae* cell lines from fat bodies and the *A. eucalypti* cell line from ovaries gave different protein patterns from those of the *Bombyx* cell lines. These results indicate that the methods described can be applied at least for the discrimination and characterization of the cell lines from different species.

Cell lines derived from insects are generally designated by their origins and characterized by their morphology, growth rate and chromosome numbers. However, the characters defined by these methods do not enable to identify the cell lines, thus, biochemical methods using antibody, isozyme and Southern blotting analysis were developed to characterize the cell lines (1, 7 and 10). However, there are no reports about the use of SDS-PAGE or O'Farrell's two dimensional electrophoresis for such purpose. Since both methods enable to distinguish various proteins in the cells, it was considered that the methods might be useful for the characterization of the cell lines. In this report attempts were made to determine whether the characterization by these methods enables to discriminate cell lines originating from different strains and species of *Bombyx* and related insects.

Materials and Methods

The cell lines used in this experiment are listed in Table 1. They were provided by Dr. H. Inoue and Dr. O. Ninaki of The Sericultural Experiment Station. These cells were grown at 25°C in IPL-41 medium with 10% fetal bovine serum, except for the *A. eucalypti* cells that were grown in Grace's insect medium with 5% fetal bovine serum and 5% *B. mori* haemolymph. SDS-PAGE was carried out as described by Laemmli(5) using 11% polyacrylamide gel. The cells for electrophoresis were harvested by centrifugation, and the cell pellets were washed three times in a physiological salt solution. The harvested cells were homogenized in SDS sample buffer, and electrophoresis was performed at 25 mA constant current until the bromophenol blue tracking dye reached the bottom of the gel. The gel was stained with Coomassie brilliant blue.

Two-dimensional gel electrophoresis was carried out as described by O'Farrell(8). The cells were homogenized in lysis buffer containing 8M urea, 2%

Nonidet P-40, 2% Ampholine pH 3.5-10(LKB), 5% 2-mercaptoethanol and 0.5% SDS. The first dimension gels contained 8M urea, 3.5% acrylamide and 0.5% Ampholine,pH3.5-10 (LKB) ,and the gels for the second dimension contained 11% polyacrylamide and 0.1% SDS.

Table 1. B. mori and related insect cell lines used in the experiment

Cell Line Designation	Origin	Confirmed By
SES-BoMo-15A	Embryos of bombyx mori	Inoue,Mitsuhashi, 1984
SES-Bm-e21	Embryos of Bombyx mori	Ninaki,Maekawa, 1986
SES-Bm-130	Embryos of Bombyx mori	Ninaki,Maekawa, 1986
SES-Bm-130R	Embryos of Bombyx mori	Ninaki,Maekawa, 1986
S.P.C. Bm36	pupal ovaries of Bombyx mori	Quiot, 1982
SES-Bma	Embryos of Bombyx mandarina	Ninaki,personal communication
A.eucalypti cells	Pupal ovaries of Antheraea eucalypti	Grace, 1962
SES-MaBr-3	Larval fat bodies of Mamestra brassicae	Inoue,Mitsuhashi, 1985
SES-MaBr-5	Larval fat bodies of Mamestra brassicae	Inoue,Mistuhashi, 1985

Results

The electrophoretic patterns of B. mori and related insect cell lines obtained by SDS-PAGE are shown in Fig. 1. In the cell lines derived from B. mori embryos, BoMo-15A, Bm-e21, Bm-130, Bm-130R, out of the 70 bands detected at least 50 bands appeared to be identical due to the similarity in their mobilities. No bands with

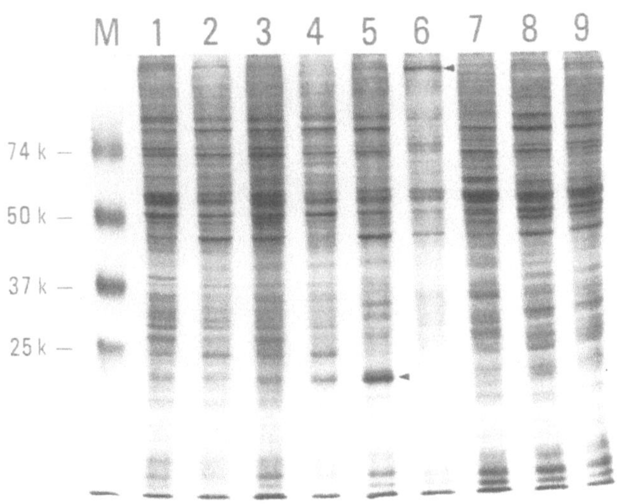

Figure 1. Separation of proteins of B. mori and related insect cell lines by SDS-PAGE. Lanes: 1, BoMo-15A; 2, Bm-e21; 3, Bm-130; 4, Bm-130R; 5, Bm36; 6, Bma; 7, A. eucalypti cells; 8, MaBr-3; 9,MaBr-5; M, Molecular weight markers. Arrows indicate the bands that characterize the cell lines.

apparently different mobilities were detected among these cell lines. In the bands of Bm36 and BoMo-15A, both derived from B. mori, about 20 out of the total number of bands were found in positions with the same mobilities. However, it was observed that a large band with a high mobility (indicated by an arrow in lane 5 of Fig.1) was present only in Bm36. On the other hand, most bands of Bma, the cell line derived from B. mandarina embryos, were the same as those of the other four B. mori cell lines, BoMo-15A, Bm-e21, Bm-130 and Bm-130R. However, one band that characterizes this cell line was found in a position with low mobility (see arrow of lane 6 of Fig.1).

The cell line from A. eucalypti ovaries gave different separation patterns of the proteins when compared with the Bombyx cell lines (Fig. 1, lane 7). Furthermore, the cell line from M. brassicae fat bodies, MaBr-3 and MaBr-5, exhibited many bands that were different from those of the Bombyx and A. eucalypti cell lines, and the bands detected were present in both cell lines.

The separation patterns of various cell line proteins by two-dimensional gel electrophoresis are shown in Fig.2. More than 200 spots of proteins were detected from each cell line. It appeared that many minor spots are specific for each cell line. However, these minor spots in each cell line cannot be easily differentiated. Therefore, only the differences between the major spots were considered in the current report.

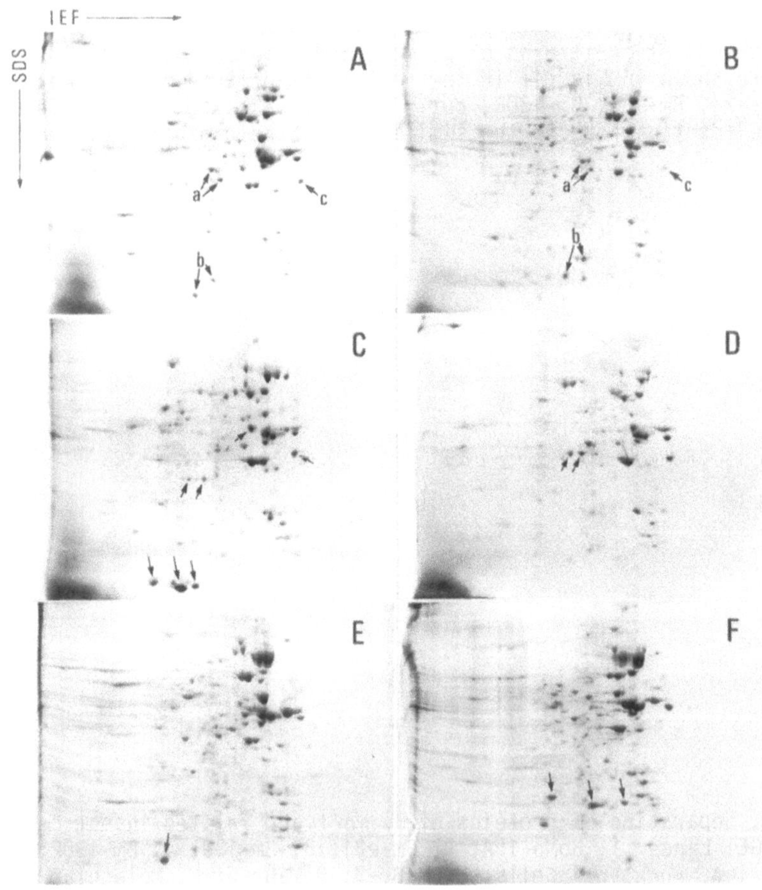

Figure 2. A-F. Separation of proteins of B. mori and related insect cell lines by O'Farrell's two-dimensional gel electrophoresis. A, BoMo-15A; B,Bm-130; C, Bm36; D, Bma; E, A. eucalypti cells; F, MaBr-3.

The B. mori embryonic lines, BoMo-15A, Bm-e21, Bm-130 and Bm-130R, showed nearly identical separation patterns of the proteins. The spots characterizing these four cell lines are indicated by arrows in Fig. 2A and B. It is, however, rather difficult to distinguish these cell lines by the spots because there were only minor differences among the spots in these B. mori cell lines. Bm36 reported as a cell line derived from B. mori ovaries gave apparently different spots when compared with those of the B. mori cell lines. The spots that were present characteristically in this cell line are indicated by arrows in Fig.2C. The Bma, B. mandarina cell line from embryos, showed a separation pattern similar to that of the four B. mori cell lines, but the two spots indicated by arrows in Fig. 2D were found to be specific to this cell line. The separation pattern of the proteins of the A. eucalypti cell line was apparently different from that of the other cell lines (Fig.2E). The patterns of MaBr-3 and MaBr-5, both cell lines derived from M. brassicae were also different from those of the other cell lines, and the three spots that could be distinguished from the other cell lines are shown by the arrows in Fig. 2F. The spot that is apparently different between these two cell lines was not detected by this method.

Discussion

Based on these experiments, it was observed that the cell lines derived from the same species gave almost the same separation patterns of proteins by SDS-PAGE and two dimensional gel electrophoresis, even if the cell lines exhibited different morphological properties or were cultured for many generations as separated strains. This phenomenon may be ascribed to the fact that most proteins that appeared in these methods are necessary to maintain the living cells as components constituting the cells. On the other hand, significant differences were observed among the cell lines from different species. Therefore, it appears that these methods enable to distinguish easily cell lines from different species.

The patterns of proteins in the Bm36 and the other four B. mori cell lines were distinctly different, presumably because they were derived from different tissues, ovaries and embryos. However, Southern bloting analysis showed that Bm36 DNA did not contain the B. mori fibroin gene and that the patterns of ribosomal DNA fragments digested with several restriction enzymes were apparently different from that of B. mori(7). Thus, based on the results of protein and DNA analysis, it was concluded that Bm36 probably did not originate from Bombyx. It was also concluded that the methods applied in this experiment enable to characterize the cell lines of B. mori and related insects and are useful for identification of cell lines introduced from various locations and of mislabelled cell lines.

References

1. Aldridge, C. A., and Knudson, D. L. In Vitro, 16: 384-391, 1980.
2. Grace, T. D. C. Nature, 195: 788-789, 1962.
3. Inoue, H., and Mitsuhashi, J. J. Seric. Sci. Jpn., 53, 108-113, 1984.
4. Inoue, H., and Mitsuhashi, J. Appl. Entmol. Zool.,20: 496-498, 1985.
5. Laemmli, U. K. Nature, 227: 680-685, 1970.
6. Ninaki, O., and Maekawa, H. Proc. 56th Meet. Jpn. Soc. Sericult. Sci;60, 1986.
7. Ninaki, O., Fujiwara, H., Ogura, H., Miyajima, N., Takada, N., and Maekawa, H. Proc.36th Meet. Jpn. Soc. Sericult. Sci. Kanto; 38, 1985.
8. O'Farrell, P. H. J. Biol. Chem., 250: 4007-4021, 1975.
9. Quiot, J. M. Sericologia, 22: 25-31, 1982.
10. Tabachnich, W. J., and Knudson, D. L. In Vitro, 16: 392-398, 1980.

Author Index

Subject Index

MIX
Papier aus verantwortungsvollen Quellen
Paper from responsible sources
FSC® C105338

If you have any concerns about our products,
you can contact us on
ProductSafety@springernature.com

In case Publisher is established outside the EU,
the EU authorized representative is:
Springer Nature Customer Service Center GmbH
Europaplatz 3, 69115 Heidelberg, Germany

Printed by Libri Plureos GmbH
in Hamburg, Germany